D1117281

# HSDPA/HSUPA for UMTS

# HSDPA/HSUPA for UMTS

High Speed Radio Access for Mobile Communications

**Edited by**

Harri Holma and Antti Toskala

*Both of Nokia Networks, Finland*

Telephone (+44) 1243 779777

Email (for orders and customer service enquiries): cs-books@wiley.co.uk
Visit our Home Page on www.wiley.com



***Other Wiley Editorial Offices***

John Wiley & Sons, Inc., 111 River Street, Hoboken, NJ 07030, USA

Jossey-Bass, 989 Market Street, San Francisco, CA 94103-1741, USA

Wiley-VCH Verlag GmbH, Boschstr. 12, D-69469 Weinheim, Germany

John Wiley & Sons Australia Ltd, 42 McDougall Street, Milton, Queensland 4064, Australia

John Wiley & Sons (Asia) Pte Ltd, 2 Clementi Loop #02-01, Jin Xing Distripark, Singapore 129809

John Wiley & Sons Canada Ltd, 22 Worcester Road, Etobicoke, Ontario, Canada M9W 1L1

Wiley also publishes its books in a variety of electronic formats. Some content that appears in print may not be available in electronic books.

***British Library Cataloguing in Publication Data***

A catalogue record for this book is available from the British Library

ISBN-13 978-0-470-01884-2 (HB)
ISBN-10 0-470-01884-4 (HB)

Project management by Originator, Gt Yarmouth, Norfolk (typeset in 10/12pt Times).
Printed and bound in Great Britain by Antony Rowe Ltd, Chippenham, Wiltshire.
This book is printed on acid-free paper responsibly manufactured from sustainable forestry in which at least two trees are planted for each one used for paper production.

# Contents

# Preface

When the first edition of *WCDMA for UMTS* was published by John Wiley & Sons, Ltd 6 years ago (in April 2000), 3GPP had just completed the first set of wideband CDMA (WCDMA) specifications, called 'Release 99'. At the same time, the Universal Mobile Telecommunication Services (UMTS) spectrum auction was taking place in Europe. UMTS was ready to go. The following years were spent on optimizing UMTS system specifications, handset and network implementations, and mobile applications. As a result, WCDMA has been able to bring tangible benefits to operators in terms of network quality, voice capacity, and new data service capabilities. WCDMA has turned out to be the most global mobile access technology with deployments covering Europe, Asia including Korea and Japan, and the USA, and it is expected to be deployed soon in large markets like China, India, and Latin America.

WCDMA radio access has evolved strongly alongside high-speed downlink packet access (HSDPA) and high-speed uplink packet access (HSUPA), together called 'high-speed packet access' (HSPA). When the International Telegraphic Union (ITU) defined the targets for IMT-2000 systems in the 1990s, the required bit rate was 2 Mbps. 3rd Generation Partnership Project (3GPP) Release 99 does support up to 2 Mbps in the specifications, but the practical peak data rate chosen for implementations is limited to 384 kbps. HSPA is now able to push practical bit rates beyond 2 Mbps and is expected to exceed 10 Mbps in the near future. In addition to the higher peak data rate, HSPA also reduces latency and improves network capacity. The new radio capabilities enable a new set of packet-based applications to go wireless in an efficient way. For operators the network upgrade from WCDMA to HSPA is straightforward as the HSPA solution builds on top of the WCDMA radio network, reusing all network elements. The first commercial HSDPA networks were launched during the last quarter of 2005.

This book was motivated by the fact that HSDPA and HSUPA are the next big steps in upgrading WCDMA networks. While the WCDMA operation has experienced some enhancements on top of dedicated channel operation, there was a clear need – it was felt – to focus just on HSDPA and HSUPA issues without having to repeat what was already presented in the different editions of *WCDMA for UMTS* for Release 99 based systems. Also, valuable feedback obtained from different lecturing events on HSDPA and HSUPA training sessions had clearly indicated a shift in the learning focus from basic WCDMA to the HSPA area. Thus, this book's principal task is to focus on HSPA specifications, optimization, and performance. The presentation concentrates on the differences that HSPA has brought to WCDMA radio access. Detailed information about WCDMA radio can be obtained from *WCDMA for UMTS*.

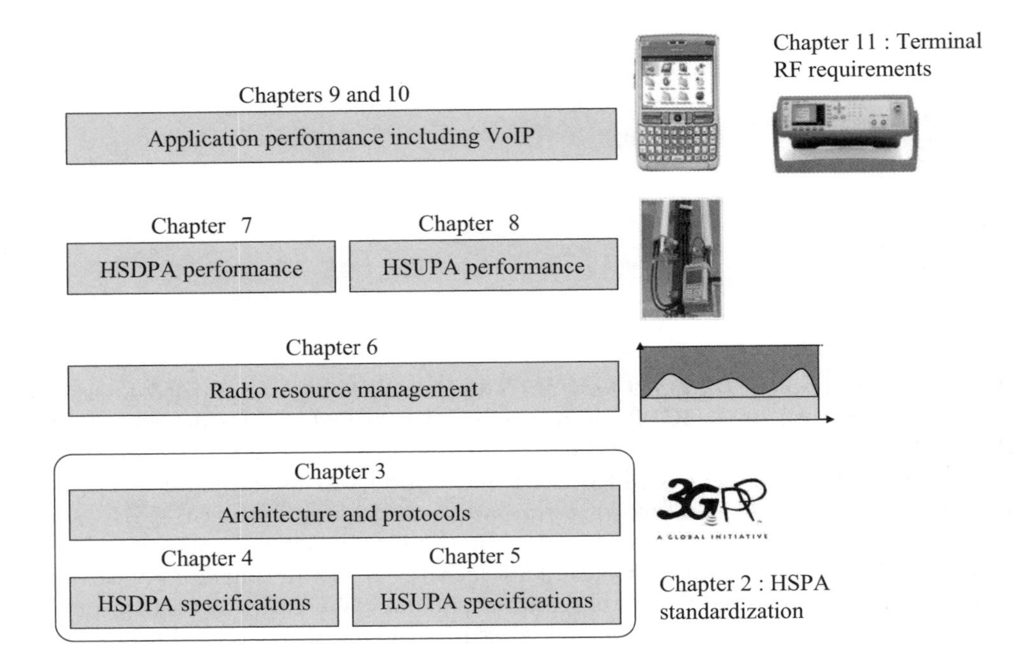


Summary of the book's contents.

The contents of this book are summarized in the above diagram. Chapter 1 gives an introduction to the status of WCDMA and HSPA capabilities. Chapter 2 provides an overview of HSPA standardization. Chapter 3 presents the HSPA network architecture and radio protocols. Chapters 4 and 5 explain the 3GPP physical layer HSDPA and HSUPA standards and the background of the selected solutions. Radio resource management algorithms are discussed in Chapter 6. Chapters 7 and 8 present HSDPA and HSUPA performance including data rates, capacity, and their coexistence with WCDMA. Application performance is presented in Chapter 9, and Voice over Internet Protocol (VoIP) performance aspects in Chapter 10. A terminal's radio frequency (RF) requirements are introduced in Chapter 11.

This book is aimed at R&D engineers, network planners, researchers, technical managers, regulators, and mobile application developers who wish to broaden their technical understanding to cover HSDPA and HSUPA as well. The views in the book are based on the authors' opinions and do not necessarily represent their employer's views.

*Harri Holma and Antti Toskala*
Nokia, Finland

# Acknowledgements

The editors would like to acknowledge the effort from their colleagues to contribute to this book. Besides the editors themselves, the other contributors to this book were: Frank Frederiksen, Sandro Grech, Jussi Jaatinen, Chris Johnson, Troels Kolding, Martin Kristensson, Esa Malkamäki, Jussi Numminen, Karri Ranta-Aho, Claudio Rosa, Klaus Pedersen, Markus Pettersson, Juho Pirskanen, and Jeroen Wigard.

In addition to their direct contribution, we would also like to acknowledge the constructive suggestions, illustrations, and comments received from Erkka Ala-Tauriala, Jorma Kaikkonen, Sami Kekki, Markku Kuusela, Svend Lauszus, Juhani Onkalo, Jussi Reunanen, Kai Sahala, Sasi Sasitharan, and Tuomas Törmänen. Further, we are grateful for the good suggestions received from the people participating in HSDPA/HSUPA training events in different locations who came up with suggestions as to what constitutes the key topics of interest and what issues deserve attention.

The team at John Wiley & Sons, Ltd deserve to be acknowledged as well for their patience and support during the production process.

We are grateful to our families, as well as the families of all contributors, for the personal time needed in the evening and weekends for writing and editing work.

Special thanks are due to our employer, Nokia Networks, for supporting and encouraging such an effort and for providing some of the illustrations in this book.

We would like to acknowledge Sierra Wireless for permission to use their product picture in the book.

Finally, it is good to remember that this book would not have been possible without the huge effort invested by our colleagues in the wireless industry within the 3rd Generation Partnership Project (3GPP) to produce the different specification releases of the global WCDMA/HSDPA/HSUPA standard and, thereby, making the writing of this book possible.

The editors and authors welcome any comments and suggestions for improvements or changes that could be implemented in forthcoming editions of this book.

*Harri Holma and Antti Toskala*
Espoo, Finland
*harri.holma@nokia.com* and *antti.toskala@nokia.com*

# Abbreviations

| | |
|---|---|
| 16QAM | 16 Quadrature Amplitude Modulation |
| 2G | Second Generation |
| 3G | Third Generation |
| 3GPP | 3rd Generation Partnership Project |
| 64QAM | 64 Quadrature Amplitude Modulation |
| 8PSK | 8 Phase Shift Keying |
| A-DPCH | Associated DPCH |
| AAL | ATM Adaptation Layer |
| AC | Admission Control |
| ACIR | Adjacent Channel Interference Ratio |
| ACK | ACKnowledgement |
| ACLR | Adjacent Channel Leakage Ratio |
| ACS | Adjacent Channel Selectivity |
| AG | Absolute Grant |
| AGC | Automatic Gain Control |
| ALCAP | Access Link Control Application Part |
| AM | Acknowledged Mode |
| AMC | Adaptive Modulation and Coding |
| AMR | Adaptive Multi-Rate |
| APN | Access Point Name |
| ARIB | Association of Radio Industries and Businesses (Japan) |
| ARP | Allocation and Retention Priority |
| ARQ | Automatic Repeat reQuest |
| ASN.1 | Abstract Syntax Notation 1 |
| ATIS | Alliance for Telecommunications Industry Solutions (US) |
| ATM | Asynchronous Transfer Mode |
| AWGN | Additive White Gaussian Noise |
| BCCH | BroadCast Control CHannel (logical channel) |
| BCFE | Broadcast Control Functional Entity |
| BCH | Broadcast CHannel (transport channel) |
| BER | Bit Error Rate |
| BLEP | BLock Error Probability |
| BLER | BLock Error Rate |
| BMC | Broadcast/Multicast Control protocol |
| BPSK | Binary Phase Shift Keying |

| | |
|---|---|
| BS | Base Station |
| BSC | Base Station Controller |
| BSS | Base Station Subsystem |
| BTS | Base Transceiver Station |
| C/I | Carrier-to-Interference ratio |
| CC | Congestion Control |
| CC | Chase Combining |
| CCSA | China Communications Standards Association |
| CCTrCH | Coded Composite Transport CHannel |
| CDMA | Code Division Multiple Access |
| CFN | Connection Frame Number |
| CLTD | Closed Loop Transmit Diversity |
| CLTD2 | Closed Loop Transmit Diversity mode-2 |
| CM | Cubic Metric |
| CN | Core Network |
| COST | COoperation Européenne dans le domaine de la recherche Scientifique et Technique |
| CP | Cyclic Prefix |
| CPICH | Common PIlot CHannel |
| CQI | Channel Quality Information |
| CRC | Cyclic Redundancy Check |
| CRNC | Controlling RNC |
| CS | Circuit Switched |
| CT | Core and Terminals |
| DAB | Digital Audio Broadcasting |
| DCCH | Dedicated Control CHannel (logical channel) |
| DCH | Dedicated CHannel (transport channel) |
| DDI | Data Description Indicator |
| DL | DownLink |
| DPCCH | Dedicated Physical Control CHannel |
| DPCH | Dedicated Physical CHannel |
| DPDCH | Dedicated Physical Data CHannel |
| DRNC | Drift RNC |
| DRX | Discontinuous Reception |
| DS-CDMA | Direct Spread Code Division Multiple Access |
| DSCH | Downlink Shared CHannel |
| DSL | Digital Subscriber Line |
| DT | Discard Timer |
| DTCH | Dedicated Traffic CHannel |
| DTX | Discontinuous Transmission |
| DVB | Digital Video Broadcasting |
| E-AGCH | E-DCH Absolute Grant CHannel |
| E-DCH | Enhanced uplink Dedicated CHannel |
| E-DPCCH | E-DCH Dedicated Physical Control CHannel |
| E-DPDCH | E-DCH Dedicated Physical Data CHannel |
| E-HICH | E-DCH Hybrid ARQ Indicator CHannel |

| | |
|---|---|
| E-RGCH | E-DCH Relative Grant CHannel |
| E-RNTI | E-DCH Radio Network Temporary Identifier |
| E-TFC | E-DCH Transport Format Combination |
| E-TFCI | E-DCH Transport Format Combination Indicator |
| ECR | Effective Code Rate |
| EDGE | Enhanced Data rates for GSM Evolution |
| EDGE | Enhanced Data Rate for Global Evolution |
| EGPRS | Enhanced GPRS |
| EGPRS | Extended GPRS |
| ETSI | European Telecommunications Standards Institute |
| EVM | Error Vector Magnitude |
| F-DCH | Fractional Dedicated CHannel |
| F-DPCH | Fractional Dedicated Physical CHannel |
| FACH | Forward Access CHannel |
| FBI | FeedBack Information |
| FCC | Federal Communications Commission |
| FCS | Fast Cell Selection |
| FDD | Frequency Division Duplex |
| FDMA | Frequency Division Multiple Access |
| FER | Frame Error Ratio |
| FER | Frame Erasure Rate |
| FFT | Fast Fourier Transform |
| FP | Frame Protocol |
| FRC | Fixed Reference Channel |
| FTP | File Transfer Protocol |
| G-factor | Geometry factor |
| GB | GigaByte |
| GBR | Guaranteed Bit Rate |
| GERAN | GSM/EDGE RAN |
| GGSN | Gateway GPRS Support Node |
| GI | Guard Interval |
| GP | Processing gain |
| GPRS | General Packet Radio Service |
| GSM | Global System for Mobile Communications |
| HARQ | Hybrid Automatic Repeat reQuest |
| HC | Handover Control |
| HLBS | Highest priority Logical channel Buffer Status |
| HLID | Highest priority Logical channel ID |
| HLR | Home Location Register |
| HS-DPCCH | Uplink High-Speed Dedicated Physical Control CHannel |
| HS-DSCH | High-Speed Downlink Shared CHannel |
| HS-PDSCH | High-Speed Physical Downlink Shared CHannel |
| HS-SCCH | High-Speed Shared Control CHannel |
| HSDPA | High-Speed Downlink Packet Access |
| HSPA | High-Speed Packet Access |
| HSUPA | High-Speed Uplink Packet Access |

| | |
|---|---|
| HTTP | Hypertext markup language |
| IFFT | Inverse Fast Fourier Transform |
| IP | Internet Protocol |
| IR | Incremental Redundancy |
| IRC | Interference Rejection Combining |
| IS-95 | Interim Standard 95 |
| ITU | International Telecommunication Union |
| ITU | International Telegraphic Union |
| LAU | Location Area Update |
| LMMSE | Linear Minimum Mean Square Error |
| LTE | Long-Term Evolution |
| MAC | Medium Access Control |
| MAC-d | dedicated MAC |
| MAC-es/s | E-DCH MAC |
| MAC-hs | high-speed MAC |
| MAI | Multiple Access Interference |
| MAP | Maximum *A Posteriori* |
| max-C/I | maximum Carrier-to-Interference ratio |
| MB | MegaByte |
| MBMS | Multimedia Broadcast and Multicast Service |
| MIMO | Multiple Input Multiple Output |
| min-GBR | minimum Guaranteed Bit Rate |
| MRC | Maximal Ratio Combining |
| MS | Mobile Station |
| MSC | Mobile Switching Centre |
| MSC/VLR | Mobile services Switching Centre/Visitor Location Register |
| MUD | MultiUser Detection |
| MUX | Multiplexing |
| NACC | Network Assisted Cell Change |
| NBAP | Node B Application Part |
| NF | Noise Figure |
| Node B | Base station |
| O&M | Operation & Maintenance |
| OFDM | Orthogonal Frequency Division Multiplexing |
| OFDMA | Orthogonal Frequency Division Multiple Access |
| OLPC | Outer Loop Power Control |
| OMA | Open Mobile Alliance |
| OSS | Operations Support System |
| OTDOA | Observed Time Difference Of Arrival |
| OVSF | Orthogonal Variable Spreading Factor |
| P-CPICH | Primary CPICH |
| PA | Power Amplifier |
| PAD | PADding |
| PAR | Peak-to-Average Ratio |
| PAS | Power Azimuth Spectrum |
| PC | Power Control |

| | |
|---|---|
| PCCC | Parallel Concatenated Convolutional Code |
| PCH | Paging CHannel |
| PCMCIA | Personal Computer Memory Card Industry Association |
| PCS | Personal Communication Services |
| PCS | Personal Communication System |
| PDCP | Packet Data Convergence Protocol |
| PDP | Packet Data Protocol |
| PDU | Protocol Data Unit |
| PDU | Payload Data Unit |
| PF | Proportional Fair |
| POC | Push-to-talk Over Cellular |
| PRACH | Physical RACH |
| PS | Packet Switched |
| PU | Payload Unit |
| QAM | Quadrature Amplitude Modulation |
| QoS | Quality of Service |
| QPSK | Quadrature Phase Shift Keying |
| RAB | Radio Access Bearer |
| RACH | Random Access CHannel |
| RAN | Radio Access Network |
| RANAP | Radio Access Network Application Part |
| RAU | Routing Area Update |
| RB | Radio Bearer |
| RF | Radio Frequency |
| RG | Relative Grant |
| RLC | Radio Link Control |
| RLL | Radio Link Layer |
| RLS | Radio Link Set |
| RM | Resource Manager |
| RNC | Radio Network Controller |
| RNTI | Radio Network Temporary Identifier |
| ROHC | RObust Header Compression |
| RR | Round Robin |
| RRC | Radio Resource Control |
| RRM | Radio Resource Management |
| RSCP | Received Signal Code Power |
| RSN | Retransmission Sequence Number |
| RSSI | Received Signal Strength Indicator |
| RTCP | Real Time Control Protocol |
| RTO | Retransmission TimeOut |
| RTP | Real Time Protocol |
| RTT | Round Trip Time |
| RTWP | Received Total Wideband Power |
| S-CCPCH | Secondary CCPCH |
| SA | Services and system Architecture |
| SC-FDMA | Single Carrier FDMA |

| | |
|---|---|
| SCCP | Signalling Connection Control Part |
| SCCPCH | Secondary Common Control Physical CHannel |
| SDU | Service Data Unit |
| SF | Spreading Factor |
| SGSN | Serving GPRS Support Node |
| SI | Scheduling Information |
| SIB | System Information Block |
| SID | Size index IDentifier |
| SINR | Signal-to-Interference-plus-Noise Ratio |
| SIR | Signal to Interference Ratio |
| SNR | Signal to Noise Ratio |
| SPI | Scheduling Priority Indicator |
| SRB | Signalling Radio Bearer |
| SRNC | Serving RNC |
| SRNS | Serving Radio Network System |
| STTD | Space Time Transmit Diversity |
| TC | Traffic Class |
| TCP | Transmission Control Protocol |
| TD-SCDMA | Time division synchronous CDMA |
| TDD | Time Division Duplex |
| TEBS | Total E-DCH Buffer Status |
| TF | Transport Format |
| TFCI | Transport Format Combination Indicator |
| TFRC | Transport Format and Resource Combination |
| THP | Traffic Handling Priority |
| TMSI | Temporary Mobile Subscriber Identity |
| TPC | Transmission Power Control |
| TR | Technical Report |
| TS | Technical Specification |
| TSG | Technical Specification Group |
| TSN | Transmission Sequence Number |
| TTA | Telecommunications Technology Association (Korea) |
| TTC | Telecommunication Technology Committee (Japan) |
| TTI | Transmission Time Interval |
| TX GAP | Transmit GAP |
| TxAA | Transmit Adaptive Antennas |
| UDP | User Datagram Protocol |
| UE | User Equipment |
| UL | UpLink |
| UM | Unacknowledged Mode |
| UM-RLC | Unacknowledged Mode RLC |
| UMTS | Universal Mobile Telecommunications System |
| UPH | UE Power Headroom |
| UPH | UE transmission Power Headroom |
| URA | UTRAN Registration Area |
| UTRA | UMTS Terrestrial Radio Access (ETSI) |

| | |
|---|---|
| UTRA | Universal Terrestrial Radio Access (3GPP) |
| UTRAN | UMTS Terrestrial Radio Access Network |
| VCC | Virtual Channel Connection |
| VF | Version Flag |
| VoIP | Voice over IP |
| VPN | Virtual Private Network |
| WAP | Wireless Application Protocol |
| WCDMA | Wideband CDMA |
| WG | Working Group |
| Wimax | Worldwide Interoperability for microwave access |
| WLAN | Wireless Local Area Network |
| WWW | World Wide Web 9 |

# 1

# Introduction

Harri Holma and Antti Toskala

## 1.1 WCDMA technology and deployment status

The first Third Generation Partnership Project (3GPP) Wideband Code Division Multiple Access (WCDMA) networks were launched during 2002. By the end of 2005 there were 100 open WCDMA networks and a total of over 150 operators having frequency licenses for WCDMA operation. Currently, the WCDMA networks are deployed in Universal Mobile Telecommunications System (UMTS) band around 2 GHz in Europe and Asia including Japan and Korea. WCDMA in America is deployed in the existing 850 and 1900 spectrum allocations while the new 3G band at 1700/2100 is expected to be available in the near future. 3GPP has defined the WCDMA operation also for several additional bands, which are expected to be taken into use during the coming years.

The number of WCDMA subscribers globally was 17 million at the end of 2004 and over 50 million by February 2006. The subscriber growth rate is illustrated in Figure 1.1. WCDMA subscribers represent currently 2% of all global mobile subscribers, while in Western Europe WCDMA's share is 5%, in the UK 8%, in Italy 14% and in Japan over 25%. The reason for the relatively high WCDMA penetrations in the UK and Italy is Three, the greenfield 3G operator, and in Japan NTT Docomo, who are pushing the technology forward. These two operators were also the ones behind the first large-scale commercial WCDMA operation that took place between 2001 and 2003.

The mobile business is driven by the availability of attractive terminals. In order to reach a major market share, terminal offering for all market segments is required. There are currently available over 200 different WCDMA terminal models from more than 30 suppliers launched by 2005. As an example, Nokia WCDMA terminal portfolio evolution is illustrated in Figure 1.2. In 2003, Nokia launched one new WCDMA handset, in 2004 two, and during 2005 more than 10 new WCDMA handsets were launched. It is expected that soon all new medium-price and high-end terminals will support WCDMA.

As WCDMA mobile penetration increases, it allows WCDMA networks to carry a larger share of voice and data traffic. WCDMA technology provides a few advantages for

*HSDPA/HSUPA for UMTS* Edited by Harri Holma and Antti Toskala

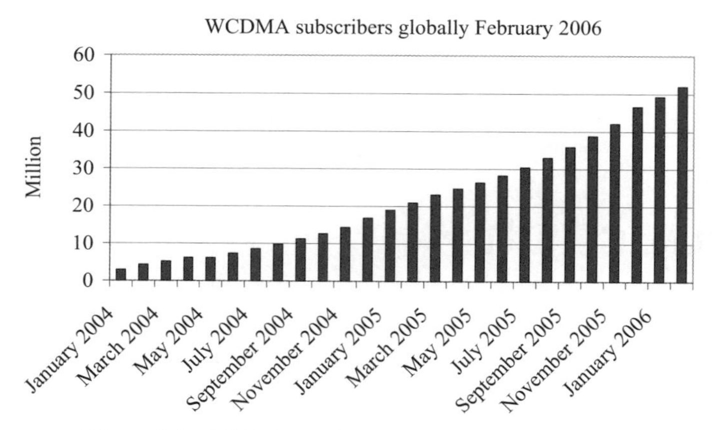


**Figure 1.1** 3G WCDMA subscriber growth monthly.

**Figure 1.2** Evolution of Nokia 3G terminal offering.
[*www.nokia.com*]

the operator in that it enables data but also improves basic voice. The offered voice capacity is very high because of interference control mechanisms including frequency reuse of 1, fast power control and soft handover. Figure 1.3 shows the estimated number of voice minutes per subscriber per month that can be supported with a two-carrier, three-sector, 2 + 2 + 2 WCDMA site depending on the number of subscribers in the site coverage area. Adaptive multi-rate (AMR) 5.9-kbps voice codec is assumed in the calculation. With 2000 subscribers in each base station coverage area, 4300 minutes per month can be offered to each subscriber, while with 4000 subscribers even more than 2100 minutes can be used. These capacities include both incoming and outgoing minutes. Global average usage today is below 300 minutes per month. This calculation shows that

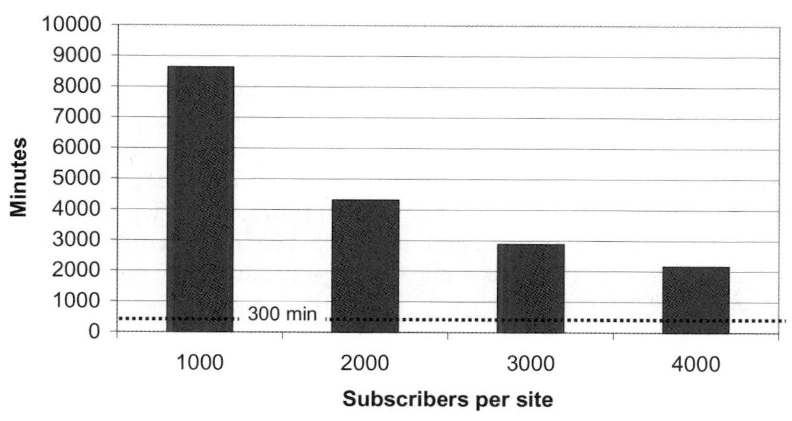


**Figure 1.3** Voice minutes per subscriber per month (minutes of usage, mou).

WCDMA makes it possible to offer substantially more voice minutes to customers. At the same time WCDMA can also enhance the voice service with wideband AMR codec, which provides clearly better voice quality than the fixed land line telephone. In short, WCDMA can offer more voice minutes with better quality.

In addition to the high spectral efficiency, third-generation (3G) WCDMA provides even more dramatic evolution in terms of base station capacity and hardware efficiency. Figure 1.4 illustrates the required base station hardware for equivalent voice capacity with the best second-generation (2G) technology from the early 2000s and with the latest 3G WCDMA base station technology. The high integration level in WCDMA is achieved because of the wideband carrier: a large number of users are supported per carrier, and fewer radio frequency (RF) carriers are required to provide the same capacity. With fewer RF parts and more digital baseband processing, WCDMA can take benefit of the fast evolution in digital signal processing capacity. The high base

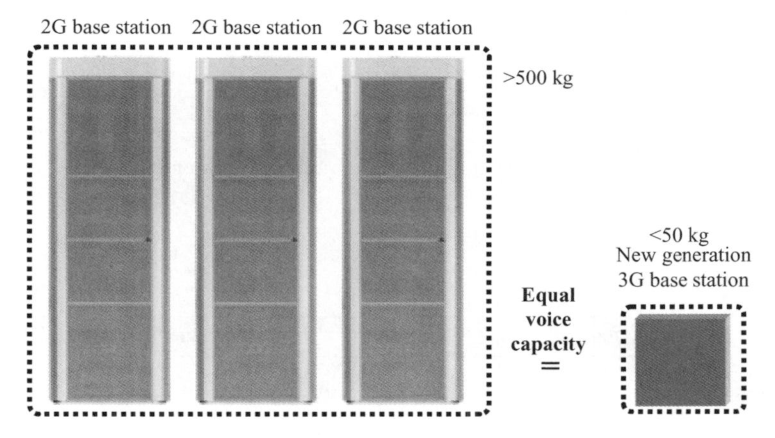


**Figure 1.4** Base station capacity evolution with 3G WCDMA.

station integration level allows efficient building of high-capacity sites since the complexity of RF combiners, extra antennas or feeder cables can be avoided.

WCDMA operators are able to provide interesting data services including browsing, person-to-person video calls, sports and news video clips and mobile-TV. WCDMA enables simultaneous voice and data which allows, for example, browsing or emailing during voice conferencing, or real time video sharing during voice calls. The operators also offer laptop connectivity to the Internet and corporate intranet with the maximum bit rate of 384 kbps both in downlink and in uplink. The initial terminals and networks were limited to 64–128 kbps in uplink while the latest products provide 384-kbps uplink.

## 1.2 HSPA standardization and deployment schedule

High-speed downlink packet access (HSDPA) was standardized as part of 3GPP Release 5 with the first specification version in March 2002. High-speed uplink packet access (HSUPA) was part of 3GPP Release 6 with the first specification version in December 2004. HSDPA and HSUPA together are called 'high-speed packet access' (HSPA). The first commercial HSDPA networks were available at the end of 2005 and the commercial HSUPA networks are expected to be available by 2007. The estimated HSPA schedule is illustrated in Figure 1.5.

The HSDPA peak data rate available in the terminals is initially 1.8 Mbps and will increase to 3.6 and 7.2 Mbps during 2006 and 2007, and potentially beyond 10 Mbps. The HSUPA peak data rate in the initial phase is expected to be 1–2 Mbps with the second phase pushing the data rate to 3–4 Mbps. The expected data rate evolution is illustrated in Figure 1.6.

HSPA is deployed on top of the WCDMA network either on the same carrier or – for a high-capacity and high bit rate solution – using another carrier, see Figure 1.7. In both cases, HSPA and WCDMA can share all the network elements in the core network and

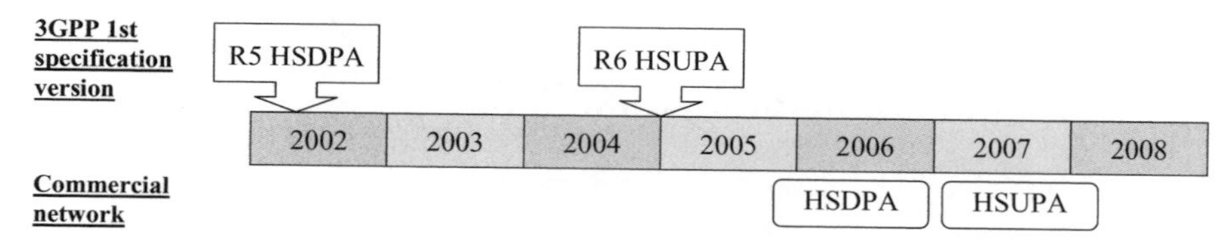


**Figure 1.5** HSPA standardization and deployment schedule.

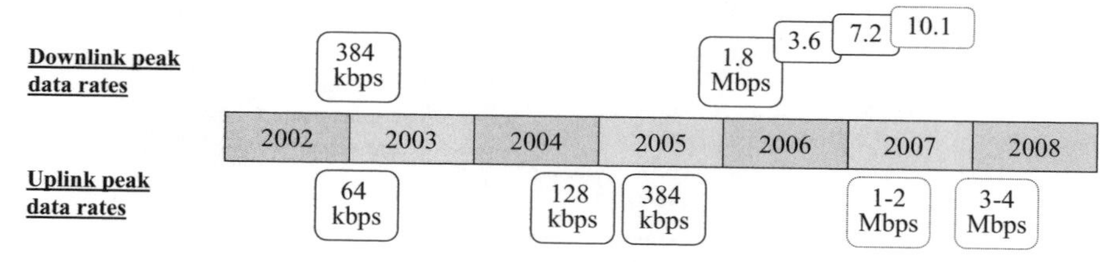


**Figure 1.6** Data rate evolution in WCDMA and HSPA.

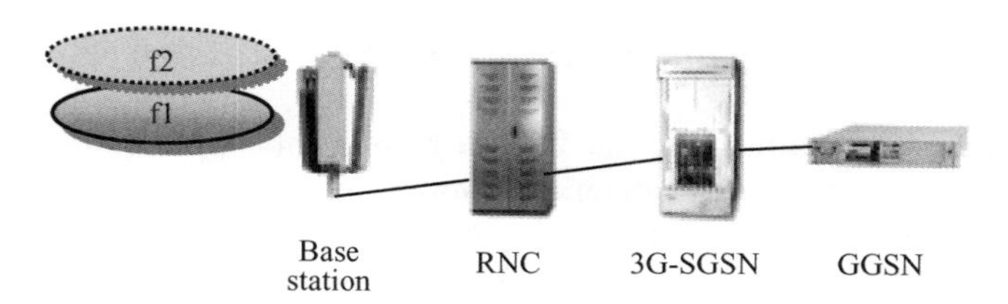


**Figure 1.7** HSPA deployment with (f2) new carrier deployed with HSPA and (f1) carrier shared between WCDMA and HSPA.

in the radio network including base stations, Radio Network Controller (RNC), Serving GPRS Support Node (SGSN) and Gateway GPRS Support Node (GGSN). WCDMA and HSPA are also sharing the base station sites, antennas and antenna lines. The upgrade from WCDMA to HSPA requires new software package and, potentially, some new pieces of hardware in the base station and in RNC to support the higher data rates and capacity. Because of the shared infrastructure between WCDMA and HSPA, the cost of upgrading from WCDMA to HSPA is very low compared with building a new standalone data network.

The first HSDPA terminals are data cards providing fast connectivity for laptops. An example terminal – Sierra Wireless AirCard 850 – is shown in Figure 1.8 providing 1.8-Mbps downlink and 384-kbps uplink peak data rates.

HSDPA terminal selection will expand beyond PCMCIA cards when integrated HSDPA mobile terminals are available during 2006. It is expected that HSPA will be a standard feature of most 3G terminals after some years in the same way as Enhanced Data Rates for GSM Evolution (EDGE) capability is included in most GSM/GPRS terminals. HSDPA will also be integrated to laptop computers in the future, as is indicated already by some of the laptop manufacturers.

**Figure 1.8** Example of first-phase HSDPA terminal.
[Courtesy of Sierra Wireless]

## 1.3 Radio capability evolution with HSPA

The performance of the radio system defines how smoothly applications can be used over the radio network. The key parameters defining application performance include data rate and network latency. There are applications that are happy with low bit rates of a few tens of kbps but require very low delay, like voice-over-IP (VoIP) and real time action games. On the other hand, the download time of a large file is only defined by the maximum data rate, and latency does not play any role. GPRS Release 99 typically provides 30–40 kbps with latency of 600 ms. EGPRS Release 4 pushes the bit rates 3–4 times higher and also reduces latency below 300 ms. The EGPRS data rate and latency allow smooth application performance for several mobile-based applications including Wireless Application Protocol (WAP) browsing and push-to-talk.

WCDMA enables peak data rates of 384 kbps with latency 100–200 ms, which makes Internet access close to low-end digital subscriber line (DSL) connections and provides good performance for most low-delay Internet Protocol (IP) applications as well.

HSPA pushes the data rates up to 1–2 Mbps in practice and even beyond 3 Mbps in good conditions. Since HSPA also reduces network latency to below 100 ms, the end user experienced performance is similar to the fixed line DSL connections. No or only little effort is required to adapt Internet applications to the mobile environment. Essentially, HSPA is a broadband access with seamless mobility and extensive coverage. Radio capability evolution from GPRS to HSPA is illustrated in Figure 1.9.

HSPA was initially designed to support high bit rate non-real time services. The simulation results show, however, that HSPA can provide attractive capacity also for low bit rate low-latency applications like VoIP. 3GPP Releases 6 and 7 further improve the efficiency of HSPA for VoIP and other similar applications.

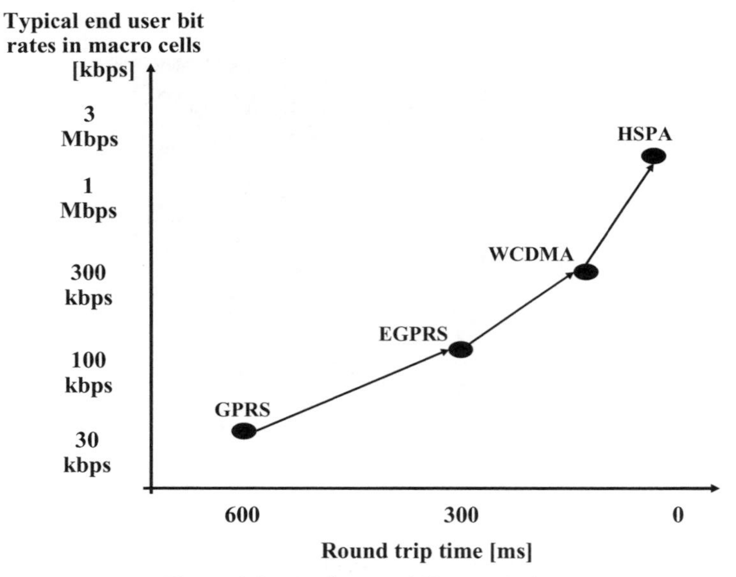


**Figure 1.9** Radio capability evolution.

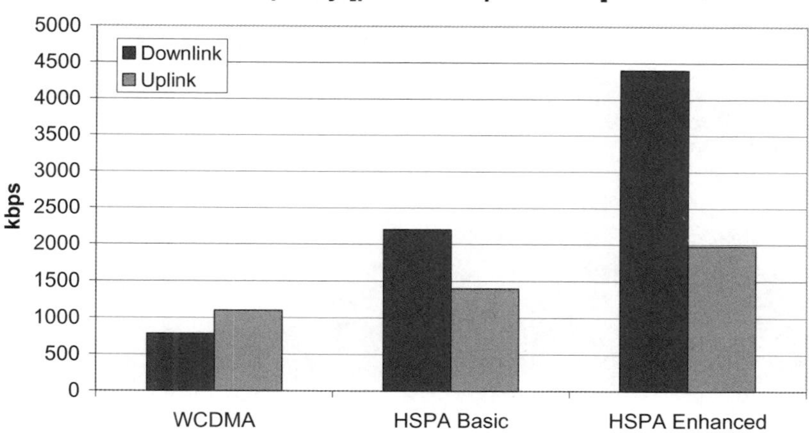


**Figure 1.10** Capacity evolution with HSPA.

Higher cell capacity and higher spectral efficiency are required to provide higher data rates and new services with the current base station sites. Figure 1.10 illustrates the estimated cell capacity per sector per 5 MHz with WCDMA, with basic HSPA and with enhanced HSPA in the macro-cell environment. Basic HSPA includes a one-antenna Rake receiver in the terminals and two-branch antenna diversity in the base stations. Enhanced HSPA includes two-antenna equalizer mobiles and interference cancellation in the base station. The simulation results show that HSPA can provide substantial capacity benefit. Basic HSDPA offers up to three times WCDMA downlink capacity, and enhanced HSDPA up to six times WCDMA. The spectral efficiency of enhanced HSDPA is close to 1 bit/s/Hz/cell. The uplink capacity improvement with HSUPA is estimated between 30% and 70%. HSPA capacity is naturally suited for supporting not only symmetric services but also asymmetric services with higher data rates and volumes in downlink.

# 2

# HSPA standardization and background

Antti Toskala and Karri Ranta-Aho

This chapter introduces the standardization framework around high-speed downlink packet access (HSDPA) and high-speed uplink packet access (HSUPA) and presents the standardization schedule and future development for HSDPA and HSUPA currently on-going. Also the developments beyond HSDPA and HSUPA are introduced at the end of this chapter.

## 2.1 3GPP

The 3rd Generation Partnership Project (3GPP) is the forum [1] where standardization is handled for HSDPA and HSUPA, as well as has been handled from the first wideband code division multiple access (WCDMA) specification release. Further, 3GPP also has responsibility for Global System for Mobile Communications (GSM)/Enhanced Data Rates for Global Evolution (EDGE) standardization. The background of 3GPP is in the days when WCDMA technology was being standardized following technology selections in different regions during 1997. Following that, WCDMA was chosen in several places as the basis for third-generation mobile communication systems and there was regional activity in several places around the same technological principles. It became evident, however, that this would not lead to a single global standard aligned down to bit level details. Thus, at the end of 1998 the US, Europe, Korea and Japan joined forces and created 3GPP. China followed a bit later. Note also that the related standardization organization, although marked as regional, usually had members from other countries/regions as well.

The first major milestone was reached at the end of 1999 when Release 99 specifications were published, containing the first full series of WCDMA specifications. Release 4 specifications followed in early 2001. The working method had been moved between Release 99 and Release 4 away from the yearly 'release' principle. The release cycle was

*HSDPA/HSUPA for UMTS* Edited by Harri Holma and Antti Toskala

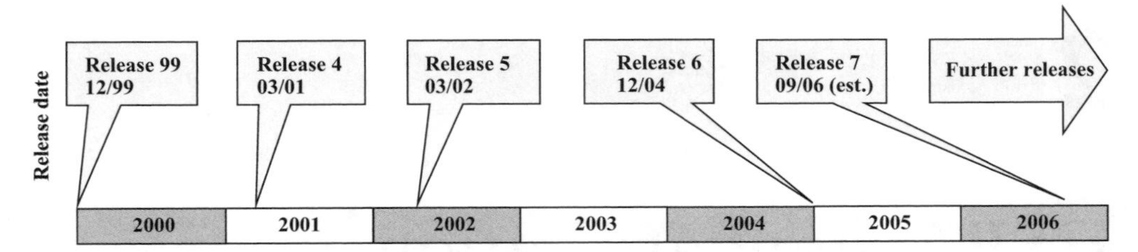


**Figure 2.1** 3GPP release timeline indicating release dates.

made longer than just 1 year, which enabled making bigger releases with less frequent intervals. This also allowed having more consideration of what is the necessary release content rather than when are release publication data needed. Release 5 followed in 2002 and Release 6, the latest full release completed during the writing of this chapter, in 2004. Release 7 specifications are expected to be ready in the second half of 2006, as shown in Figure 2.1, though the first versions were available at the beginning of 2006.

3GPP originally had four different technical specification groups (TSGs), and later five following the move of GSM/EDGE activities to 3GPP. Following restructuring in 2005, there are once again four TSGs, as shown in Figure 2.2:

- TSG RAN (Radio Access Network). TSG RAN focuses on the radio interface and internal interfaces between base transceiver stations (BTSs)/radio network controllers (RNCs) as well as the interface from RNC to the core network. HSDPA and HSUPA standards were under TSG RAN responsibility.
- TSG CT (core and terminals). TSG CT focuses on the core network issues as well as covering, for example, signalling between the core network and terminals.
- TSG SA (services and system architecture). TSG SA focuses on the services and overall system architecture.
- TSG GERAN (GSM/EDGE RAN). TSG GERAN covers similar issues like TSG RAN but for the GSM/GPRS/EDGE-based radio interface.

Under each TSG there are further working groups where the actual technical work is done. For example, under TSG RAN, where HSDPA and HSUPA have been done, there

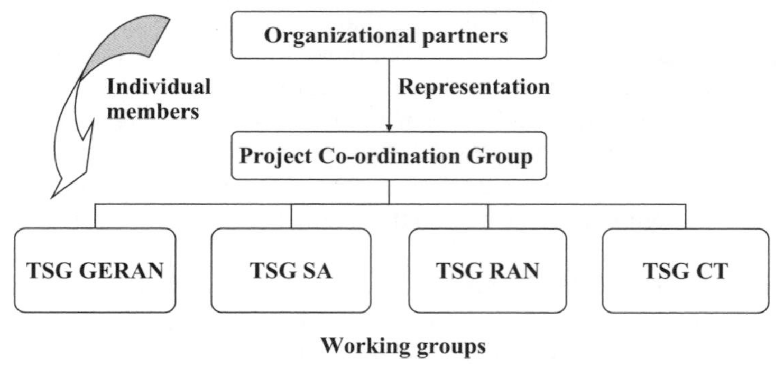


**Figure 2.2** 3GPP structure.

are five working groups (WGs) as follows:

- TSG RAN WG1: responsible for physical layer aspects;
- TSG RAN WG2: responsible for layer 2 and 3 aspects;
- TSG RAN WG3: responsible for RAN internal interfaces;
- TSG RAN WG4: responsible for performance and radio frequency (RF) requirements;
- TSG RAN WG5: responsible for terminal testing.

The 3GPP membership is organized through organizational partners. Individual companies need to be members of one of the organizational partners, and based on this membership there is the right to participate in 3GPP activity. The following are the current organizational partners:

- Alliance for Telecommunications Industry Solutions (ATIS) from the US.
- European Telecommunications Standards Institute (ETSI) from Europe.
- China Communications Standards Association (CCSA) from China.
- Association of Radio Industries and Businesses (ARIB) from Japan.
- Telecommunication Technology Committee (TTC) from Japan.
- Telecommunications Technology Association (TTA) from South Korea.

3GPP creates the technical content of the specifications, but it is the organizational partners that actually publish the work. This enables having identical sets of specifications in all regions, thus ensuring roaming across continents. In addition to the organizational partners, there are also so-called market representation partners, such as the UMTS Forum, part of 3GPP.

The work in 3GPP is based around work items, though small changes can be introduced directly as 'change requests' against specification. With bigger items a feasibility study is done usually before rushing in to making actual changes to the specifications.

### 2.1.1 HSDPA standardization in 3GPP

When Release 99 was completed, HSDPA or HSUPA were not yet on the agenda. During 2000, while also making corrections to WCDMA Release 99 and working on Release 4 to include, for example, TD-SCDMA, it became obvious that some improvements for packet access would be needed. To enable such an evolution, a feasibility study (study item) for HSDPA was started in March 2000. As shown in the study item proposal [2], the work was initiated in line with 3GPP principles, having at least four supporting companies. The companies supporting the start of work on HSDPA were Motorola and Nokia from the vendor side and BT/Cellnet, T-Mobile and NTT DoCoMo from the operator side, as indicated in [2].

The feasibility study was finalized for the TSG RAN plenary for March 2001 and the conclusions reported in [3] were such that there was clear benefit to be seen with the solutions studied. In the HSDPA study item there were issues studied to improve the downlink packet data transmission over Release 99 specifications. Topics such as physical layer retransmissions and BTS-based scheduling were studied as well as adaptive coding and modulation. The study also included some investigations for multi-antenna transmission and reception technology, titled 'multiple input multiple output' (MIMO), as well as on fast cell selection (FCS).

As the feasibility study clearly demonstrated that significant improvement was possible to achieve and with reasonable complexity, it was clear to proceed for an actual work item to develop specifications. When the work item was set up [4] the scope of the work was in line with the study item, but MIMO was taken as a separate work item and a separate feasibility study on FCS was started. For the HSDPA work item there was wider vendor support and the actual work item was supported from the vendor side by Motorola, Nokia and Ericsson. During the course of work, obviously much larger numbers of companies contributed technically to the progress.

When Release 5 specifications with HSDPA were released 1 year later – in March 2002 – there were clearly still corrections to do in HSDPA, but the core functionality was already in the physical layer specifications. The work was partly slowed down by the parallel corrections activity needed for the Release 99 terminals and networks being rolled out. Especially with protocol aspects, intensive testing tends to reveal details that need corrections and clarifications in the specifications and this was the case with Release 99 devices preceding the start of commercial operations in Europe in the second half of 2002. The longest time was taken for HSDPA protocol parts, on which backward compatibility was started in March 2004.

From the other topics that relate to HSDPA, the MIMO work item did not complete in the Release 5 or Release 6 time frame, and it is still under discussion whether there is sufficient merit for the introduction of it – as covered in the Release 7 topic listing. The feasibility study on FCS concluded that the benefits were limited to the extent that additional complexity would not be justified, thus no work item was created around FCS after the study was closed. While focus was on the frequency division duplex (FDD) side, time division duplex (TDD) was also covered in the HSDPA work item to include similar solutions in both TDD modes (narrowband and wideband TDD).

### *2.1.2 HSUPA standardization in 3GPP*

Although HSUPA is a term used broadly in the market, in 3GPP the standardization for HSUPA was done under the name 'enhanced uplink dedicated channel' (E-DCH) work item. Work started during the corrections phase for HSDPA, beginning with the study item on '*uplink enhancements for dedicated transport channels*' in September 2002. From the vendor side, Motorola, Nokia and Ericsson were the supporting companies to initiate the study in 3GPP.

The techniques investigated in the study for HSUPA (E-DCH) were, as shown in Figure 2.3:

- fast physical layer Hybrid-ARQ (HARQ) for the uplink;
- fast Node B based uplink scheduling;
- shorter uplink transmission time interval (TTI) length;
- higher order modulation;
- fast DCH setup.

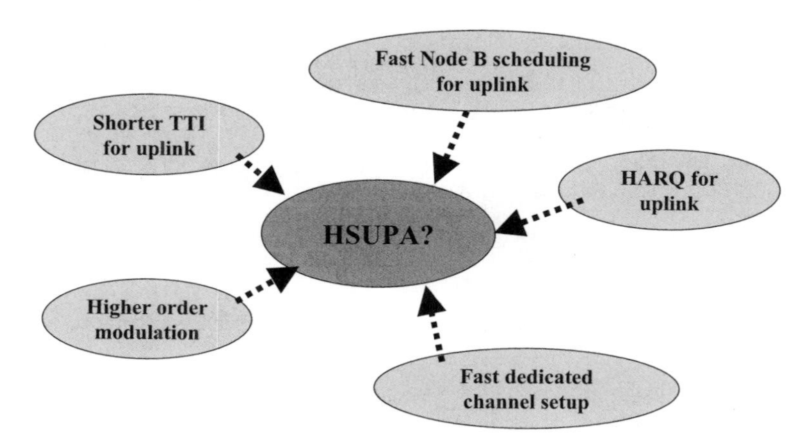


**Figure 2.3** Techniques investigated for HSUPA.

After long and detailed study work that produced the study item report [5], clear benefits were shown with the techniques investigated. The report showed no potential gains from using higher order modulation in the uplink direction, which resulted that adaptive modulation was not included in the actual work item.

As a conclusion, the study item, finalized in March 2004, recommended commencement of a work item in the 3GPP for specifying a fast physical layer HARQ and a Node B based scheduling mechanism for the uplink as well as a shorter uplink TTI length. Also, the faster setup mechanisms for DCHs were left outside the recommendation for a 3GPP work, but those issues were (partly) covered under different work items for 3GPP Release 6 specification, based on the findings during the study item phase.

3GPP started a work item titled '*FDD enhanced uplink*' to specify the HSUPA features according to the recommendations of the study report. The TDD part was not progressed with the same timing, but is being worked on for the Release 7 schedule.

Due to good and detailed background work that was carried out during the 18-month study phase, as well as not having major burden from the correction work with earlier releases, the work towards specifications was fast and the first version of the feature was introduced to the core specifications in December 2004. This version was not yet a final and complete one, but it did contain all the key functionality around which the more detailed finalization and correction work continued.

In March 2005 the work item was officially completed for the functional specifications, which meant that the feature was moved to correction and maintenance. During the rest of 2005 the open issues as well as performance requirements were finalized. The 3GPP standardization process with HSUPA as an example is shown in Figure 2.5. The final step with HSUPA is completion of protocol backward compatibility, which will enable establishment of a release baseline for devices to be introduced in the market. This is scheduled to take place in March 2006 when the ASN.1 review is scheduled to be finalized (ASN.1 is the signalling protocol message-encoding language used in 3GPP in several protocols).

Completion of backward compatibility impacts how corrections are made. When such a functionality is not totally broken, the correction to the protocol message is done in

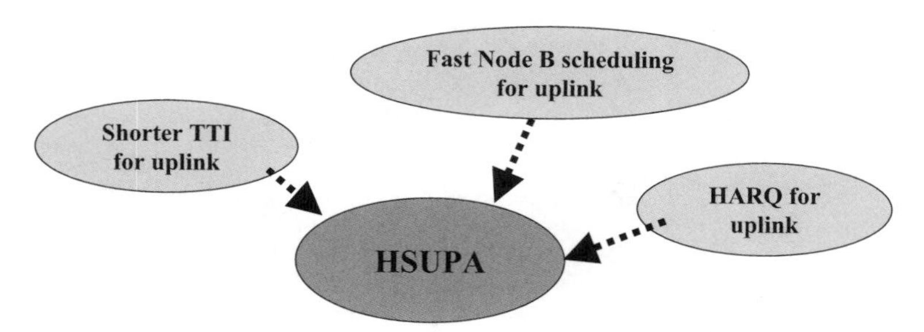


**Figure 2.4** The techniques chosen for the HSUPA work item.

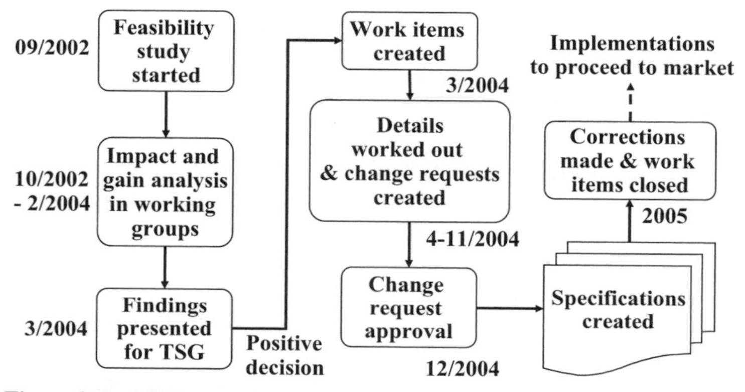


**Figure 2.5** 3GPP standardization process with HSUPA as an example.

such a way that the original message is kept and protocol extensions are used instead. This enables terminals based on the baseline version to operate though they do not obviously react to the correction/additions in the message extension. If the error was such that no proper operation is possible then the correction is non-backward compatible and needs to be taken into account when preparing products for the market. Thus, typically, terminals are based on some baseline version but accommodate critical corrections that have been done later as well, on top of the baseline specification version.

### 2.1.3 *Further development of HSUPA and HSDPA*

While HSUPA was being specified, there were developments for improving Release 6 HSDPA as well in several areas, such as:

- performance specifications for more advanced terminals with receiver diversity and/or advanced receivers, as covered later in Chapter 11;
- improved uplink range with optimized uplink feedback signalling, as covered in Chapter 4;
- improvements in the area of HSDPA mobility by faster signalling and shorter processing times, as described in Chapter 4.

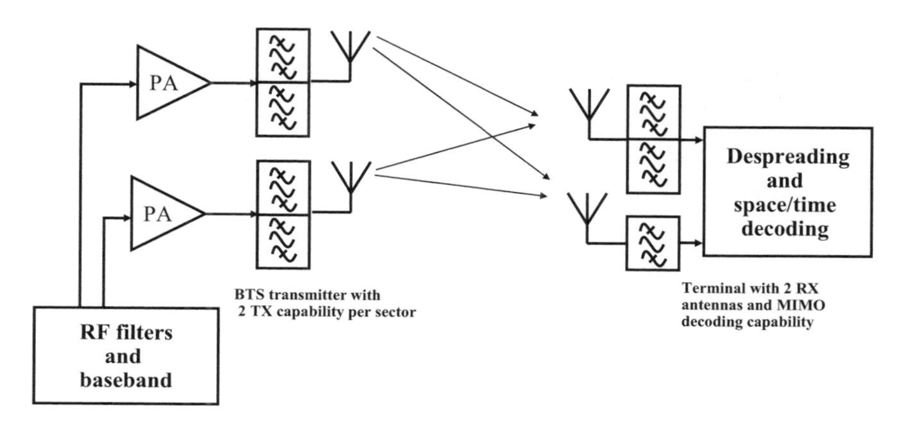


**Figure 2.6** MIMO principle with 2 transmit and 2 receive antennas.

For Release 7 a work item has been defined, titled '*continuous connectivity for packet data users*', aiming for reduced overhead during services that require maintaining the link but do not have the necessary continuous data flow. An example of such a service would be the packet-based voice service, often known as 'Voice over IP' (VoIP). What is being worked with currently can be found in [6], but conclusions are yet to be drawn as to what actually will be included in Release 7 specifications.

The MIMO work item is on-going as discussed previously, with current proposals captured in [7]. The key principle is to have two (or more) transmit antennas with at least partly different information streams and then to use two or more antennas and advanced signal processing in the terminal to separate the different sub-streams, with the principle illustrated in Figure 2.6.

The key challenge is to demonstrate whether there is a sufficient incremental gain still available when taking into account the receiver performance improvements done for Release 6 or other existing alternatives to improve capacity by adding more transmitters – such as going from a three-sector to a six-sector configuration. The conclusions so far in 3GPP indicate that in the macro-cell environment HSDPA with MIMO does not seem to bring any capacity benefit over the case with receiver diversity and an advanced receiver at the terminal end. Thus, it remains to be seen whether there will be anything on MIMO in Release 7 or later releases. The study will look towards smaller (micro-cells), and it is then to be discussed further in 3GPP whether adding MIMO to the specifications is reasonable or not, with conclusions expected by mid-2006.

Additional on-going items relevant for HSDPA or HSUPA operation include a work item on reduced circuit-switched (CS) and packet-switched (PS) call setup delay, which is aiming to shorten the time it takes to move from the idle state to the active (Cell_DCH) state. As most of the steps in WCDMA will remain the same, regardless of whether one is setting up a CS or PS call, the improvements will benefit both HSDPA/HSUPA use as well as the normal speech call setup. The work has first focused on identifying how to improve setting up a Release 99 speech call and coming up as well with methods that could be applied even for existing devices. Now the focus has shifted to bigger adjustments that do not work with existing devices, but have more potential for improvements because changes to the terminals can also be done. This means that further improvements would be obtained with Release 7 capable devices in most cases. Further details of the solutions identified so far and those being studied can be found in [8].

### *2.1.4 Beyond HSDPA and HSUPA*

While WCDMA-based evolution will continue in 3GPP with further releases, something totally new has been started as well. 3GPP has started a feasibility study on the UMTS Terrestrial Radio Access Network (UTRAN) long-term evolution (LTE), to ensure the competitiveness of the 3GPP technology family for the long term as well. The work has been started with the following targets defined [9]:

- Radio network user plane latency below 5 ms with 5-MHz or higher spectrum allocation. With smaller spectrum allocation below, latency below 10 ms should be facilitated.
- Reduced control plane latency.
- Scalable bandwidth up to 20 MHz, with smaller bandwidths covering 1.25 MHz, 2.5 MHz, 5 MHz, 10 MHz and 15 MHz for narrow allocations.
- Downlink peak data rates up to 100 Mbps.
- Uplink peak data rates up to 50 Mbps.
- Two to three times the capacity of existing Release 6 reference scenarios with HSDPA or HSUPA.
- Improved end user data rates at the cell edge.
- Support for the PS domain only.

There are additional targets for the reduced operator system cost which are being taken into account in the system architecture discussion. Naturally, inter-working with WCDMA/HSDPA/HSUPA and GSM/GPRS/EDGE is part of the system design, to prevent degradation of end user experience when moving between the system and ensuring moving to WCDMA or GSM when meeting the end of coverage area.

The work in 3GPP has already progressed, especially in the area of multiple access selection. The decision was reached about how to achieve multiple access. Further work for long-term evolution is based on pursuing single-carrier frequency division multiple access (SC-FDMA) for the uplink transmission and orthogonal frequency division multiplexing (OFDM) in the downlink direction.

SC-FDMA means having frequency resource allocated to one user at a time (sharing is also possible) which achieves similar orthogonal uplink as in the OFDM principle. The reason for not taking OFDM in the uplink as well was the resulting poor signal waveform properties for high-performance terminal amplifiers. With SC-FDMA there is a cyclic prefix added to the transmission to enable low-complexity frequency domain equalization in the base station receiver. The example transmitter/receiver chain in Figure 2.7 is just one possible way of generating an SC-FDMA signal, with the exact details still to be decided in 3GPP by 2007.

The OFDM is known on local area technologies as the 'wireless local area network' WLAN and on the broadcast systems as 'digital audio broadcasting' (DAB) or as a different version of the digital video broadcasting (DVB) system, with one of them

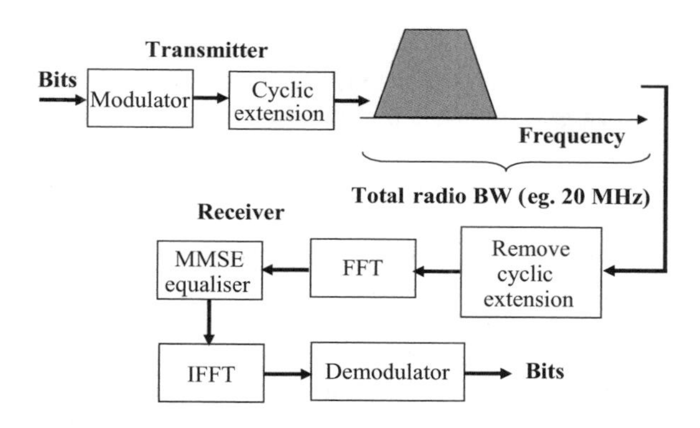


Figure 2.7 SC-FDMA principle.

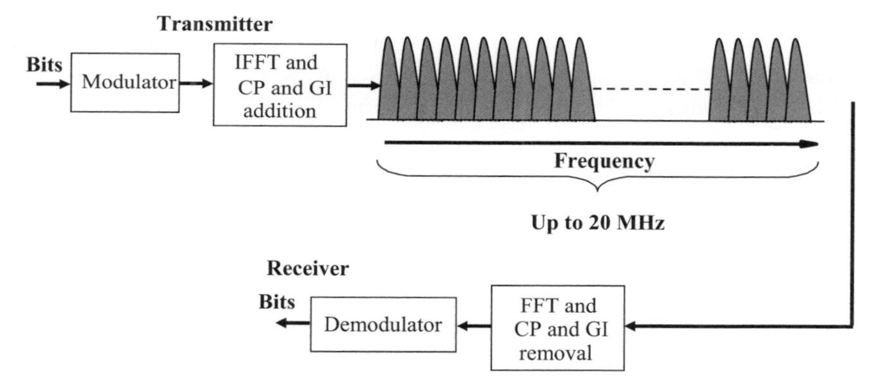


Figure 2.8 Classical OFDM principle with cyclic prefix (CP) and guard interval (GI).

(DVB-H) intended for mobile use [10]. Each user is allocated a set of sub-carriers, with the OFDM principle as shown in Figure 2.8. Unlike with SC-FDMA, the Inverse Fast Fourier Transform (IFFT) is now based at the transmitter end. With this structure there is the resulting situation that parallel sub-carriers carry different information, thus the envelope characteristics of the signal suffer from that. This is bad for terminal power amplifiers and was the key motivation not to adopt OFDM for the uplink direction.

The motivation for new radio access is, on the other hand, the long-term capacity needs and, on the other hand, the resulting complexity of implementing high data rates up to 100 Mbps. Of the proposals considered, multi-carrier WCDMA was seen as too complicated especially from the terminal point of view. On the other hand, frequency flexibility, as shown in Figure 2.9, enabled with the new access technology, is attaining a

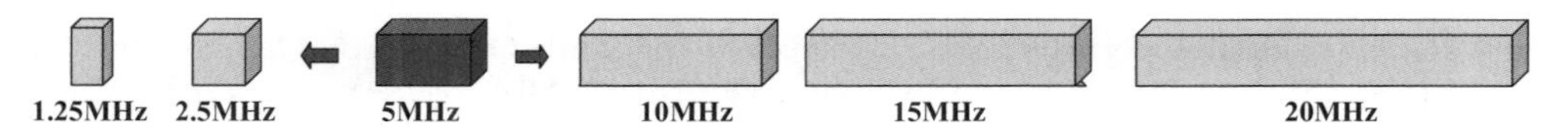


Figure 2.9 Spectrum flexibility with evolved UTRAN.

lot of interest when considering the possibility of re-farming lower frequency bands in the future. While this chapter was being written, the work in 3GPP entailed the feasibility study phase, but the work item phase is planned to start in mid-2006 and the first set of specifications, expected to be part of Release 8, should be available during 2007.

The other interesting area of the evolution is the network architecture. The trend seems to be to go towards such an architecture where there are fewer nodes involved for user plane processing compared with the current functionality. To achieve low latency it has been agreed to go for a two-node approach, where one is naturally equivalent to the base station and the other, upper, node handles core network functionalities and possibly some radio aspects as well. As radio does not support uplink or downlink macro-diversity (soft handover) there are some additional degrees of freedom in the architecture for the control plane, though macro-diversity has been applied in some systems with very flat architecture as well. Also, more control is being suggested to be moved to the base station, to make the base station even more intelligent than with HSDPA and HSUPA, adding more radio-related signalling responsibilities for the base station, even radio resource control (RRC) signalling. Until actual specifications start to become available some time in 2007 more details about the physical layer aspects can be obtained from [11] and on the protocol issues from [12].

## 2.2 References

[1] *www.3gpp.org*

[2] RP-000032, Work Item, Description Sheet for High Speed Downlink Packet Access, 13–15 March 2000, 3GPP TSG RAN7, Madrid, Spain. Available at *www.3gpp.org*

[3] 3GPP Technical Report, TR 25.848, Technical Specification Group RAN: Physical Layer Aspects of UTRA High Speed Downlink Packet Access (Release 4) 3GPP TR 25.848, Version 4.0.0, March 2001. Available at *www.3gpp.org*

[4] RP-010262, Proposal for Rel-5 Work Item on HSDPA, 13–16 March 2001, 3GPP TSG RAN11, Palm Springs, California, USA. Available at *www.3gpp.org*

[5] 3GPP Technical Report, TR 25.896, Technical Specification Group RAN: Feasibility Study for Enhanced Uplink for UTRA FDD, Release 6, Version 6.0.0, March 2004. Available at *www.3gpp.org*

[6] 3GPP Technical Report, TR 25.903, Technical Specification Group RAN: Continuous Connectivity for Packet Data Users, Release 7, Version 0.2.0, November, 2005. Available at *www.3gpp.org*

[7] 3GPP Technical Report, TR 25.876, Technical Specification Group RAN: Multiple-Input Multiple-Output in UTRA, Version 1.7.1, October 2005. Available at *www.3gpp.org*

[8] 3GPP Technical Report, TR 25.815, 3GPP Technical Specification Group RAN: Signaling enhancements for Circuit-Switched (CS) and Packet-Switched (PS) Connections; Analyses and Recommendations, Release 7, Version 0.4.2, November 2005. Available at *www.3gpp.org*

[9] 3GPP Technical Report, TR 25.913, Technical Specification Group RAN: Requirements for Evolved UTRA (E-UTRA) and Evolved UTRAN (E-UTRAN), Release 7, Version 2.1.0, September 2005. Available at *www.3gpp.org*

[10] *www.dvb-h-online.org*

[11] 3GPP Technical Report, TR 25.814, Technical Specification Group RAN: Physical Layer Aspects for Evolved UTRA (Release 7). Available at *www.3gpp.org*
[12] 3GPP Technical Report, TR 25.813, Technical Specification Group RAN: Evolved Universal Terrestrial Radio Access (UTRA) and Universal Terrestrial Radio Access Network (UTRAN); Radio interface protocol aspects. Available at *www.3gpp.org*

# 3

# HSPA architecture and protocols

Antti Toskala and Juho Pirskanen

This chapter covers the high-speed downlink packet access (HSDPA) and high-speed uplink packet access (HSUPA) impacts on the radio network and protocol architecture as well as on network element functionalities and interfaces. At the end of the chapter the radio resource control (RRC) states are covered.

## 3.1 Radio resource management architecture

The radio resource management (RRM) functionality with HSDPA and HSUPA has experienced changes compared with Release 99. In Release 99 the scheduling control was purely based in the radio network controller (RNC) while in the base station (BTS or Node B in 3GPP terminology) there was mainly a power control related functionality (fast closed loop power control). In Release 99 if there were two RNCs involved for connection, the scheduling was distributed. The serving RNC (SRNC) – the one being connected to the core network for that connection – would handle the scheduling for dedicated channels (DCHs) and the one actually being connected to the base transceiver station (BTS) would handle the common channel (like FACH). Release 99's RRM distribution is shown in Figure 3.1, and is covered in more detail in [1].

As scheduling has been moved to the BTS, there is now a change in the overall RRM architecture. The SRNC will still retain control of handovers and is the one which will decide the suitable mapping for quality of service (QoS) parameters. With HSDPA the situation is simplified in the sense that as there are no soft handovers for HSDPA data, then there is no need to run user data over multiple Iub and Iur interfaces and, even though HSDPA is supported over Iur in the specifications, the utilization of the Iur interface can be completely avoided by performing SRNC relocation, when the serving high-speed downlink shared channel (HS-DSCH) cell is under a different controlling RNC (CRNC). With Release 99 this cannot be avoided at RNC area boundaries when soft handover is used between two base stations under different RNCs. Thus, the typical HSDPA scenario could be presented by just showing a single RNC. Please note that the

*HSDPA/HSUPA for UMTS* Edited by Harri Holma and Antti Toskala

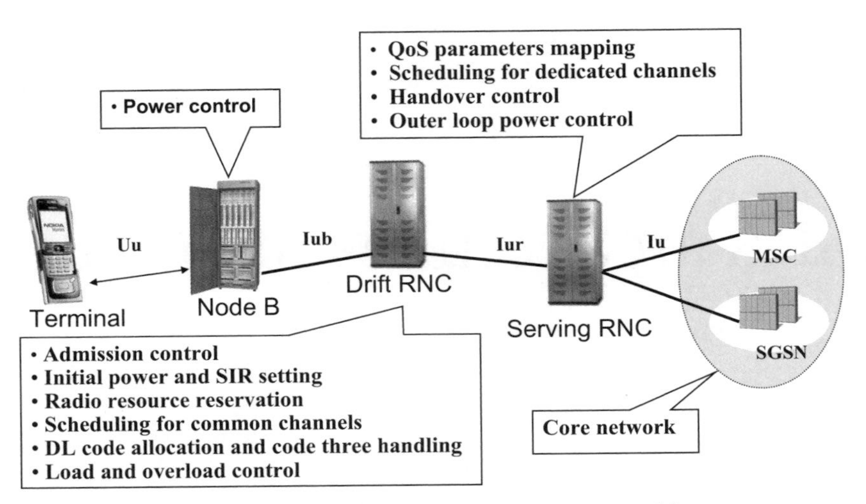


**Figure 3.1** Release 99 radio resource management architecture.

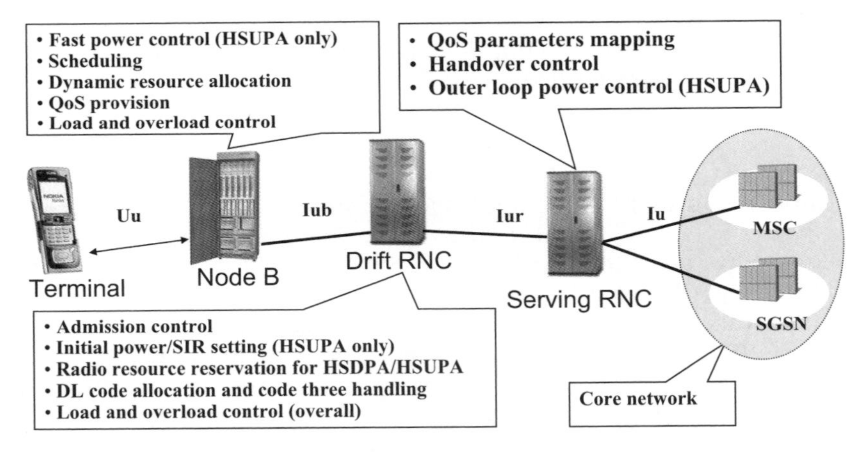


**Figure 3.2** HSDPA and HSUPA RRM architecture in Release 6.

associated DCH may still be in soft handover, while there is always only one serving cell for HSDPA use.

### *3.1.1 HSDPA and HSUPA user plane protocol architecture*

The basic functionality of the different protocol layers is valid with HSDPA and HSUPA and is similar to Release 99. The architecture can be defined as the user plane part, handling user data, and the control plane part. The RRC layer in the control plane part handles all the signalling related to configuring the channels, mobility management, etc. that is hidden from the end user, as is shown in the overall protocol architecture in Figure 3.3.

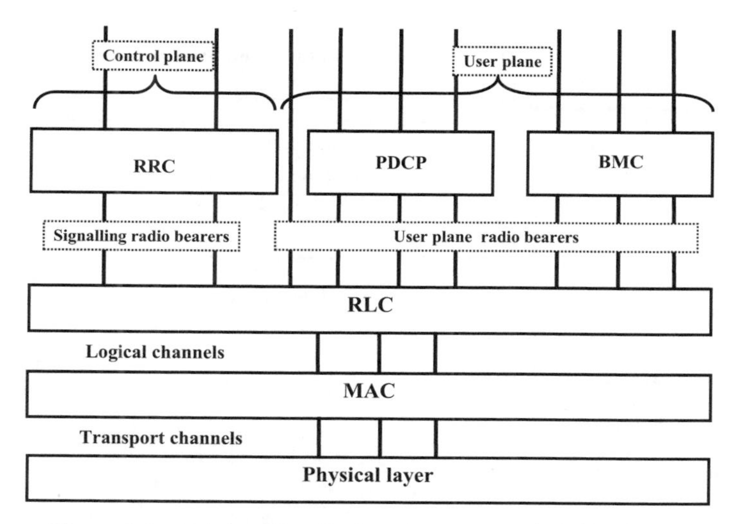


**Figure 3.3** Release 99 radio interface protocol architecture.

The Packet Data Convergence Protocol (PDCP) has as its main functionality header compression which is not relevant for circuit-switched services. The importance of header compression, as discussed in connection with Chapter 10, is easy to understand since a non-compressed Internet Protocol (IP) header can be two to three times the size of the actual voice packet payload itself.

Radio link control (RLC) handles the segmentation and retransmission for both the user and control data. RLC may be operated on three different modes:

- Transparent mode, when no overhead is added due to the RLC layer, such as with AMR speech, and is not applicable when the transport channels of HSDPA and HSUPA are used.
- Unacknowledged mode, when no RLC layer retransmission will take place. This is used with applications that can tolerate some packet loss, as is the case with VoIP, and cannot allow delay variation due to RLC level retransmission.
- Acknowledged mode operation, when data delivery is ensured with RLC layer retransmissions with applications that require all packets to be delivered.

The medium access control (MAC) layer in Release 99 focuses on mapping between the logical channels and handling the priorities, as well as selection of the data rates being used – i.e., selection of the transport format (TF) being applied. Transport channel switching is also a MAC layer functionality.

Both HSDPA and HSUPA introduce new elements in the architecture. The MAC layer functionalities for HSDPA and HSUPA can operate independently of Release 99's DCH operation, but takes into account the overall resource limitations of the air interface. Figure 3.4 illustrates the overall radio interface architecture for HSDPA and HSUPA user data, highlighting the new protocol entities dealing with user data. Control plane signalling – omitted in Figure 3.4 – would simply connect to the RLC and then have

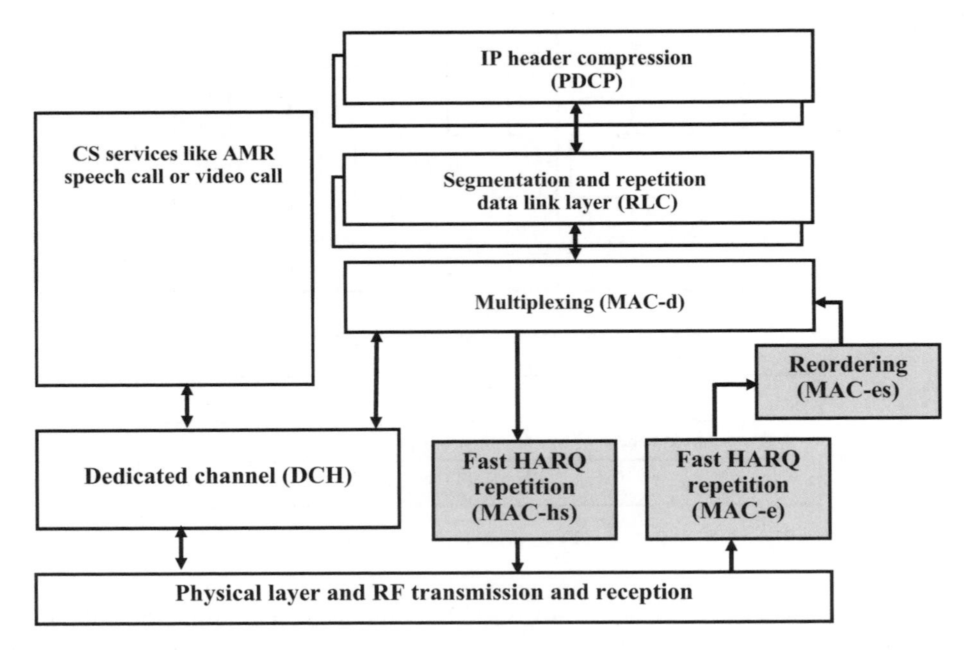


**Figure 3.4** HSDPA and HSUPA radio interface architecture for the user data.

the signalling carried either over the DCH or over the HSDPA/HSUPA. For user data, PDCP handles IP header compression. There are several PDCP and RLC entities shown in the figure to indicate the possibility of running parallel services.

The BTS-based (fast) scheduling functionality is a MAC layer functionality, and thus there is now a new protocol entity, MAC-hs (hs for high speed), in the BTS. This is shown as part of the user plane protocol architecture in Figure 3.5, which covers HSDPA-specific additions and their location in the network elements. The RNC retains the MAC-d (d for dedicated), but the only remaining functionality is transport channel switching as all other functionalities, such as scheduling and priority handling, are moved to MAC-hs. It is worth noting that the layer above the MAC layer – namely, the RLC layer – stays mainly unchanged, but some optimizations for RT services such as VoIP

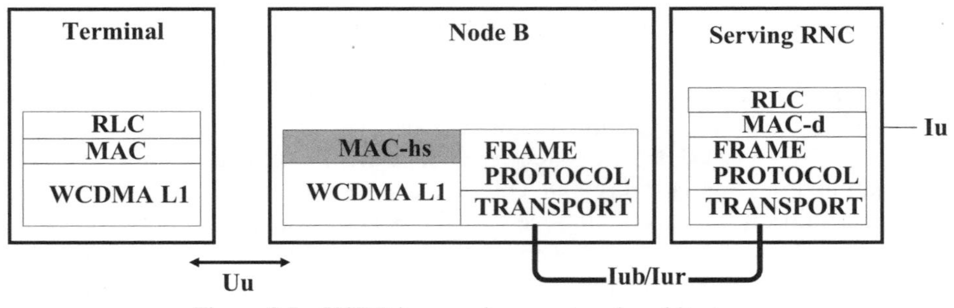


**Figure 3.5** HSDPA user plane protocol architecture.

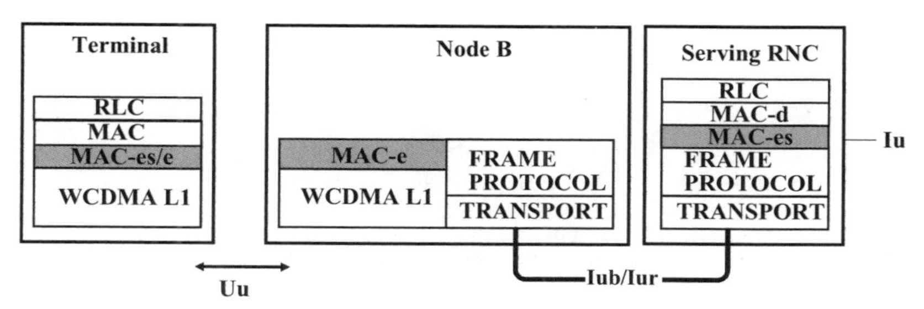


**Figure 3.6** HSUPA user plane protocol architecture.

were introduced for unacknowledged mode (UM) RLC in Release 6. As covered in the introduction, even if HSDPA introduced physical layer retransmissions, there is still the RLC layer handling the retransmissions should physical layer operation fail or, especially, in connection with the different mobility events like serving HS-DSCH cell change. This is assuming acknowledged mode (AM) RLC operation. In the case of UM-RLC, physical year retransmissions are the only ones available. An example would be a VoIP call where the RLC layer retransmissions from the RNC would be too slow.

With HSUPA there is in like manner a new MAC entity added to the BTS, as shown in Figure 3.6. This is, however, not the only place where additions were made to the protocol architecture. The terminal has a new MAC entity as well (MAC-es/s), to cover the fact that part of the scheduling functionality is now moved to Node B, though it is based on control information from the RNC and a direct capacity request from the user equipment (UE) to the Node B. There is a new protocol entity for the RNC as well. This is due to HSUPA soft handover and the fact that the physical layer retransmission introduced in HSUPA has the effect of placing packets out of order. When data are received in several BTS sites, there is a possibility when operating in the soft handover region that packets from different BTSs arrive such that the order of packets is not retained, and to allow reordering to be done for single packet streams the reordering functionality needs to be combined with macro-diversity combining in MAC-es. Thus, the new MAC-es 'in-sequence delivery' functionality has as its main task to ensure that for the layers above the packets are provided in the order they were transmitted from the terminal. Should such ordering be handled at the BTS, then an unnecessary delay would be introduced as the BTS would have to wait for missing packets until they could be determined to have been correctly received by another BTS in the active set. Further details of the MAC layer architecture can be found in [2].

In like manner to HSDPA, the RLC layer in HSUPA is involved with the retransmission of packets if the physical layer fails to correctly deliver them after the maximum number of retransmissions is exceeded or in connection with mobility events.

### *3.1.2 Impact of HSDPA and HSUPA on UTRAN interfaces*

While the impacts of HSDPA and HSUPA in terms of data rates over the air interface are well known and often the focus of discussion, the impact on the operation of the other interfaces requires attention as well. For the interface between the base station and RNC, the Iub interface, there are now larger data rates expected than with Release 99 terminals.

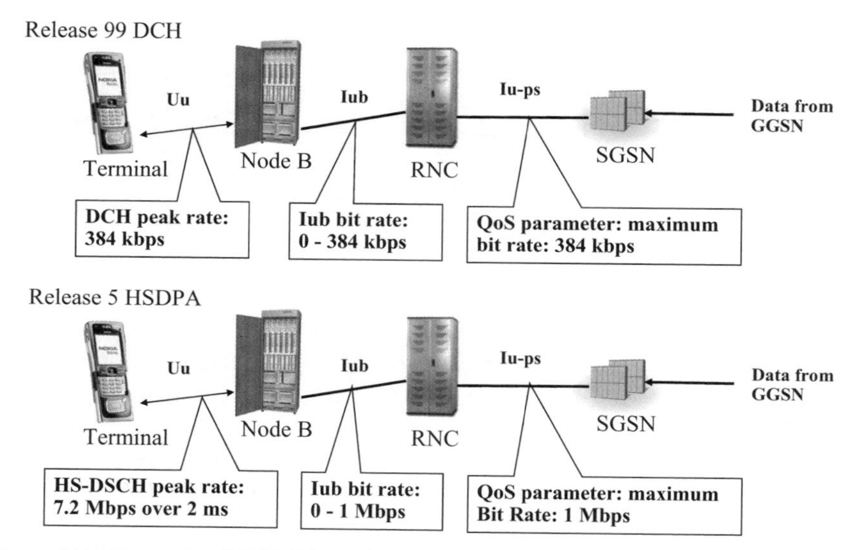


**Figure 3.7** Example of HSDPA and Release 99 data rates on different interfaces.

While Release 99 terminals are at most 384 kbps, the data rate on different interfaces, including the Iu-ps interface to the packet core network Serving GPRS Support Node (SGSN), is equal to the one used for radio. With HSDPA the situation has changed, however. Over the air interface, denoted as Uu in Third Generation Partnership Project (3GPP) terminology, there can be data rates up to 14.4 Mbps over short (2-ms) periods. This, however, does not mean the same data rate being used on the Iub and Iu-ps interface for that particular user. From a single user point of view, the radio resource is time and code shared with other users in the cell. Thus, the average bit rate for a user in a loaded cell is clearly lower. Further, the peak rates of 10 Mbps require extremely favourable radio conditions and it is unlikely that all the users in the cell would be able to get such a high data rate. Thus, the average traffic on the Iub interface is also less than the peak rate for radio. An example is illustrated in Figure 3.7, which shows as a comparison the Release 99 384-kbps downlink and HSDPA for the 7.2-Mbps case. The 384-kbps downlink will have equal data rate booking on all interfaces and will not exceed the 384-kbps limit. With HSDPA the radio interface peak rate supported by the terminal in the example is 7.2 Mbps. The service data rate over the Iu-ps and Iub interface could be limited, for example, to 1 Mbps. The use of buffering in the BTS makes it possible to have the peak rate for the connection as high as terminal and BTS capabilities allow, while keeping the maximum bit rate over the Iub and Iu-ps in line with the QoS parameters received from the packet core.

The buffer in the BTS – together with a scheduler that time-shares the resources – enables having a higher peak rate (over a short period of time) for radio than the average rate on Iub/Iu-ps. Having the transmission buffer in the BTS also requires flow control to be applied to avoid buffer overflow. In this way the user under better radio conditions may get more of the Iub resources, as the situation exists to get a lot of data through. The principle of flow control operation is shown in Figure 3.8. The user under good radio

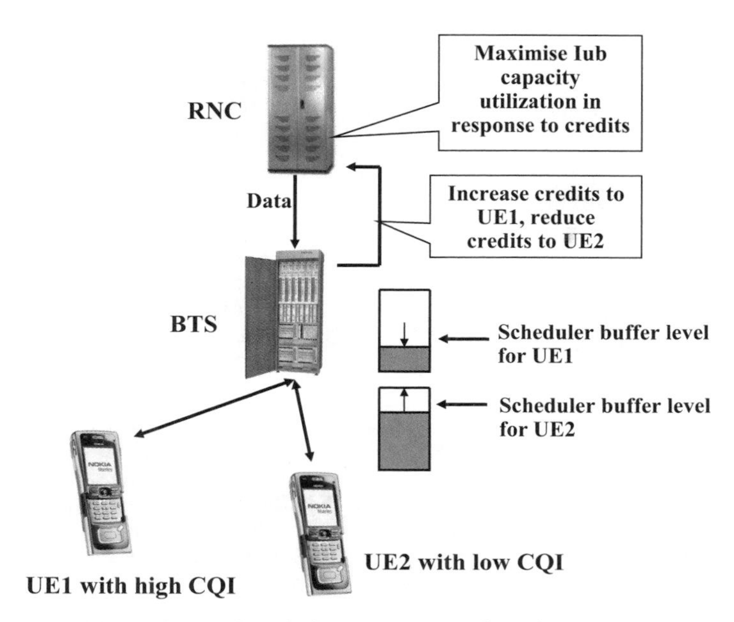


**Figure 3.8** HSDPA flow control on the Iub interface.

conditions will get more Iub allocation (credits) as the data move fast from the radio point of view. On the other hand, when the buffer starts to get filled due to poor radio conditions (and resulting low data rate) the flow control will slow the data stream down for that kind of user, as shown in Figure 3.8.

For the network elements and the terminal itself the use of HSDPA and HSUPA will cause a lot of changes, especially on the physical layer as well as on the MAC and RLC layers. From the RNC point of view there is an impact which is due not only to the already mentioned flow control but also, on the other hand, to the changes in RRM. Mobility events cause some changes in addition to the opportunity to share Iub resources dynamically between all users. Further, the data rates increase from the practical maximum of 384 kbps for Release 99 devices up to 10 Mbps and, theoretically, even up to 14 Mbps.

For the BTS and the terminal, the key changes with HSDPA are related to the addition of scheduling and retransmission functionalities at the BTS and, respectively, the needed packet combining functionality at the terminal. Additional channels are then needed for actual data transmission itself, including new modulation, as well as for signalling purposes to facilitate the new functionality. The key new functionalites due to HSDPA are illustrated in Figure 3.9. Further details of the new functionalities are covered in Chapter 4.

Similarly, with HSUPA new functionalities are needed as shown in Figure 3.10. Here, scheduling is controlled from the BTS and there are needs for data flow handling in the other (uplink direction) as well as new signalling channels for HSUPA. The combining functionality is now at the base station and the new re-ordering functionality is added to the RNC. Uplink scheduling is required in addition to downlink scheduling for HSDPA.

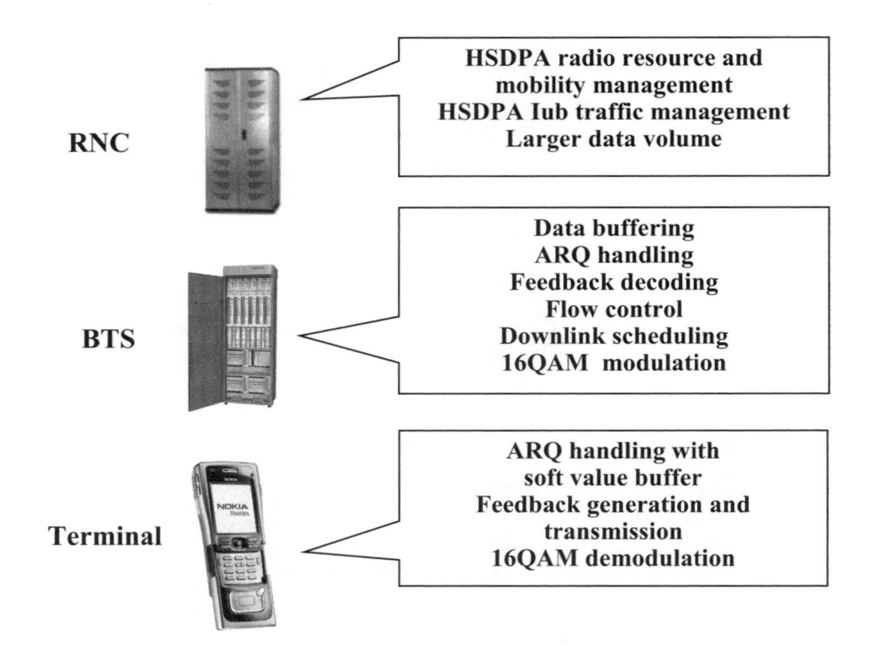


**Figure 3.9** New functionality on different elements due to HSDPA.

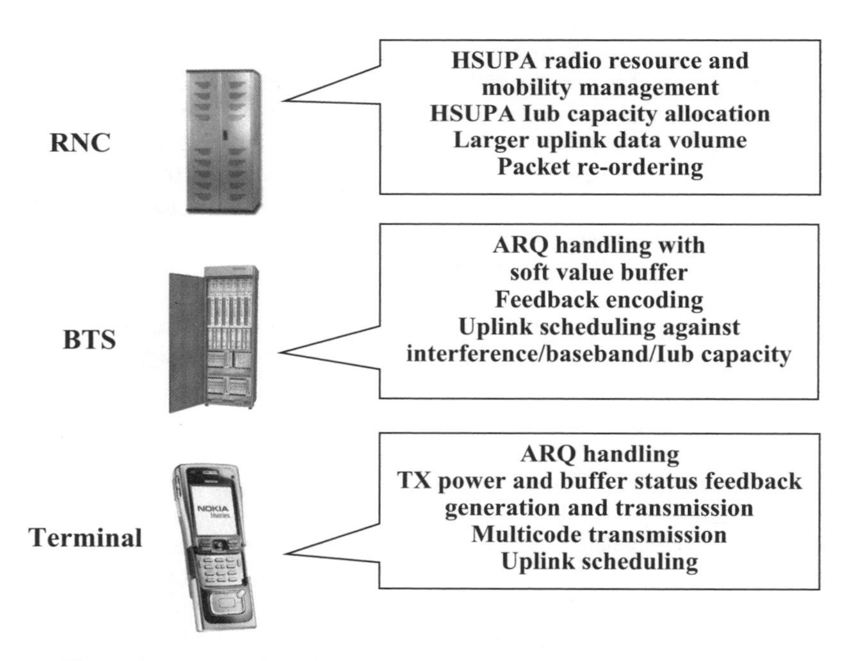


**Figure 3.10** New functionality on different elements due to HSUPA.

The terminal will now be able to accept multi-code transmission, which was the case in Release 99 as well, but not implemented in the actual market place. Further details of HSUPA-related new functionalities are covered in Chapter 5.

### *3.1.3 Protocol states with HSDPA and HSUPA*

RRC states [1], [3] are the same with HSDPA and HSUPA as in Release 99. The Cell_DCH is the state which is used when active data transmission to and from the terminal on the DCH or HSDPA/HSUPA is possible. From Cell_DCH state the terminal will be moved to Cell_FACH or further states, either directly from Cell_DCH state or via Cell_FACH state if there are no data in the buffers. This happens depending on the network timer settings after some seconds. There needs to be an obvious trade-off with network response time for the first packet transmission instant after idle period and the timer valve applied. The transitions take time due to the reconfiguration/setup processes needed. Keeping a user booking HSDPA/HSUPA resource when there are no data to transfer is not efficient from the system capacity or from the BTS resource use point of view. Terminal battery life would also suffer, as keeping a terminal active with no data passing through it will cause the battery to run out quickly.

Data can be transmitted in Cell_FACH state as well, but only using the forward access channel (FACH) for the downlink and the random access channel (RACH) in the uplink, which means limited data rates as these channels do not offer any of the performance enhancement features of HSDPA and HSUPA. A terminal in Cell_FACH state continuously decodes FACH channels and then initiates the response downlink data (or initiates transmission of data due in the uplink buffer) on the RACH. Depending on the data volume (based on the reporting), the terminal may be moved back to Cell_DCH state. As estimated in Chapter 10, the terminal battery power consumption ratio between Cell_FACH state and Cell_DCH state is approximately 1 to 2, thus Cell_FACH state should be avoided for lengthy time periods as well.

If the idle period in data transmission continues for lengthy periods, it makes sense to move the terminal onwards to Cell_PCH or URA_PCH state. These states are the most efficient ones from the terminal battery consumption point of view. Naturally, the use of the discontinuous reception (DRX) cycle with the paging operation will cause some additional delay to resume actual data transmission as the terminal needs to be paged first. This happens in either one cell (location known at cell level in Cell_PCH state) or in the registration area level (URA_PCH state). This is the benefit of the URA_PCH state for terminals with high mobility. Terminals moving fast in a dense network would necessitate a lot of cell updates, increasing the RACH load, but would then need to be paged in multiple cells in case of downlink-originated activity. RRC states are illustrated in Figure 3.11.

In idle mode there are no HSDPA/HSUPA-specific issues, the operation again uses the DCH. Release 6 specifications indicate the position in the broadcast information at which the network will inform whether there is HSUPA/HSDPA available, but this is only for user information and not to influence the existing network and cell selection behaviour. This has been defined in such a way in the specifications that early implementation (on top of Release 5) is possible as well. Obviously, the RRM reason will dictate whether a user gets allocated HSDPA/HSUPA or not. Similarly to GPRS, it is

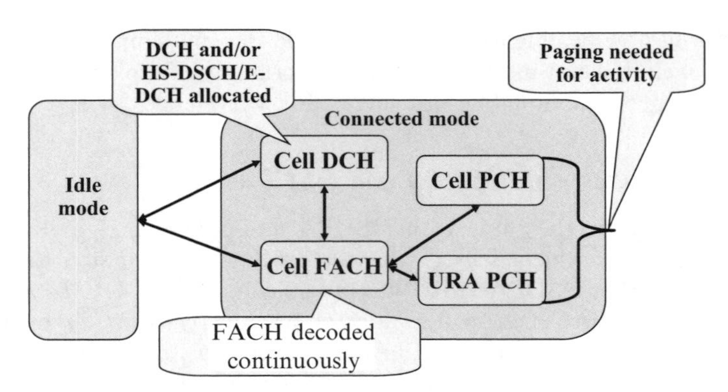


**Figure 3.11** RRC states with HSDPA/HSUPA.

unknown before actual connection establishment how many slots – i.e., what practical data rates – are available.

## 3.2 References

[1] H. Holma and A. Toskala (eds) (2004). *WCDMA for UMTS* (3rd edn). John Wiley & Sons.
[2] 3GPP, Technical Specification Group RAN: Medium Access Control (MAC) Protocol Specification, 3GPP, TS 25.321, Version 6.7.0, Release 6. Available at *www.3gpp.org*
[3] 3GPP, Technical Specification Group RAN: Radio Resource Control (RRC) Protocol Specification, 3GPP, TS 25.331, Version 6.8.0, Release 6. Available at *www.3gpp.org*

# 4

# HSDPA principles

Juho Pirskanen and Antti Toskala

This chapter covers high-speed downlink packet access (HSDPA) principles for wideband code division multiple access (WCDMA) – the key new feature included in Release 5 specifications and enhanced further in Release 6 specifications. This chapter focuses on describing how HSDPA works and what features are included in Release 5 and Release 6 specifications. HSDPA has been designed to increase downlink packet data throughput by means of fast physical layer (L1) retransmission and transmission combining as well as fast link adaptation controlled by the Node B – i.e., base transceiver station (BTS). The specification work has resulted in quite different structures and solutions when compared with Release 99 based WCDMA. This chapter is organized as follows. First, HSDPA key aspects are presented and a comparison with Release 99 dedicated channel (DCH)-based downlink packet access is made. Second, the different transport and physical channels needed to actually implement HSDPA functionality are covered. Third, the functionalities that are the basis of HSDPA operation, retransmission and link adaptation are covered in detail. The chapter then continues with the mobility principles and moves on to the terminal capability classes. Toward the end, the chapter addresses the theoretical data rates achievable with HSDPA on different layers and the HSDPA-specific parameters on the interface between the radio network controller (RNC) and BTS – Iub – are covered. The chapter concludes with the HSDPA medium access control (MAC) layer description.

## 4.1 HSDPA vs Release 99 DCH

In Release 99 there basically exists in the specifications three different methods for downlink packet data operation: DCH, forward access channel (FACH) and downlink shared channel (DSCH). Since the DSCH has been *de facto* replaced with the high-speed DSCH of HSDPA, it is not covered in more depth here. This has been recognized in the Third Generation Partnership Project (3GPP) as well and the DSCH has been removed from the specifications from Release 5 onwards, simply due to the lack of interest for

*HSDPA/HSUPA for UMTS* Edited by Harri Holma and Antti Toskala

actual implementation. For those interested in the historical reasons for DSCH use please refer to [1] and the references therein for further details.

The FACH is used either for small data volumes or when setting up the connection and during state transfers. In connection with HSDPA, the FACH is used to carry the signalling when the terminal has moved, due to inactivity from Cell_DCH state to Cell_FACH, Cell_PCH or URA_PCH state. The FACH is operated on its own and, depending on the state the terminal is in, FACH is decoded either continuously (Cell_FACH) or based on the paging message. For the FACH there is neither fast power control nor soft handover. For the specific case of the Multimedia Broadcast Multicast Service (MBMS), selection combining or soft combining may be used on the FACH, but in any case there is no physical layer feedback for any activity on the FACH and thus power control is not feasible. The FACH is carried on the secondary common control physical channel (S-CCPCH) with fixed spreading factor. If there is a need for service mix, like speech and data, then FACH cannot be used.

The Release 99 based DCH is the key part of the system – despite the introduction of HSDPA – and Release 5 HSDPA is always operated with the DCH running in parallel (as shown in Figure 4.1). If the service is only for packet data, then at least the signalling radio bearer (SRB) is carried on the DCH. In case the service is circuit-switched – like AMR speech call or video call parallel to PS data – then the service always runs on the DCH. With Release 6 signalling can also be carried without the DCH, as explained in connection with the fractional DCH (F-DCH). In Release 5, uplink user data always go on the DCH (when HSDPA is active), whereas in Release 6 an alternative is provided by the Enhanced DCH (E-DCH) with the introduction of high-speed uplink packet access (HSUPA), as covered in Chapter 5.

The DCH can be used for any kind of service and it has fixed a spreading factor in the downlink with fixed allocation during the connection. Thus, it reserves the code resources corresponding to the peak data rate of the connection. Of course, higher layer signalling could be used to reconfigure the code used for the DCH, but that is too slow to react to changes in the traffic. In the case of multiple services, the reserved capacity is equal to the sum of the peak data rate of the services. The theoretical peak rate is 2 Mbps but it seems that 384 kbps is the highest implemented DCH data rate on the market so far. Any

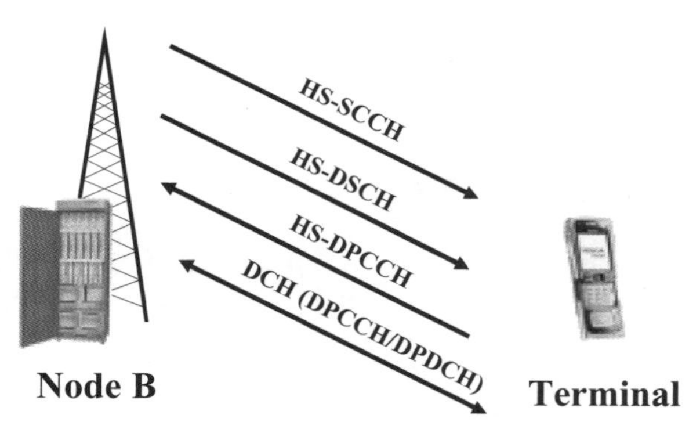


**Figure 4.1** Channels needed for HSDPA operation in Release 5.

Table 4.1 Comparison of fundamental properties of the DCH and HS-DSCH.

| *Feature* | *DCH* | *HS-DSCH* |
|---|---|---|
| Variable spreading factor | No | No |
| Fast power control | Yes | No |
| Adaptive modulation and coding | No | Yes |
| Multi-code operation | Yes | Yes, extended |
| Physical layer retransmissions | No | Yes |
| BTS-based scheduling and link adaptation | No | Yes |

scheduling and retransmission handling is done in the RNC. At the base station end, fast power control is the key functionality for the DCH in addition to encoding the data packet provided by the RNC. Soft handover (macro-diversity) is supported for the DCH, and in such a case all base stations need to transmit exactly the same content (except for power control commands) with identical timing to facilitate combining in the Rake receiver.

HSDPA's fundamental operation is based on the use of link adaptation, fast scheduling and physical layer retransmission. All these methods have the aim of improving downlink packet data performance both in terms of capacity and practical bit rates. HSDPA does not support DCH features like fast power control or soft handover (as summarized in Table 4.1).

## 4.2 Key technologies with HSDPA

Several new channels have been introduced for HSDPA operation. For user data there is the high-speed downlink shared channel (HS-DSCH) and the corresponding physical channel. For the associated signalling needs there are two channels: high-speed shared control channel (HS-SCCH) in the downlink and high-speed dedicated physical control channel (HS-DPCCH) in the uplink direction. In addition to the basic HSDPA channel covered in Release 5 specifications, there is now a new channel in Release 6 specifications – the fractional dedicated physical channel (F-DPCH) – to cover for operation when all downlink traffic is carried on the HS-DSCH. The channels needed for HSDPA operation are shown in Figure 4.1. The channels missing from Figure 4.1 are the broadcast channels from Release 99 for channel estimation, system information broadcast, cell search, paging message transmission, etc., as covered in [1] in detail.

The general HSDPA operation principle is shown in Figure 4.2, where the Node B estimates the channel quality of each active HSDPA user on the basis of the physical layer feedback received in the uplink. Scheduling and link adaptation are then conducted at a fast pace depending on the scheduling algorithm and the user prioritization scheme.

The other key new technology is physical layer retransmission. Whereas in Release 99 once data are not received correctly, there is a need for retransmission to be sent again from the RNC. In Release 99 there is no difference in physical layer operation, regardless of whether the packet is a retransmission or a new packet. With HSDPA the packet is first received in the buffer in the BTS (as illustrated in Figure 4.3). The BTS keeps the

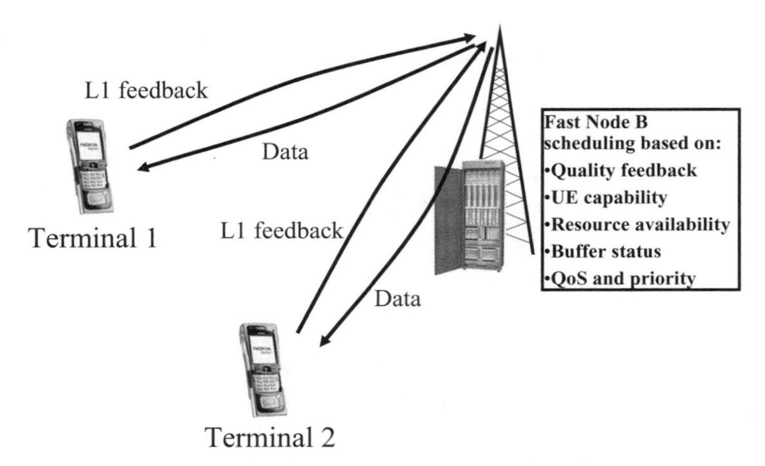


**Figure 4.2** HSDPA Node B scheduling principle.

packet in the buffer even if has sent it to the user and, in case of packet decoding failure, retransmission automatically takes places from the base station without RNC involvement. So, the terminal has respectively combined the transmissions, capturing the energy of both. Should physical layer operation fail due to, say, signalling error, then RNC-based retransmission may still be applied on top. As shown in the example in Figure 4.3, using a radio link control (RLC)-acknowledged mode of operation, RLC layer acknowledgement is provided in the RLC layer as would be done for Release 99 based operation. Now, the use of RLC retransmissions is not a very frequent event as, for example, with

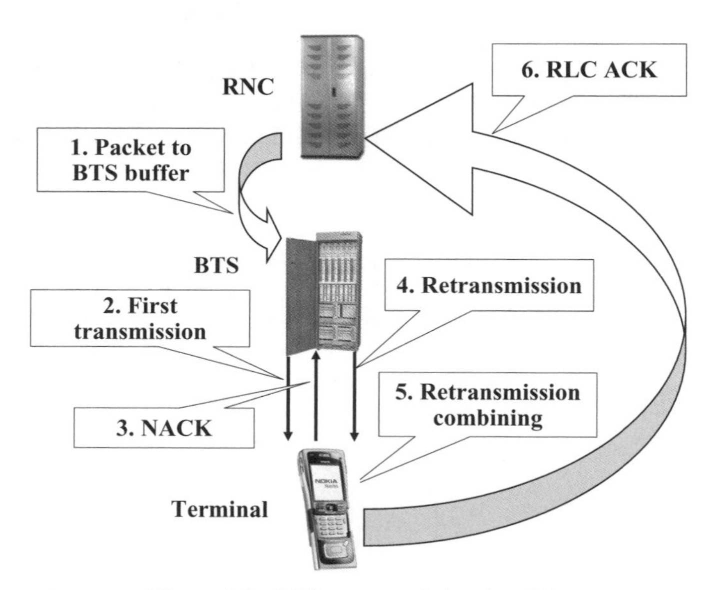


**Figure 4.3** BTS retransmission handling.

four retransmissions and an initial block error rate (BLER) of 10%, the probability for retransmissions should be less than 0.1%. RLC layer retransmissions are more likely in connection with mobility operations, as is discussed in the HSDPA mobility section.

### *4.2.1 High-speed downlink shared channel*

The HS-DSCH is the transport channel that carries the actual user data with HSDPA. In the physical layer the HS-DSCH is mapped on the high-speed physical downlink shared channel (HS-PDSCH). The key differences from the Release 99 DCH-based packet data operation are as follows:

- Lack of fast power control. Instead, link adaptation selects the suitable combination of codes, coding rates and modulation to be used.
- Support of higher order modulation than the DCH. With 16-quadrature amplitude modulation (16QAM) the number of bits carried per symbol is doubled in favourable conditions compared with the quadrature phase shift keying (QPSK) in Release 99.
- User allocation with base station based scheduling every 2 ms, with fast physical layer signalling. With DCH the higher layer signalling from the RNC allocates semi-permanent code (and a spreading factor) to be used. The transmission time interval (TTI) is also longer with the DCH, allowing values such as 10, 20, 40 or 80 ms. (The longest is limited in the specific case of small data rates that have a spreading factor of 512).
- Use of physical layer retransmissions and retransmission combining, while with the DCH – if retransmissions are used – they are based on RLC level retransmissions.
- Lack of soft handover. Data are sent from one serving HS-DSCH cell only.
- Lack of physical layer control information on the HS-PDSCH. This is carried instead on the HS-SCCH for HSDPA use and on the associated DCH (uplink power control, etc.)
- Multicode operation with a fixed spreading factor. Only spreading factor 16 is used, while with the DCH the spreading factor could be a static parameter between 4 and 512.
- With HSDPA only turbo-coding is used, while with the DCH convolutional coding may also be used.
- No discontinuous transmission (DTX) on the slot level. The HS-PDSCH is either fully transmitted or not transmitted at all during the 2-ms TTI.

An important property of the HS-DSCH is the dynamic nature of resource sharing enabled by the short 2-ms allocation period. When there are data for a user allocated on the HS-PDSCH, they are sent continuously during the 2-ms TTI. There is no discontinuous transmission (DTX) on the slot level as with the DCH, while code resource (recall allocation of a downlink code with a fixed spreading factor) with the DCH is partially filled with lower data rates. In line with Figure 4.4, the DTX positions in the slots (and divided over the TTI) do reduce the downlink interference generated but keep the code resource occupied according to the highest data rate possible on the DCH. For example, with 384 kbps downlink, the code resource reservation is not changed when moving to a lower data rate. Thus, if enabling a peak rate of 384 kbps, then as the application goes

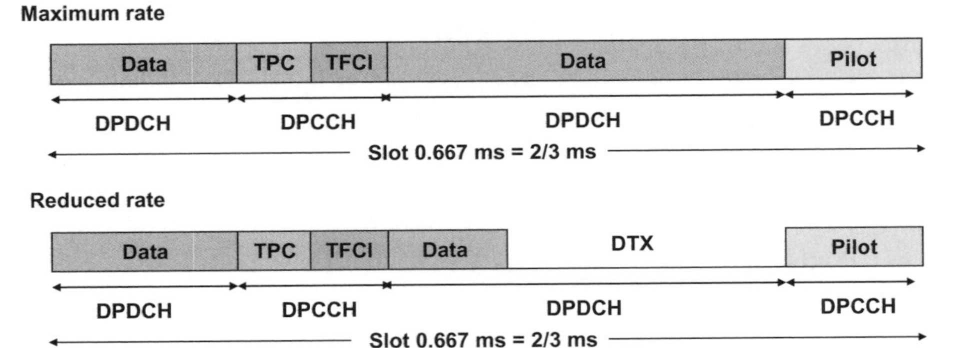


**Figure 4.4** Discontinuous downlink transmission with Release 99 DCH.

down to, say, 16 kbps, the only way to reduce resource consumption would be to reconfigure the radio link. This again takes time both when reconfiguring and 'locking' the data rate to a new smaller value until another reconfiguration would take place to upgrade the data rate again. With the HS-PDSCH, once there are no data to be transmitted, there is no transmission at all on the the HS-DSCH for the user in question but the resource for the 2 ms is allocated to another user instead.

The HS-PDSCH is always transmitted on connection with the HS-SCCH and, additionally, the terminal also always receives the DCH which carries services like circuit-switched AMR speech or video as well as the SRB. Release 6 enhancements enable operating with the SRB mapped on the HS-DSCH as well, as covered later in connection with the section on the fractional dedicated physical channel (DPCH). HS-PDSCH details are specified in [2].

#### 4.2.1.1 HS-DSCH coding

As covered in the introduction on the HS-DSCH, only turbo-coding is used for the HS-DSCH. This was motivated by the fact that turbo-coding outperforms convolutional coding otherwise expected with the very small data rates. The channel coding chain is simplified from the corresponding DCH one, as there is no need to handle issues like DTX or compressed mode with the HS-DSCH. Also, there is only one transport channel active at a time, thus fewer steps in multiplexing/de-multiplexing are needed. An additional new issue is the handling of 16QAM and the resulting varying number of bits carried by the physical channel even when the number of codes used remains fixed. Another new functionality is bit scrambling on the physical layer for the HS-DSCH. The HS-DSCH channel coding chain is illustrated in Figure 4.5.

For the 16QAM there is the specific function of constellation rearrangement, which maps the bits to different symbols depending on the transmission numbers. This is beneficial as with 16QAM all the symbols do not have equal error probability in the constellation. This is due to different symbols having different numbers of 'neighbouring' symbols which places the symbols closer to the axis, with a greater number of neighbour-

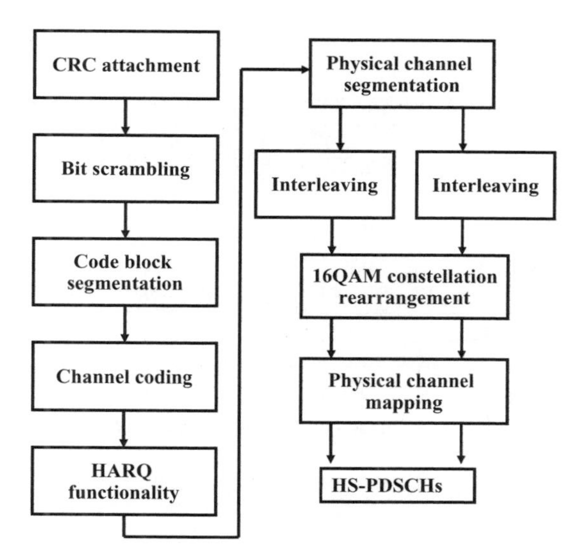


**Figure 4.5** HS-DSCH channel coding chain.

ing symbols (as in Figure 4.9) more likely to be decoded incorrectly than the other symbols further away from the axis.

The bit scrambling functionality in Figure 4.5 was introduced to avoid having sequences which end up repeating the same symbol (long sequences of '1s' or '0s'). These could occur with some type of content, and especially when not using ciphering at higher layers. In such a case the terminal would have difficulties with HS-DSCH power level estimation (see the 16QAM description given later) and, thus, the reason that the physical layer scrambling operation was introduced. Operation is the same for all users and is thus purely for ensuring good signal properties for demodulation.

The Hybrid-ARQ (HARQ) functionality shown as a single block in the coding chain in Figure 4.5 can be further split into different elements (as shown in Figure 4.6). The HARQ functionality consists of a two-stage rate matching functionality which allows

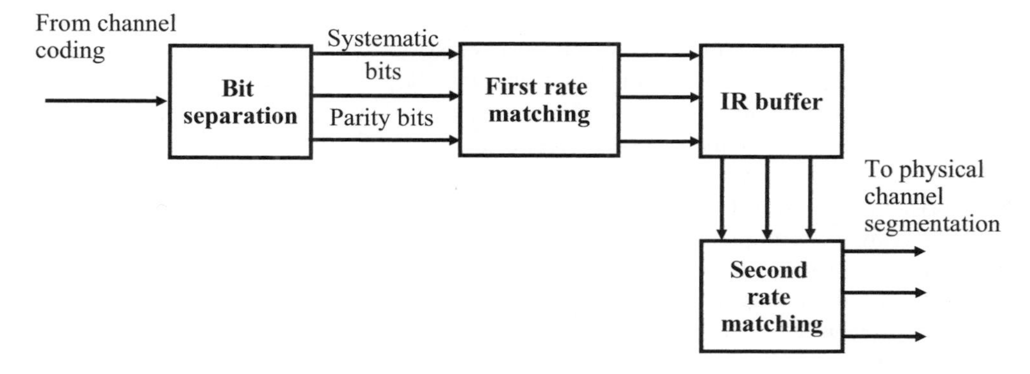


**Figure 4.6** HARQ functionality.

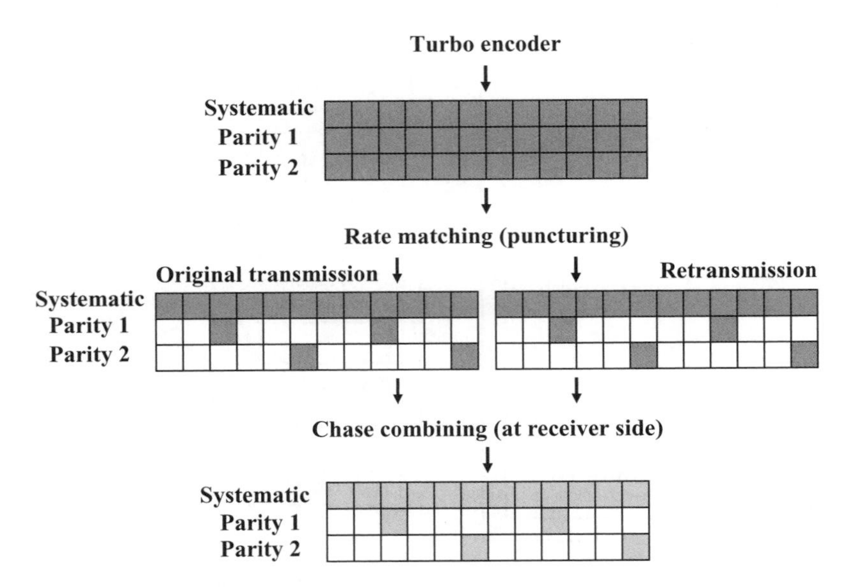


**Figure 4.7** Chase (soft) combining principle with identical retransmissions.

tuning the redundancy version of different retransmissions when using non-identical retransmissions. The buffer shown should be considered as a virtual buffer only, as actual implementation can consist of a single rate matching block as well. HARQ can be operated in two different ways, with identical or with non-identical retransmissions.

The use of identical retransmissions is often referred in the literature as Chase or soft combining. In this case the rate matching functionality is identical between transmissions and always the same bits remain after rate matching operation are sent. This principle is illustrated in Figure 4.7. Regardless of the number of retransmissions, the rate matching operation is always unchanged for every transmission of the same packet.

The terminal receiver has to store received samples as soft values. This needs more memory than buffering just the output of the turbo-decoder. As this is also fast memory there are limits to how much of such memory is reasonable to have in a terminal. The fact that retransmissions are handled by the base station makes the whole solution feasible from the terminal point of view. With retransmission handling from the RNC there would be too much delay and soft combining with large data rates would simply have required too much memory.

Non-identical retransmissions or so-called 'incremental redundancy' uses a different rate matching between retransmissions. The relative number of parity bits to systematic bits varies between retransmissions. This solution requires more memory in the receiver and has been accommodated in user equipment (UE) capabilities. The terminal – with identical parameters but larger soft memory capability – can otherwise manage with incremental redundancy even at the maximum data rate. The principle is illustrated in Figure 4.8. The rate matching function is varied between different retransmissions and in the actual implementation channel encoding can be done for each transmission or the data can then be kept in the virtual buffer (as shown in Figure 4.6).

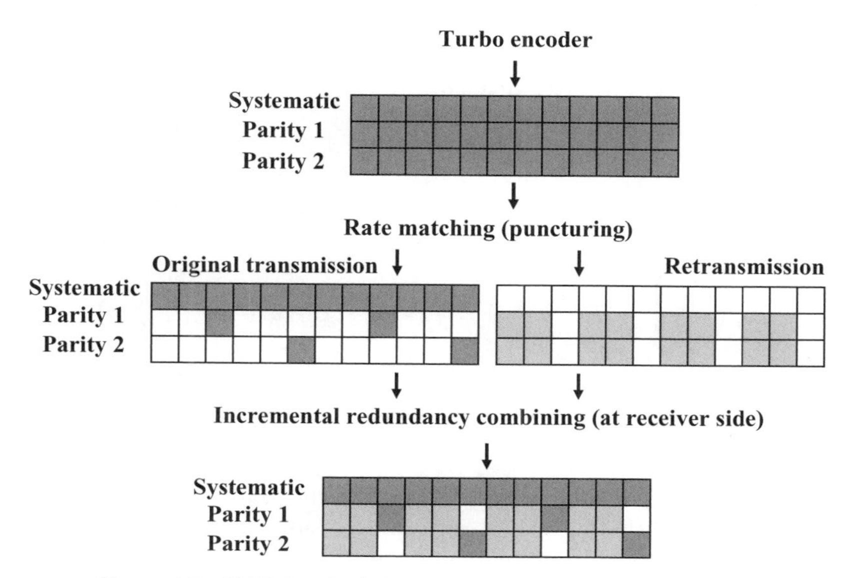


**Figure 4.8** HARQ principle with non-identical retransmissions.

If physical layer retransmissions fail or exceed the maximum number of retransmissions then the radio link layer (RLC) will handle further retransmissions. This typically happens with serving HS-DSCH cell change or sometimes due to poor coverage or due to a signalling error that could fill the buffer with undesired data. The latter rare event is due to the error checks in the signalling, as covered in connection with the HS-SCCH details.

The physical channel segmentation in Figure 4.5 maps the data to physical channel interleavers. The two interleavers are identical to Release 99 interleavers in the QPSK case and in the 16QAM case when two interleavers are used. Note that as part of turbo-coding there is a separate turbo-code internal interleaver in use. Details of the one-third rate turbo-coder in use are unchanged from Release 99, based on the use of two parallel concatenated convolutional code (PCCC) with two eight-state constituent encoders and one turbo-code internal interleaver.

HS-DSCH channel coding is specified in Release 5 and in newer versions of [3].

#### 4.2.1.2 HS-DSCH modulation

While the DCH only uses QPSK modulation, the HS-DSCH may additionally use the higher order modulation: 16QAM. During the original HSDPA feasibility study other alternatives – such as 8-PSK or 64QAM – were also discussed but not considered worth adding to the system in addition to the link adaptation range already available with QPSK and 16QAM and the different repetition/puncturing rates for turbo-coding. The 16QAM and QPSK constellations are shown in Figure 4.9. By having more constellation points – 16 instead of 4 – now 4 bits can be carried per symbol instead of 2 bits per symbol with QPSK.

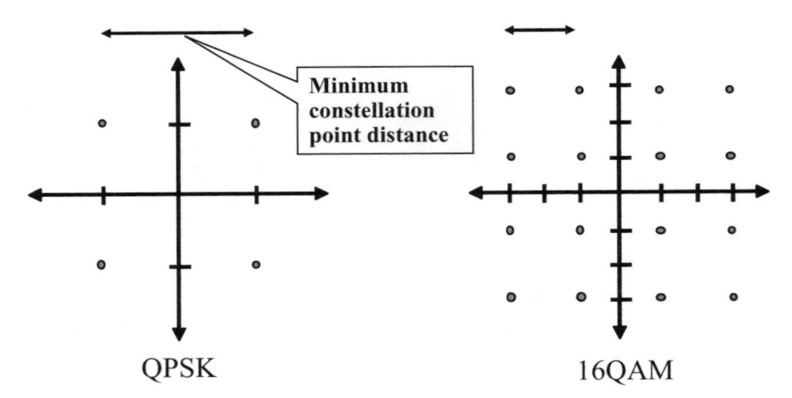


**Figure 4.9** QPSK and 16QAM constellations.

As shown in Figure 4.9, the use of higher order modulation introduces additional decision boundaries. With 16QAM, it is no longer sufficient to not only have phase figured out correctly but also amplitude needs to be estimated for more accurate phase estimate. This explains why signal quality needs to be better when using 16QAM instead of QPSK. In the downlink, a good-quality common pilot channel (CPICH) allows estimation of the optimum channel without excessive user-specific pilot overhead. The CPICH offers the phase information directly, but there is the need to estimate the power difference between the CPICH and HS-DSCH power level to estimate the amplitude information accurately as well. This suggests that at the base station end there are also power changes that – during the 2-ms transmission – should be avoided.

The HS-DSCH can use a number of multicodes, with a spreading factor of 16. The theoretical maximum number of codes available in a code tree with such a spreading factor is 16, but as the common channels and associated DCHs need some room, the maximum feasible number is 15. Whether a single terminal can receive up to 15 codes during the 2-ms TTI depends on the terminal's capabilities, as described in Section 4.5. In the system there can be other traffic that are consuming code space as well – such as CS speech or video calls – which cannot be mapped on HSDPA. Thus, radio resource management will then determine the available code space for the scheduler at the BTS.

In principle, one could create more code space with secondary scrambling codes, but as they are not orthogonal to the codes under the primary scrambling codes, the resulting total capacity is not expected to increase. Thus, the use of other scrambling codes has been restricted to compressed mode. HS-DSCH spreading and modulation is specified in [4].

### *4.2.2 High-speed shared control channel*

The HS-SCCH has two slots offset compared with the HS-DSCH (as shown in Figure 4.10). This enables the HS-SCCH to carry time-critical signalling information which allows the terminal to demodulate the correct codes. A spreading factor of 128 allows 40 bits per slot to be carried (with QPSK modulation). There are no pilots or power control bits on the HS-SCCH and, thus, the phase reference is always the same as

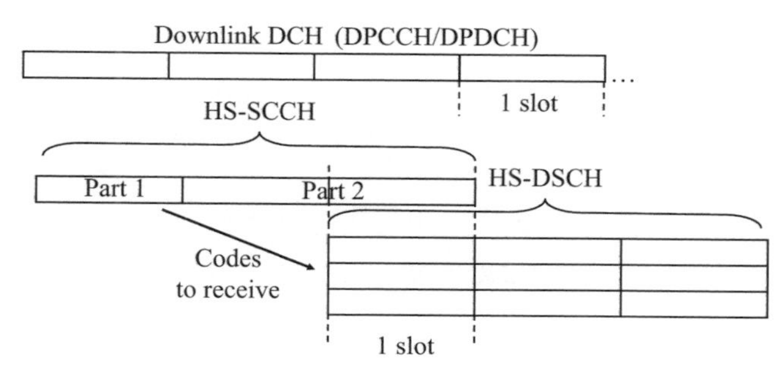


**Figure 4.10** Relative timing between HS-SCCH and HS-DSCH.

for the HS-DSCH. As part of the information – like which codes to de-spread in the HS-DSCH – needs to be available before beginning to use the HS-DSCH, the HS-SCCH is divided into two parts:

- The first part carries the information needed to be available to enable de-spreading of correct codes and contains the modulation information as well (as indicated in Figure 4.10). This enables having terminals supporting just 5 or 10 codes, even with a code space allocation up to 15 codes. Thus, with a terminal supporting a smaller number of codes, the codes in use are decoded from the HS-SCCH and the need for de-spreading is limited to just those codes intended for the terminal. The scheduler will not exceed the limits set by the terminal's capability.
- The second part contains less urgent information – such as which ARQ process is being transmitted. An indication whether the transmission is new or related to an earlier transmitted packet is also included. Even if the previous transmissions for a packet have not been received correctly, the new data indicator tells the terminal that the older transmissions can be flushed from the buffer. This means – depending on the service – that either there will be an RLC level retransmission (acknowledged mode RLC) later, or that the data are simply to be discarded and the application is expected to have error tolerance (RLC unacknowledged mode). Information on the redundancy version and constellation is also carried in the second part, as is transport block size.

The timing between the HS-SCCH and the HS-DSCH allows the terminal to have one slot time to figure out which codes to de-spread and ascertain the modulation. For the remaining parameters, slot processing time has to be considered before the transmission is over and a new 2-ms TTI may start.

When HSDPA is operated using the time multiplexing principle, then only one HS-SCCH can be configured. In this case only one user receives data at a time. Figure 4.11 illustrates the case with a single HS-SCCH.

When there is a need to have code multiplexing, then more than one HS-SCCH needs to be included, as shown in Figure 4.12. A single terminal may consider at most four HS-SCCHs; the system itself could configure even more. In practical scenarios it is not likely that more than four would be needed. One of the motivations for code multiplexing is the case when more than five codes can be devoted to HSDPA use. Then, especially when considering the expected first-phase devices, there are going to be terminals that can

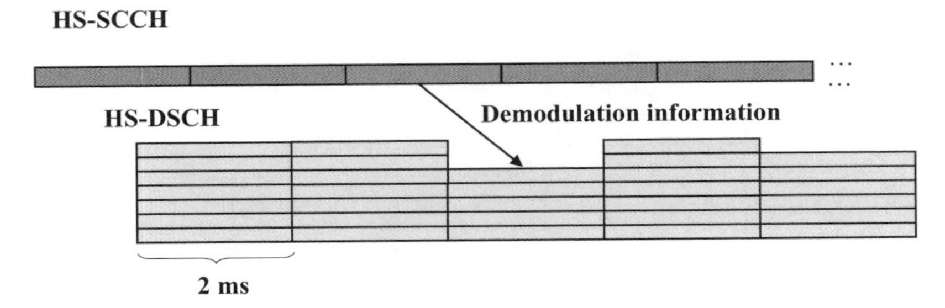


**Figure 4.11** Operation in the case of single HS-SCCH and time multiplexing.

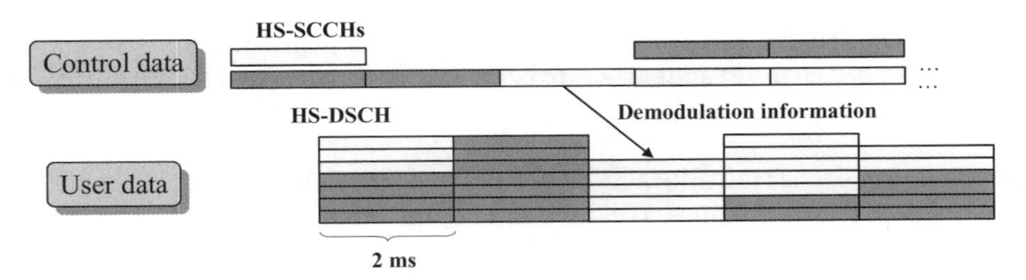


**Figure 4.12** Operation in case of two HS-SCCHs and code multiplexing

receive only five (or ten) codes of the maximum – i.e., 15. The higher the data rates and terminal capabilities the smaller the need for code multiplexing. The use of code multiplexing is not necessarily needed either when the carrier is shared with DCH traffic, or when there is a desire to have HSDPA data users operating with reasonable data rates – in the order of 384 kbps or more. In general, the data rate available for each user in different cases will depend both on power allocation and the environment and the type of terminal being used, as discussed in Chapter 7

The channel coding is one-third convolutional coding (as turbo-coding does not make sense with such a small amount of information). In the second part there is a cyclic redundancy check (CRC) to make sure that there is no corruption of the information. A signalling error with, say, an HARQ process number would cause problems as it would cause buffer corruption; thus, a 16-bit CRC is used to ensure sufficient reliability. For the first part there is no CRC and the additional challenge is to isolate from the multiple different HS-SCCHs which one was intended for the UE in question. This is enabled by a terminal-specific masking operation which allows detecting the HS-SCCH that was intended for the terminal in question while still keeping the time-critical information in compact form in one slot.

## 4.3 High-speed dedicated physical control channel

HSDPA operation needs uplink physical layer feedback information from the terminal to the base station to enable the link adaptation and physical layer retransmissions. The

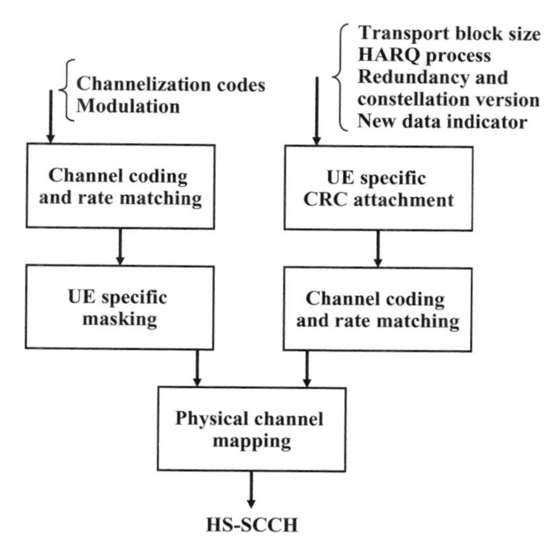


**Figure 4.13** HS-SCCH channel coding and multiplexing chain.

signalling is provided on a parallel code channel, leaving the DCH operation unchanged; this enables operation in soft handover as well in cases when all BTSs in the active set have not been upgraded to support HSDPA. This is feasible, as only the the serving HS-DSCH cell will send the HS-SCCH and HS-DSCH channels and, respectively, only the serving HS-DSCH cell needs to decode the uplink feedback. By keeping the existing uplink DPCCH and DPDCH unchanged the active set can also accommodate Release 99 based base stations. Obviously, when such a BTS becomes dominant one must switch to the DCH instead. The resulting penalty for adding a parallel code channel is the increased peak-to-average ratio (PAR) of the uplink signal waveform, which results in lower total transmission power in certain cases, as discussed in greater depth in Chapter 11.

The uplink feedback information is carried on the HS-DPCCH. The HARQ feedback informs the base station whether the packet was decoded correctly or not. Channel quality information (CQI), respectively, tells the base station scheduler the data rate the terminal expects to be able to receive at a given point in time.

The HS-DPCCH uses a fixed spreading factor of 256 and has a 2-ms/three-slot structure. The first slot is used for the HARQ information (as shown in Figure 4.14). The two remaining slots are for CQI use. The HARQ information is always sent when there has been a correctly decoded HS-SCCH received in the downlink direction while the QCI transmission frequency is controlled by the system parameter $k$. For both slots there is also a separate parameter to control repetition. Repetition over multiple 2-ms periods is needed in some cases – for example, for cell edge operation when the available power would not ensure sufficient quality for feedback reception. The power control from non-serving HSDPA cells may also reduce the received HS-DPCCH power level in the soft handover region as the terminal has to reduce the uplink transmission power level if any of the cells in the active set sends a power-down command.

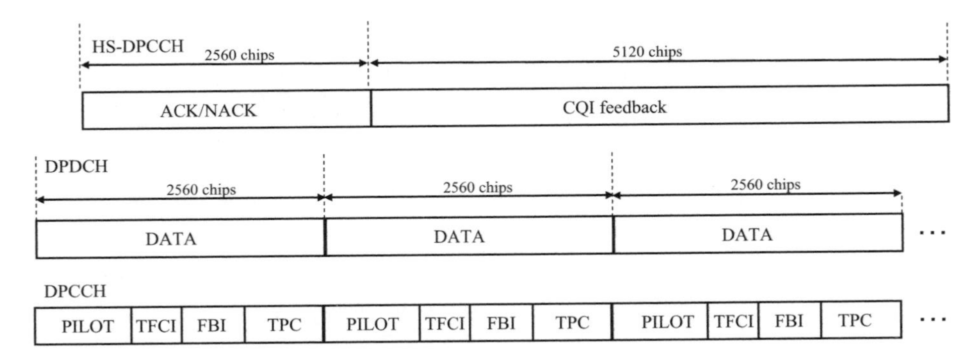


**Figure 4.14** HS-DPCCH structure.

The HS-DPCCH is only symbol aligned with DPCCH/DPDCH. This was considered necessary in order to avoid too large variations in the timing as both terminal and Node B processing is dimensioned for the most demanding case. The relative timing is illustrated in Figure 4.14. The downlink HS-SCCH/HS-DSCH timing will determine the HS-DPCCH transmission instant, as covered in Section 4.3.2.1.

HS-DPCCH operation has been enhanced in Release 6 to improve cell edge operation. Improvement is effected by introducing pre-/post-ambles for the DPCCH channel. When the HS-SCCH has been received, the terminal will send a sequence in place of the ACK/NACK signalling in the previous 2-ms HS-PDCCH frame, unless there has been a packet in the previous TTI. Doing this will cause the terminal not only to transmit more often, but it also allows the BTS receiver to have prior knowledge as to whether ACK or NACK will be transmitted. Thus, by avoiding detection of 'no transmission' – often denoted as 'DTX' – there is no need to choose between the three values – but only between ACK and NACK. The benefit is biggest when there is a continuous flow of packets, but even in the case of infrequent packets there is a reduced peak power requirement. Thus, the gain factor settings for the HS-DPCCH does not need to use such high values, and the transmission power consumed for DPCCH/DPDCH operation is reduced.

The coding for the HARQ is simple. In Release 5 there is either a sequence of '1s' sent for the ACK and '0s' for the NACK. In Release 6 different sequences are added to the pre-amble and post-amble, but ACK/NACK remain unchanged.

For CQI, (20.5) coding is applied – similar to TFCI coding – which carries CQI information from the terminal to the base station. The CQI value that the terminal reports does not just correspond to the $E_c/N_0$ or the signal to interference ratio (SIR) the terminal is experiencing. Instead, the value reported is the function of the multipath environment, terminal receiver type, ratio of the interference of the own base station compared with others and expected BTS HSDPA power availability.

The clear benefit of the approach is that the solution defined will automatically accommodate the various possible receiver implementations and environment variations and, thus, gives an indication of the best data rates needed by the terminal to cope with the environment in question. This removes the need from the network end to have to consider, say, the delay profile characteristics of the cell/sector in question. The only input value from the network is the HS-DSCH power allocation value the terminal may

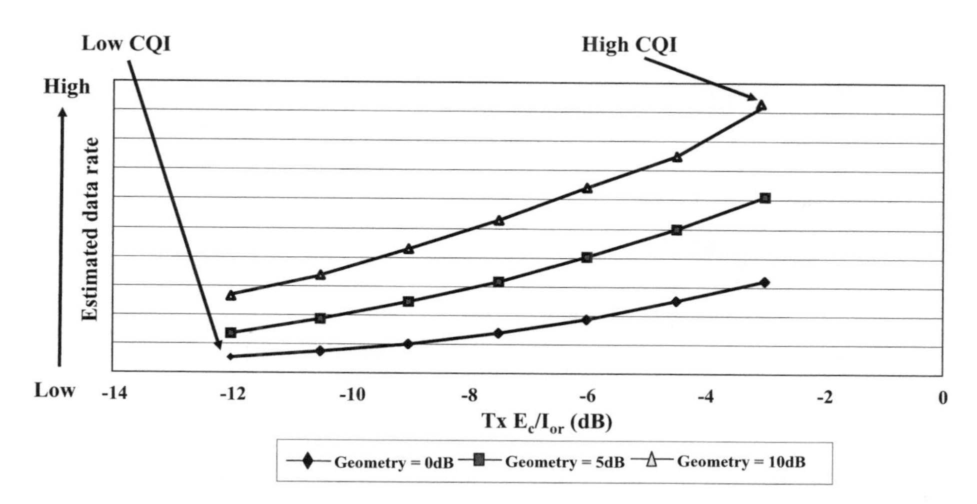


**Figure 4.15** Example CQI curves for different geometry values.

assume to be used in the network. Should this change, then the network can easily compensate for this as the terminal's assumption is known to the base station. CQI reporting is illustrated in Figure 4.15, showing that when the terminal is close to the base station and assumes high HS-DSCH power allocation (based on the value given by the network), a high CQI value is reported. Respectively, when the terminal is closer to the cell edge (the lowest curve) then the reported CQI is much lower, especially if the expected HSDPA Node B power allocation is low as well. When at or close to the cell edge, most of the interference comes from other cells and, thus, the representative geometry value is low – around 0 dB – or in some cases the value is negative.

### *4.3.1 Fractional DPCH*

For Release 6, further optimization took place for the situation where only packet services are active in the downlink other than the SRB. In such a case, especially with lower data rates, it was felt that the downlink DCH introduces too much overhead and can also consume too much code space if looking for a large number of users using a low data rate service – like VoIP. The solution was to use in such cases an F-DPCH, which is basically a stripped-down version of DPCH that handles the power control. When comparing the F-DPCH with the Release 99 DPCH, only the transmission power control (TPC) field is kept (as shown in Figure 4.16). In addition, the original timing is retained which avoids the need to adjust timings based on Release 99 power control loop implementation.

The code resource is time-shared, thus several users can share the same code space for power control information. Each user sees only the channel which has one symbol per slot for transmission power control (TPC) information and assumes there is no transmission in the rest of the symbols. With several users, the network configures each user having the same code but different frame timing and, thus, users can be transmitted on the single code source. Up to ten users can share one SF 256 code, thus reducing

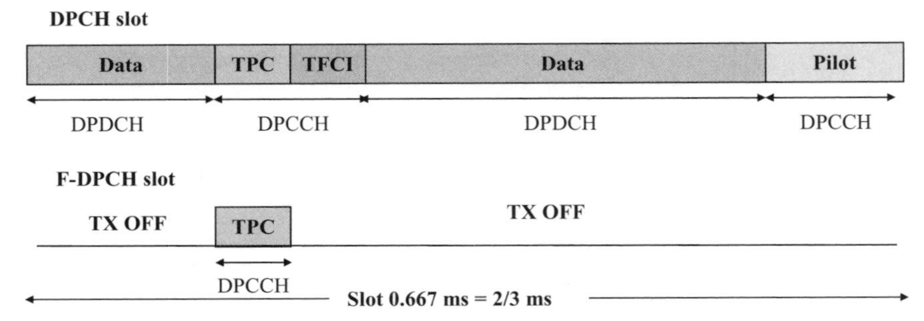


**Figure 4.16** F-DPCH slot structure compared with DPCH structure.

code space utilization for the associated DCH for users with all services mapped to the HS-DSCH. The principle is as illustrated in Figure 4.17, where two users share the same code space in the downlink direction. The uplink DCH is not impacted. In this case the SRB must be on the HS-DSCH as there is no room for data bits (no DPDCH) on the F-DPCH.

The F-DPCH has some restrictions for use. Obviously it is not usable with services requiring data to be mapped to the DCH, such as AMR speech calls and CS video. Further, the lack of pilot information means that a method like feedback-based transmit diversity (closed loop mode) is not usable. The use of closed loop diversity is based on user-specific phase modification, wherein pilot symbols would be needed for verification of the phase rotation applied. On the other hand, when utilizing the F-DPCH, SRBs can benefit from high data rates of HSDPA and reduce service setup times remarkably (as discussed in Chapter 9).

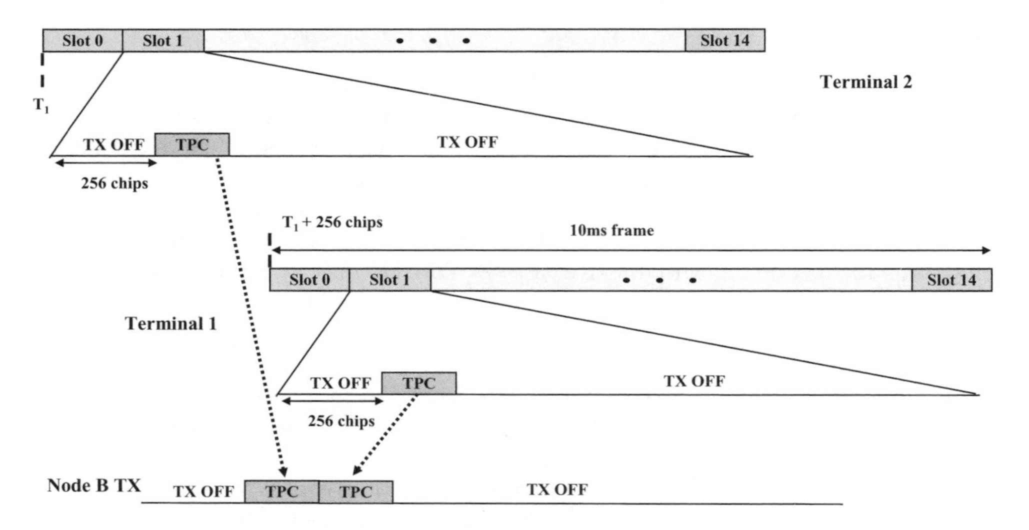


**Figure 4.17** Fractional DPCH principle for Node B transmission.

As the SRBs are mapped to the HS-DSCH using the F-DPCH, the link quality from the serving HS-DSCH cell is most critical. To ensure that there are appropriate criteria to detect radio link failure, the terminal behaviour for setting the downlink power control commands in the uplink DPCCH was modified so that the F-DPCH can be reliably detected from the serving HS-DSCH cell. Thus, radio link failure is detected in the terminal only from the F-DPCH of the serving HS-DSCH cell, and not from the soft combined DPCCH as in Release 99.

### *4.3.2 HS-DSCH link adaptation*

Link adaptation is very dynamic as it operates with 2-ms granularity with the HS-DSCH. In addition to the scheduling decision, the MAC-hs in the BTS will also decide every 2 ms which coding and modulation combination to transmit. Link adaptation is based on physical layer CQI being provided by the terminal.

Using link adaptation, the network will also gain from the limitation of power control dynamics in the downlink. As signals in the downlink cannot use too large a dynamic range to avoid the near–far problem between signals from the same source, the downlink power control dynamics is much more limited. While in the uplink a 71-dB or more dynamic range is used, in the downlink only around 10 to 15 dBs can be utilized.

The exact number depends on the implementation, channel environment and spreading factors applied. This means that for users close to the base station the power level transmitted is higher than necessary for reliable signal detection. Link adaptation takes this extra margin into use by selecting transmission parameters in such a way that the required symbol energy corresponds more accurately to the available symbol power. This is illustrated in Figure 4.18, where link adaptation as a function of the carrier-to-interference ratio (C/I) is illustrated. As discussed previously, link adaptation itself is based on CQI information that also takes other aspects into account besides just the

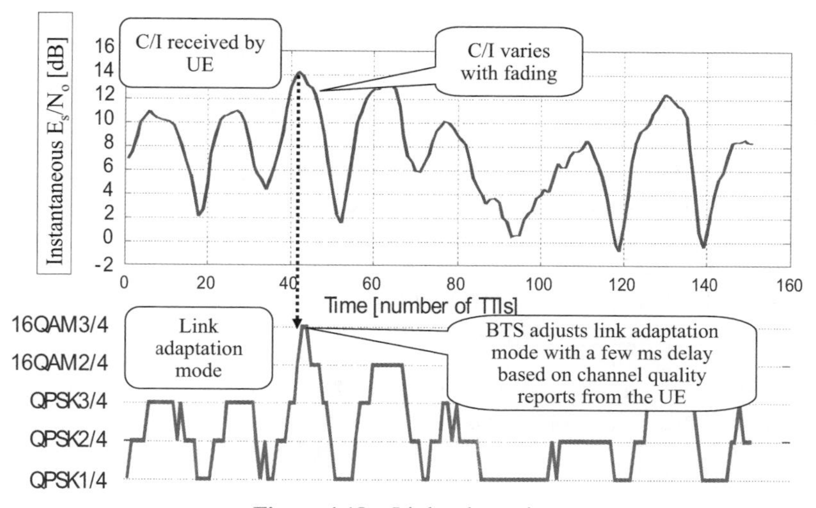


**Figure 4.18** Link adaptation.

16QAM terminals with 10 codes

| CQI value | Transport block size | Number of HS-PDSCH codes | Modulation | Reference power adjustment Δ |
|---|---|---|---|---|
| 0 | N/A | N/A | | |
| 1 | 137 | 1 | QPSK | 0 |
| 2 | 173 | 1 | QPSK | 0 |
| : | | | | |
| 14 | 2583 | 4 | QPSK | 0 |
| 15 | 3319 | 5 | QPSK | 0 |
| 16 | 3565 | 5 | 16QAM | 0 |
| : | | | | |
| 24 | 11418 | 8 | 16QAM | 0 |
| 25 | 14411 | 10 | 16QAM | 0 |
| 26 | 14411 | 10 | 16QAM | -1 |
| 27 | 14411 | 10 | 16QAM | -2 |
| 28 | 14411 | 10 | 16QAM | -3 |
| 29 | 14411 | 10 | 16QAM | -4 |
| 30 | 14411 | 10 | 16QAM | -5 |

QPSK-only terminals with 5 codes

| CQI value | Transport block size | Number of HS-PDSCH codes | Modulation | Reference power adjustment Δ |
|---|---|---|---|---|
| 0 | N/A | N/A | | |
| 1 | 137 | 1 | QPSK | 0 |
| 2 | 173 | 1 | QPSK | 0 |
| 3 | 233 | 1 | QPSK | 0 |
| : | | | | |
| 14 | 2583 | 4 | QPSK | 0 |
| 15 | 3319 | 5 | QPSK | 0 |
| 16 | 3319 | 5 | QPSK | -1 |
| 17 | 3319 | 5 | QPSK | -2 |
| 18 | 3319 | 5 | QPSK | -3 |
| : | | | | |
| 29 | 3319 | 5 | QPSK | -14 |
| 30 | 3319 | 5 | QPSK | -15 |

**Figure 4.19** Example CQI tables for categories 7/8 and 11/12.

signal strength or C/I. By just changing from QPSK to 16QAM there is a difference of a few decibels – depending on the environment – and by playing with the coding rates and the number of codes the total dynamic range can reach 30 dB.

The terminal capability impacts the reporting as all devices, especially in the early phase, will not necessarily support, say, 16QAM modulation or more than five parallel codes. When exceeding the range of data rates the terminal can support, the terminal will report only an offset rather than the highest code/modulation point it can support. The CQI table has a more or less evenly spaced combination of modulation, number of codes, coding and transport block sizes. The slightly simplified CQI example tables in Figure 4.19 represent different terminal capabilities. The rightmost table in Figure 4.19 is for terminal categories 11 and 12 (QPSK only) and, thus, after five codes only a 1.8-Mbps data rate offset for the situation is signalled. Respectively, the CQI table for category 7 (7.2 Mbps) goes all the way to ten codes (with 16QAM) before using the offset on top of that. For all other categories the offset is used after some point, except category 10 which goes all the way to 14.4 Mbps and thus does not need offsets on top of that. The inter-TTI interval parameter in the terminal's capabilities does not have an impact on which QCI table to use, only the number of codes and supported modulation makes the difference. The exception is the 14.4-Mbps class where the only difference is the maximum transport block size supported when compared with a terminal capability of 10 Mbps. As can be seen from Chapter 7, the CQI values corresponding to a case when 14.4 Mbps can be received are not expected to occur very frequently in real networks.

#### 4.3.2.1 HSDPA physical layer operation procedure

HSDPA physical layer operation goes through the following steps once one or more users have been configured as using the HS-DSCH and data start to reach the buffer in the Node B.

- The scheduler in the Node B evaluates – every 2 ms – for each user with data in the buffer: the channel condition, buffer status, time from last transmission, retransmissions pending and so forth. The exact criteria in the scheduler are naturally a vendor-specific implementation issue and not part of 3GPP specifications. Parameters to control scheduler behaviour have been specified and are covered in Section 4.5.2.
- Once a terminal has been determined as serving in a particular TTI, the Node B identifies the necessary HS-DSCH parameters, including number of codes, 16QAM usability and terminal capability limitations.
- The Node B starts to transmit the HS-SCCH two slots before the corresponding HS-DSCH TTI. HS-SCCH selection is free (from a set of at most four channels) assuming there were no data for the terminal in the previous HS-DSCH frame. Had there been data in the previous frame, then the same HS-SCCH needs to be used.
- The terminal monitors the terminal-specific set of at most four HS-SCCHs given by the network. Once the terminal has decoded Part 1 from an HS-SCCH intended for that terminal, it will start to decode the remaining parts of that HS-SCCH and will buffer the necessary codes from the HS-DSCH.
- After decoding the HS-SCCH parameters from Part 2, the terminal can determine to which ARQ process the data belong and whether they need to be combined with data already in the soft buffer.
- In the Release 6 version, a pre-amble is sent in the ACK/NACK field if the feature is configured to be used by the network (and if there was no packet in the previous TTI). Sending of the pre-amble is based on HS-SCCH decoding, not on the HS-DSCH itself.
- Upon decoding the potentially combined data, the terminal sends in the uplink direction an ACK/NACK indicator, depending on the outcome of the CRC conducted on the HS-DSCH data (an ACK follows the correct CRC outcome).
- If the network continues to transmit data for the same terminal in consecutive TTIs, the terminal will stay on the same HS-SCCH that was used during the previous TTI.
- In Release 6, when the data flows end, the terminal sends a post-amble in the ACK/NACK field, assuming the feature was activated, of course.

HSDPA operation is synchronous in terms of the terminal response for a packet transmitted in the downlink. The network side, however, is asynchronous in terms of when a packet or a retransmission for an earlier transmission is sent. This allows the necessary freedom for BTS scheduler implementation and is enabled by ARQ process information on the HS-SCCH. The physical layer operation procedure for HSDPA is covered in [5].

The terminal operation times between different events are specified accurately from the HS-SCCH reception, followed by HS-DSCH decoding and ending with the uplink ACK/NACK transmission. As indicated in Figure 4.20, there is a 7.5-slot reaction time from the end of the HS-DSCH TTI to the start of ACK/NACK transmission in the

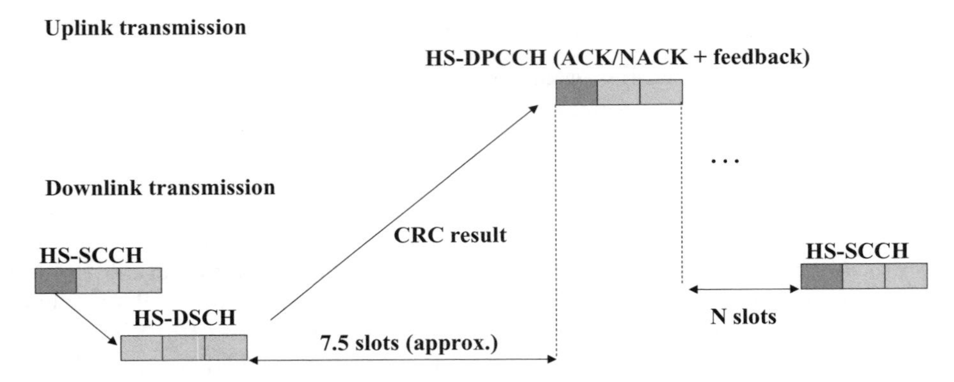


**Figure 4.20** Terminal timing with respect to one HARQ process.

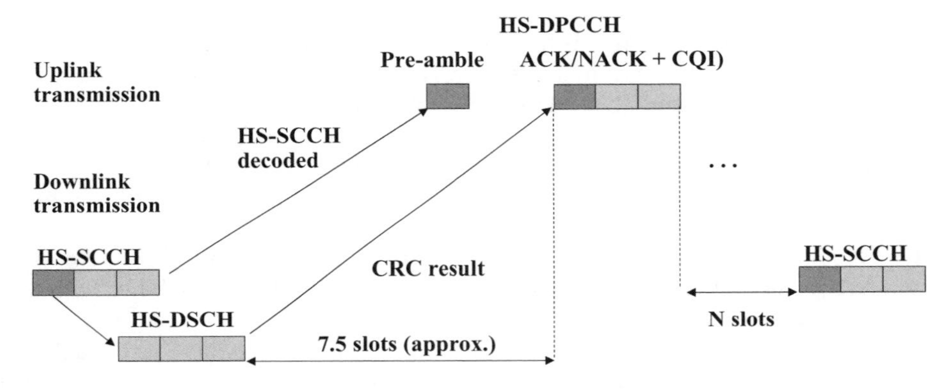


**Figure 4.21** Operation with pre/post scheme of Release 6.

HS-DPCCH in the uplink. The value of 7.5 slots is accurate – any variation is due to the need for symbol alignment between the uplink HS-DPCCH and the uplink DPCCH/DPDCH, thus the timing value is within the 256-chip window.

In Release 6, with the pre-/post-amble in use the timing is otherwise unchanged, but for the first packet the previous ACK/NACK slot is used for the pre-amble, as shown in Figure 4.21. Respectively, when the transmission ends (or at least the terminal does not detect the HS-SCCH), the post-amble is transmitted in the position where an ACK/NACK would normally occur.

### *4.3.3 Mobility*

When compared wth the DCH the big difference in mobility handling is the lack of soft handover for the HS-DSCH. While there is only one serving HS-DSCH cell, the associated DCH itself can be in soft handover and maintain the active set as in Release 99. The requirement for the terminal is to be able to cope with up to six cells in the DCH active set. When operating in soft handover with HSDPA in use, there was a need to

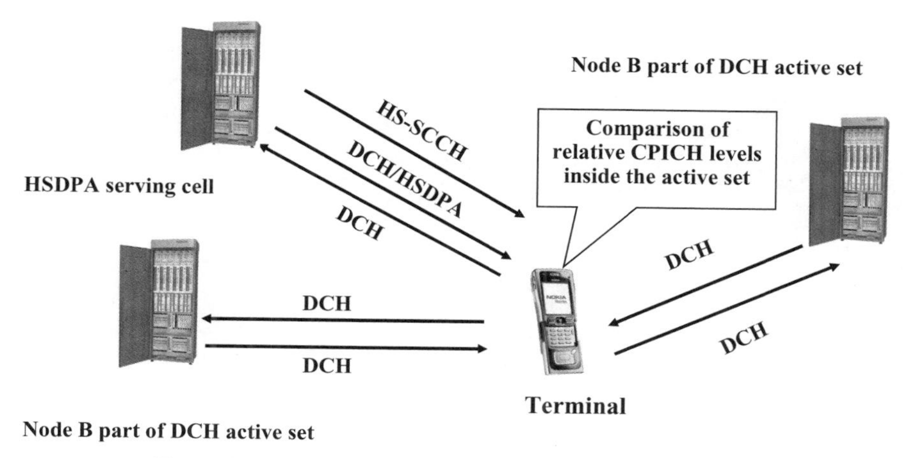


**Figure 4.22** HSDPA operation with DCH active set of 3.

modify measurement events to get information about changes in the relative strengths of the cells in the active set (as illustrated in Figure 4.22). This information (when held within the active set window) does not trigger actions for the DCH but may trigger a change in the serving HS-DSCH cell with HSDPA operation. To enable this information to be available for the RNC, the measurement event 1D was modified to enable event-based reporting when the relative strength of the cells in the active set changes, or more exactly when the cell with the optimum CPICH strength in the active set changes. When the criteria for reporting are fulfilled, the terminal will send – as part of uplink radio resource control (RRC) signalling – a measurement report to the RNC.

In Release 5 the serving HS-DSCH cell can only be changed inside the terminal's active set by using a physical channel reconfiguration procedure. Therefore, when the terminal detects that a neighbouring cell fulfills the criteria for adding the cell to the active set (measurement event 1A), the cell first needs to be added to the active set before the serving HS-DSCH operation can be changed to that cell. In Release 6 the situation was changed so that the active set update procedure could also carry out a serving HS-DSCH cell change (as illustrated in Figure 4.23).

When a change of serving HS-DSCH cell happens, the terminal will flush all the buffers at the handover time and move on to listen to the new base station as instructed in the downlink RRC signalling. Respectively, at the moment the handover takes place the Node B also flushes packets still in its buffers, including possibly unfinished HARQ processes. The network will have to sort out the unsent packets (when using RLC acknowledged mode). In unacknowledged mode the packets are not to be transmitted if some of them are lost during the cell change operation, but packet loss can be minimized by the RNC calculating carefully when the handover will take place and not sending – at the very last moment – packets to the serving HS-DSCH cell being replaced. During synchronized serving HS-DSCH cell change, any interruption to packet transmission can be minimized as both the terminal and network elements know exactly when the handover takes place. To minimize possible data loss further, the support of out-of-sequence payload data units (PDUs) was introduced in unacknowledged mode

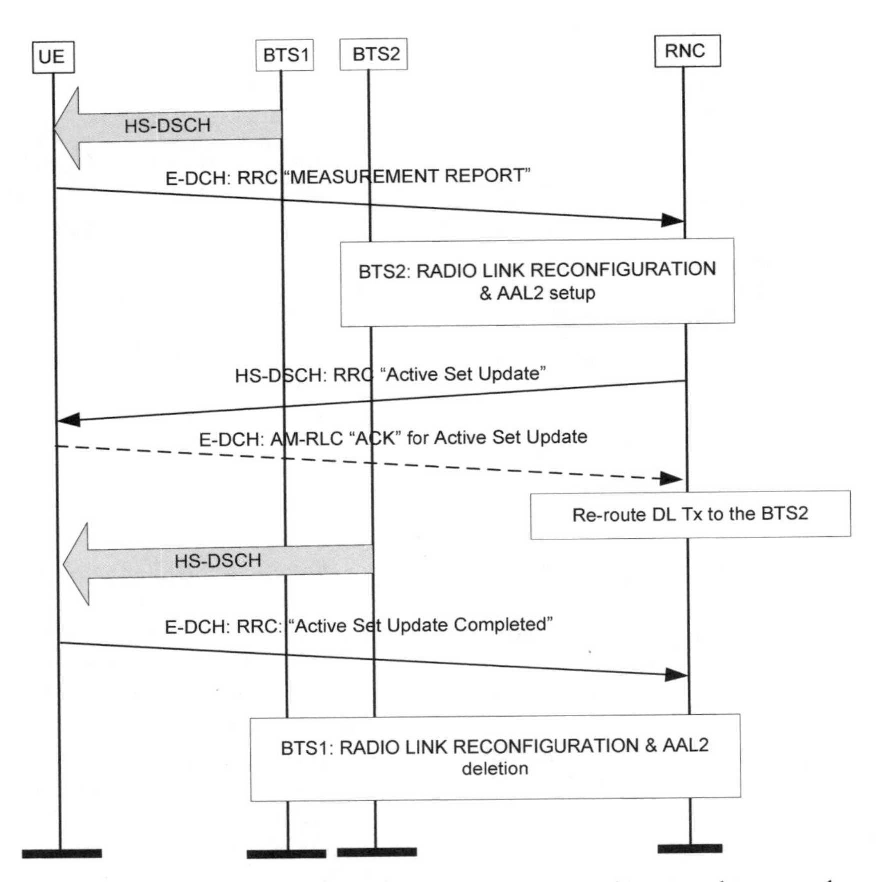


**Figure 4.23** Serving HS-DSCH cell change by using active set update procedure.

RLC in Release 6, allowing the UMTS Terrestrial Radio Access Network (UTRAN) to use bi-casting during serving HS-DSCH cell change, which ensures that packets potentially lost at the source Node B are available at the target Node B.

The HS-DSCH supports the use of compressed mode for terminal inter-system and inter-frequency measurements. Although the HS-DSCH itself will not implement any compressed mode, actual data transmission and control signalling are suspended (as shown in Figure 4.24) when there is a gap in the downlink that overlaps any part of HS-SCCH or HS-DSCH transmission. In such a case the terminal will ignore the whole transmission. Respectively, uplink signalling will obviously not take place when there is a gap scheduled in the uplink. As both compressed mode pattern information and the scheduler are located in the Node B, it is possible to combine the information. An alternative is to use DCH transmission for the data while having compressed mode active. When parameterized correctly, compressed mode is not expected to be used in any case close to the base station, but does so when WCDMA coverage is getting poor or when the HSDPA cell layer – for example, in the micro-cell layer – needs to change when entering the area of macro-cell only coverage. Changing to the Global System for Mobile

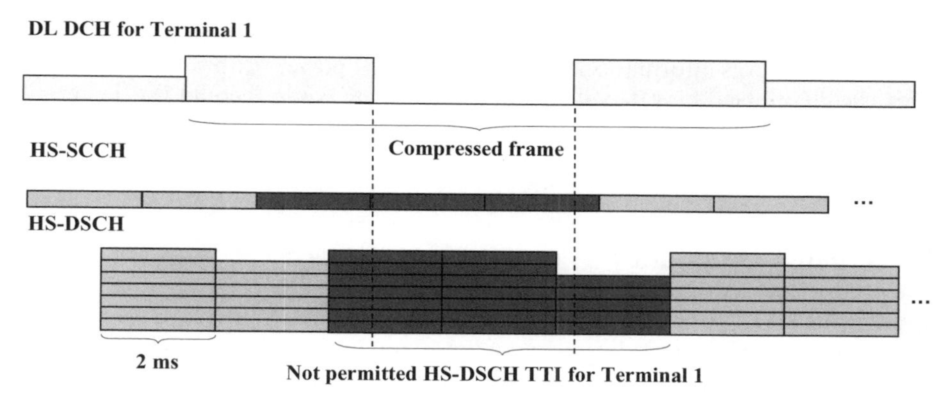


**Figure 4.24** HSDPA operation with compressed mode.

Communications (GSM) obviously causes a drop in the user data rate, the actual difference depending on GSM network capabilities. The available data rate varies theoretically up to 384 kbps depending whether there is only a basic (GPRS) service available or whether the enhanced data rate for global evolution (EDGE) is also supported. The channel conditions and number of slots supported in the multimode terminal limit the practical data rates to around 200 to 300 kbps (as discussed in Chapter 10).

For the handover to/from GSM networks, the Release 6 specifications contain support for packet handover. This allows reducing the handover interruption time to similar level achieved by Release 99 based voice call handovers, where the end user in a properly parameterized network does not detect the change of system. In Release 5 there is network-assisted cell change when changing to GSM to speed up the process, while in Release 99 based networks only cell reselection from WCDMA is available to GSM.

## 4.4 BTS measurements for HSDPA operation

There are three new BTS measurements specified in Release 5 that facilitate receipt of the necessary information about HSDPA operation in the RNC. In the physical layer there is the following measurement (defined in [6]):

- Non-HSPDA power, which basically reveals the power being used for all channels other than HSDPA (HS-DSCH and HS-SCCH). In the Release 6 version this measurement covers all downlink channels not used for HSDPA or HSUPA purposes. There is no point in measuring actual HSDPA power as that is either the difference between non-HSDPA power and the maximum BTS transmission power, or HSDPA power allocation has been provided by the RNC.

It is in the scheduler – this part of the MAC layer is covered in [7] – that the HS-DSCH provides a bit rate measurement. This measurement gives information on the average data rate per each priority class over the measurement period.

Further, a common measurement in Iub [8] is definition of the required power for the HS-DSCH. This gives information about the estimated power per priority class required to meet the guaranteed bit rate value. Additionally, the Node B could list the terminals which require very high power for meeting their guaranteed bit rate for connection.

## 4.5 Terminal capabilities

Support of the HSDPA feature itself is optional for terminals. When supporting HSDPA operation, the terminal will indicate one of the 12 different categories specified. Depending on the category supported, the resulting maximum downlink data rates vary between 0.9 and 14.4 Mbps. HSDPA capability is otherwise independent of Release 99 based capabilities, but if the HS-DSCH has been configured for the terminal, then DCH capability in the downlink is limited to the value given by the terminal. A terminal can indicate a 32, 64, 128 or 384-kbps DCH capability; thus, a terminal normally capable of running at 384 kbps on the DCH may indicate that – once HSDPA is configured – the DCH should be reconfigured down to 64 kbps.

The first ten HSDPA terminal capability categories in Table 4.2 need to support 16QAM, but the last two, categories 11 and 12, only support QPSK modulation. The other key difference between the classes is the maximum number of parallel codes that must be supported. Another value indicating the capability to sustain the peak rate over multiple continuous TTIs is the inter-TTI parameter. Categories with a value of 1 correspond to devices that can also sustain the peak rate during 2 ms over multiple TTIs, while terminals with an inter-TTI value greater than 1 must 'rest' for 2 or 4 ms after each received TTI. Additionally, there is a soft buffer capability which uses two principles for determining its value.

**Table 4.2** HSDPA terminal capability categories.

| *Category* | *Maximum number of parallel codes per HS-DSCH* | *Minimum inter-TTI interval* | *Transport channel bits per TTI* | *ARQ type at maximum data rate* | *Achievable maximum data rate* (Mbps) |
|---|---|---|---|---|---|
| 1 | 5 | 3 | 7298 | Soft | 1.2 |
| 2 | 5 | 3 | 7298 | IR | 1.2 |
| 3 | 5 | 2 | 7298 | Soft | 1.8 |
| 4 | 5 | 2 | 7298 | IR | 1.8 |
| 5 | 5 | 1 | 7298 | Soft | 3.6 |
| 6 | 5 | 1 | 7298 | IR | 3.6 |
| 7 | 10 | 1 | 14411 | Soft | 7.2 |
| 8 | 10 | 1 | 14411 | IR | 7.2 |
| 9 | 15 | 1 | 20251 | Soft | 10.2 |
| 10 | 15 | 1 | 27952 | IR | 14.4 |
| 11 | 5 | 2 | 3630 | Soft | 0.9 |
| 12 | 5 | 1 | 3630 | Soft | 1.8 |

**Table 4.3** RLC parameters for different UE categories.

| *UE category* | *Maximum number AM RLC entities* | *Minimum total RLC AM and MAC-hs memory* |
|---|---|---|
| 1–6, 11 and 12 | 6 | 50 kbytes |
| 7–8 | 8 | 100 kbytes |
| 9–10 | 8 | 150 kbytes |

AM = acknowledged mode.

Terminals with a smaller soft buffer value – for example, category 5 – cannot deal with incremental redundancy at the higher data rates (closer to the 3.6-Mbps peak rate), so the network has to use soft/chase combining instead. Whether this is the case depends also on the total number of ARQ processes configured and on relative memory partitioning.

The highest capability is provided with category 10, which allows the theoretical maximum data rate of 14.4 Mbps. This rate is achievable with one-third rate turbo-coding and significant puncturing resulting in a code rate close to 1 – that is, hardly any coding. For category 9, the maximum turbo-encoding block size (from Release 99) has been taken into account when calculating the values, and thus results in the 10.2-Mbps maximum user data rate value with four turbo-encoding blocks.

Besides the physical category to be reported to the network there are other parameters important for HSDPA operation. The size of the RLC reordering buffer determines the window length of the packets that can be 'in the pipeline' to ensure in-sequence delivery of data to higher layers in the terminal. The minimum values range from 50 to 150 kB depending on the UE category (as can be seen from Table 4.3). The buffer size has been derived so that there should be no limitations to the data rate due to this, assuming UTRAN end delays (including RLC retransmission handling) are reasonable and, thus, the memory needed for the RLC buffer does not get too large.

There is a link in the terminal capability with Release 6 HSUPA. The support of HSDPA is mandatory for a terminal that supports high-speed uplink packet access (HSUPA) in Release 6, but the HSDPA category can be chosen freely. Release 6 HSDPA additions – like F-DPCH or the pre/post scheme – are mandatory for Release 6 HSDPA capable terminals, while support for features like advanced receivers or receiver antenna diversity depends on a particular implementation and is not part of terminal capability signalling. The terminal capabilities for HSDPA operation are specified in [9].

### *4.5.1 L1 and RLC throughputs*

The physical layer data rate is constructed from the following parameters:

- number of codes in use;
- modulation;
- effective coding rate – that is, the amount of repetition or puncturing on top of the turbo-encoder output.

Table 4.4 Theoretical bit rates with 15 multicodes for different TFRCs.

| *TFRC* | *Modulation* | *Effective code rate* | *Max. throughput* (Mbps) |
|---|---|---|---|
| 1 | QPSK | $\frac{1}{4}$ | 1.8 |
| 2 | QPSK | $\frac{2}{4}$ | 3.6 |
| 3 | QPSK | $\frac{3}{4}$ | 5.3 |
| 4 | 16QAM | $\frac{2}{4}$ | 7.2 |
| 5 | 16QAM | $\frac{3}{4}$ | 10.7 |

Table 4.5 RLC level data rates for different UE categories and 320-bit PDU size.

| *UE category* | *No. of codes* | *Modulation* | *RLC blocks per 2 ms TTI* | *Transport block* | *Max. RLC data rate* |
|---|---|---|---|---|---|
| 12 | 5 | QPSK | $10 \times 320$ | 3 440 | 1.6 Mbps |
| 5/6 | 5 | 16QAM | $21 \times 320$ | 7 168 | 3.36 Mbps |
| 7/8 | 10 | 16QAM | $42 \times 320$ | 14 155 | 6.72 Mbps |
| 9 | 15 | 16QAM | $60 \times 320$ | 20 251 | 9.6 Mbps |
| 10 | 15 | 16QAM | $83 \times 320$ | 27 952 | 13.3 Mbps |

The example bit rates for different parameter combinations without overhead considerations for different transport format and resource combinations (TFRCs) is shown in Table 4.4.

The BTS scheduler can allocate the data rates for a single user to obtain a continuous high bit rate connection or to multiple users when the average user data rate is divided between the number of users sharing the resources. Obviously, at cell edge the data rates are more restricted than locations close to the base station site.

It is also important to understand that these data rates are not typically end-to-end data rates. The core network may ask for, say, a 1-Mbps peak rate which then momentarily over a 2-ms TTI may be, say, 3.6 Mbps. This is hidden from the core network by resulting data rate averaging when sharing HS-DSCH resources among several users (as was discussed in Chapter 3).

When we take into account RLC/MAC headers and fix the PDU size to be 320 bits, then we get the RLC layer data rates for different terminal categories (see Table 4.5). For the application level the difference is then less than 5% due to the Internet Protocol (IP) header overhead.

### 4.5.2 *Iub parameters*

HSDPA operation needs a large number of parameters that have to be aligned between the terminal and the Node B. These are provided (as shown in Figure 4.25) to the Node B from the RNC based on vendor-specific algorithms and on the terminal capability signalled by the terminal to the RNC using RRC signalling during connection setup.

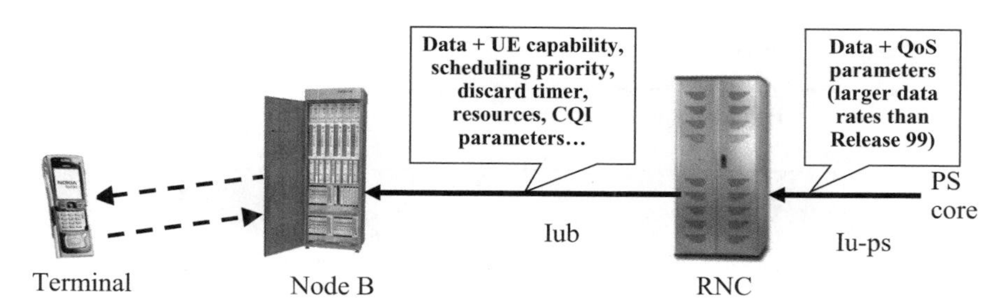


**Figure 4.25** HSDPA parameters over the Iub interface.

These parameters can be divided into the following categories:

- Parameters for Node B resource allocation – to indicate, for example, which HS-SCCH to use and which codes are available for the HS-DSCH as well as how much power to use for HSDPA.
- Scheduler parameters – to control scheduler behaviour like scheduling priority indicator or guaranteed bit rate.
- Terminal specific parameters – such as terminal capability and terminal-specific HSDPA parameters (like the set of HS-SCCH codes the terminal is monitoring).

The Node B needs to be aware of the terminal's capability. This means taking into account the limitations in terms of number of supported codes and modulation as well as the potential inter-TTI restriction. Were there a need for the use of a larger set of HS-SCCH codes than four, then the different terminals would monitor partly different set of HS-SCCHs.

Scheduler control allows quality of service (QoS) management in the Node B scheduler. Different scheduling priority indications can be used per user and per service. For example, the Voice-over-IP (VoIP) service can have a higher scheduling priority than the background-type data service regardless of whether these services are for the same user or not. Further, scheduler buffer management can be controlled by a discard timer which indicates the maximum time packets should be kept in the buffer and – upon expiration of the timer – how many are to be discarded.

Terminal-specific parameters become part of the Iur interface as well were there a need to run HSDPA over the Iur. Normally, that is not needed since there is no soft handover and, thus, relocation can always be made to the RNC directly connected to the serving HS-DSCH cell.

## 4.6 HSDPA MAC layer operation

The HSDPA MAC layer has as key functionalities the scheduling in the BTS and the handling of the HARQ process between the terminal and the BTS. The specifications do not contain details for scheduler operation as the scheduler is left for individual network vendor implementation (different scheduler types are covered in Chapter 7). As was

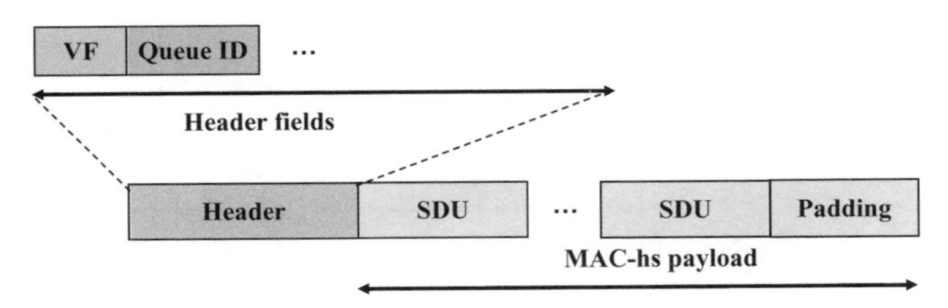


**Figure 4.26** MAC-hs PDU.

discussed in Section 4.5.2, the specifications do contain parameters to control scheduler behaviour; but, to reiterate, with most parameters the exact reaction to them will depend on the actual scheduler implemented.

The terminal is also given some MAC layer parameters. As the packets are not arriving in sequence after HSDPA HARQ operation, the terminal MAC layer has to cope with packet reordering as part of MAC layer operation. For this purpose, the terminal is given a parameter that reorders packet release. If the timer expires, the terminal gives the packets to the upper layer (meaning RLC in this case) even if there are packets missing in the sequence. This is needed to avoid stalling the process if one packet ends not being successfully delivered by the physical layer operation. Depending on the RLC configuration, further retransmission may occur on the RLC layer.

The data sent on the HS-DSCH also have a MAC layer header (as shown in Figure 4.26). The MAC-hs PDU consists of the MAC-hs header and MAC-hs payload. The payload consists of one or more MAC-hs service data units (SDUs) and potential padding if they do not fit the sizes available.

The MAC header has the following fields:

- A version flag (VF) to enable future protocol extension – this has not been utilized so far in 3GPP.
- Queue ID to allow different reordering queues at the terminal end – this is used, for example, when mapping both the SRB (in Release 6) and user data to the HS-DSCH. Note that only one transport channel may exist in a single TTI and, thus, in a single MAC-hs PDU.
- A transmission sequence number (TSN) which allows the reordering operation at the terminal end.
- A size index identifier (SID) reveals the MAC-d PDU size, while the parameter N reveals the number of MAC PDUs of the size indicated in the SID.
- The F field indicates whether there are additional SID and N fields in the MAC-hs header or not.

In the MAC layer one can multiplex different services together into a single transport channel. This requires that both services have similar QoS characteristics, as it is not possible then for the BTS scheduler to separate them for different handling. In this case the multiplexing takes place in the RNC in MAC-d (as shown in Figure 4.27) – this is done by multiplexing several logical channels into a single MAC-d flow. In this case a

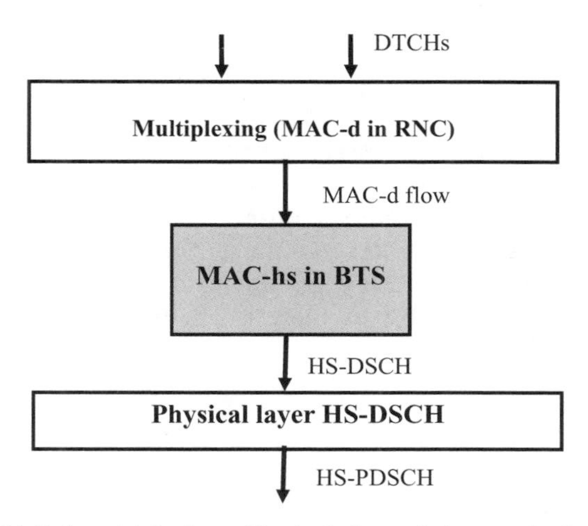


**Figure 4.27** MAC-d multiplexing of logical channels into a single MAC-d flow.

separate MAC-d header is added; otherwise, there is no need to have a MAC-d header with HSDPA and, thus, no actual operation takes place in MAC-d.

There are two logical channels relevant for HSDPA operation. These channels are mapped in the MAC layer to transport channels and then further in the physical layer to physical channels. The dedicated traffic channel (DTCH) carries user data, while the dedicated control channel (DCCH) carries the control information – like RRC signalling. The DCCH cannot be mapped to the MAC-d flow in Release 5, but in Release 6 the additional functionality is defined to ensure proper terminal behaviour just in case the connection to the serving HS-DSCH cell is lost. HSPDA MAC layer operation is specified in [7].

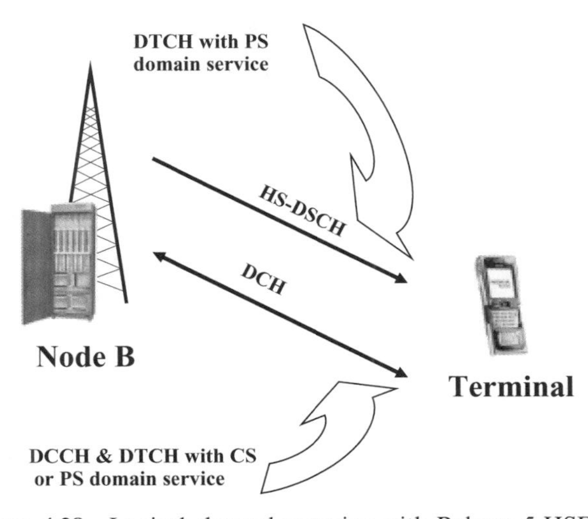


**Figure 4.28** Logical channel mapping with Release 5 HSDPA.

Also, the earlier described F-DPCH enables benefiting from reduced code resource needs if there are no services requiring a Release 99 type DCH to be allocated. For the DTCH, allocation depends on the type of service in question. For the circuit-switched (CS) domain service, obviously the DCH is always to be used (as shown in Figure 4.28), while for packet-switched (PS) domain allocation the choice is influenced by several service parameters as well as the radio resource situation.

The properties of the scheduler have an impact on what kind of service should be carried on HS-DSCH. Also if the delay requirements are very tight, there is less gain in having those allocated on HSDPA as there is not much the scheduler can do. The HSDPA radio resource management issues are covered in Chapter 6.

## 4.7 References

[1] H. Holma and A. Toskala (eds) (2004), *WCDMA for UMTS* (3rd edn), John Wiley & Sons, Chichester, UK.

[2] 3GPP, Technical Specification Group RAN, Physical channels and mapping of transport channels onto physical channels (FDD), 3GPP TS 25.211 version 6.7.0, Release 6, available at *www.3gpp.org*

[3] 3GPP, Technical Specification Group RAN, Multiplexing and channel coding (FDD), 3GPP TS 25.212 version 6.7.0, Release 6, available at *www.3gpp.org*

[4] 3GPP, Technical Specification Group RAN, Spreading and modulation (FDD), 3GPP TS 25.213 version 6.4.0, Release 6, available at *www.3gpp.org*

[5] 3GPP, Technical Specification Group RAN, Physical layer procedures (FDD), 3GPP TS 25.214 version 6.7.1, Release 6, available at *www.3gpp.org*

[6] 3GPP, Technical Specification Group RAN, Physical layer – measurements (FDD), 3GPP TS 25.215 version 6.4.0, Release 6, available at *www.3gpp.org*

[7] 3GPP, Technical Specification Group RAN, Medium Access Control (MAC) protocol specification, 3GPP TS 25.321 version 6.7.0, Release 6, available at *www.3gpp.org*

[8] 3GPP, Technical Specification Group RAN, Iub, 3GPP TS 25.433 version 6.8.0, Release 6, available at *www.3gpp.org*

[9] 3GPP, Technical Specification Group RAN, UE radio access capabilities definition, 3GPP TS 25.306 version 6.7.0, Release 6, available at *www.3gpp.org*

# 5

# HSUPA principles

Karri Ranta-Aho and Antti Toskala

This chapter covers the high-speed uplink packet access (HSUPA) principles for wideband code division multiple access (WCDMA) for Release 6 specifications. As discussed in Chapter 2, the Third Generation Partnership Project (3GPP) term was not HSUPA, but enhanced dedicated channel (E-DCH). Usage of the term 'HSUPA' instead of E-DCH follows the trend that began with employment of the term 'HSDPA'; and using the term 'HSUPA' synonymously for corresponding 'uplink improvement' has been widely adopted in the wireless industry, though not officially covered in 3GPP specifications.

This chapter introduces the key technologies as well as the channels and key procedures needed to implement HSUPA functionality. Towards the end of the chapter the medium access control (MAC) layer and mobility management as aspects are also covered. This chapter concludes with discussions on terminal capabilities and resulting data rates for the application layer.

## 5.1 HSUPA vs Release 99 DCH

HSUPA specification work started after the successful finalization of the first release of the high-speed downlink packet access (HSDPA) specifications for 3GPP Release 5 in mid-2002. HSDPA had enhanced downlink capabilities but the uplink capabilities did not match the HSDPA downlink. Thus, work on improving uplink performance started in 3GPP. The obvious choices were to investigate the techniques used for HSDPA and, if possible, adopt them for the uplink as well. The concept of HSUPA was born and in the course of 3 years the concept materialized into specifications.

HSUPA is not a standalone feature, but uses most of the basic features of the WCDMA Release 99 in order to work. Cell selection and synchronization, random access, basic mobility procedures, etc., are needed and remain unchanged with HSUPA operation. The only change is a new way of delivering user data from the user equipment (UE) to the Node B, all other parts of the specifications remain untouched. For example,

*HSDPA/HSUPA for UMTS* Edited by Harri Holma and Antti Toskala

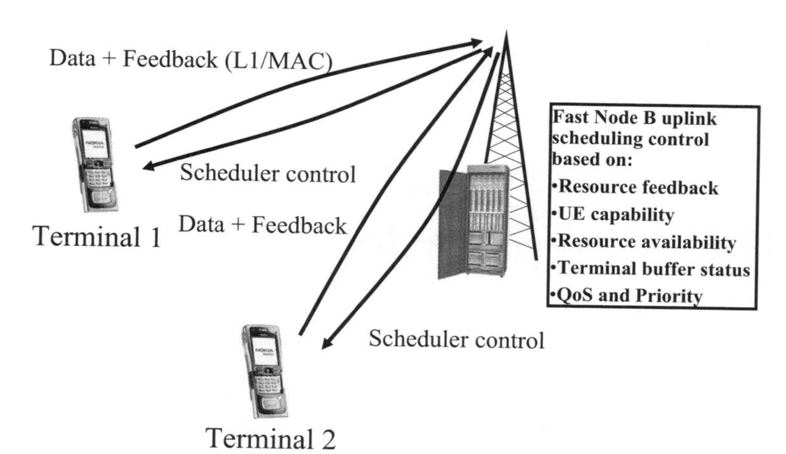


**Figure 5.1** HSUPA Node B (BTS) scheduling principle.

the basic power control loop functions in Release 99 are just as essential for HSUPA operation.

HSUPA provides a flexible path beyond the 384-kbps uplink which can be seen as the realistic maximum for WCDMA before HSUPA. A similar technology to that of HSDPA is being used by introducing fast uplink hybrid-ARQ (HARQ), Node B based uplink scheduling (as shown in Figure 5.1) and easier multicode transmission than with Release 99.

The focus in this chapter is on the additions to the Release 6 specifications that the HSUPA feature entails, and the functionality of the Release 99 uplink is described as a reference in the most relevant cases. Anyway, it is worth keeping in mind that none of the old features were replaced by HSUPA and that it was more of an add-on than a replacement. The complete description of all the features needed for, say, synchronization and cell search can be found in [1].

## 5.2 Key technologies with HSUPA

### *5.2.1 Introduction*

The HSUPA feature of the 3GPP WCDMA system is in fact a new uplink transport channel – E-DCH – that brought some of the same features to the uplink as the HSDPA with its new transport channel – high-speed downlink shared channel (HS-DSCH) – provided for the downlink. The E-DCH transport channel supports fast Node B based scheduling, fast physical layer HARQ with incremental redundancy and, optionally, a shorter 2-ms transmission time interval (TTI). Though – unlike HSDPA – HSUPA is not a shared channel, but a dedicated one, by structure the E-DCH is more like the DCH of Release 99 but with fast scheduling and HARQ than an uplink HSDPA: that is, each UE has its own dedicated E-DCH data path to the Node B that is continuous and independent from the DCHs and E-DCHs of other UEs. Table 5.1 lists the applicability of the key features for DCH, HSDPA and HSUPA.

**Table 5.1** HSDPA, HSUPA and DCH comparison table.

| *Feature* | *DCH* | *HSDPA (HS-DSCH)* | *HSUPA (E-DCH)* |
|---|---|---|---|
| Variable spreading factor | Yes | No | Yes |
| Fast power control | Yes | No | Yes |
| Adaptive modulation | No | Yes | No |
| BTS based scheduling | No | Yes | Yes |
| Fast L1 HARQ | No | Yes | Yes |
| Soft handover | Yes | No | Yes |
| TTI length [ms] | 80, 40, 20, 10 | 2 | 10, 2 |

Respectively, new signalling channels are needed (as shown in Figure 5.2); all the channels (excluding broadcast ones) shown in the figure are necessary for HSUPA operation. In Figure 5.2 it is assumed that downlink is on the DCH while in most cases it is foreseen that HSDPA could be used, but for clarity only the downlink DCH is shown in addition to HSUPA-related channels.

The channels for scheduling control – E-DCH absolute grant channel (E-AGCH) and E-DCH relative grant channel (E-RGCH) – as well as retransmission support on the E-DCH HARQ indicator channel (E-HICH) are covered in detail in later sections. The user data is carried on the enhanced dedicated physical data channel (E-DPDCH) while new control information is on the E-DPCCH, as will be covered in later sections. From the Release 99 DCH, the dedicated physical control channel (DPCCH) is unchanged and the need for the DPDCH depends on possible uplink services mapped to the DCH.

Unlike HSDPA, HSUPA does not support adaptive modulation because it does not support any higher order modulation schemes. This is due to the fact that more complex modulation schemes require more energy per bit to be transmitted than simply going for transmission with multiple parallel code channels using simple binary phase shift keying (BPSK) modulation.

In the downlink – due to the smaller dynamic range of transmitted channel power – there are cases where the downlink signal is transmitted in any case with more energy

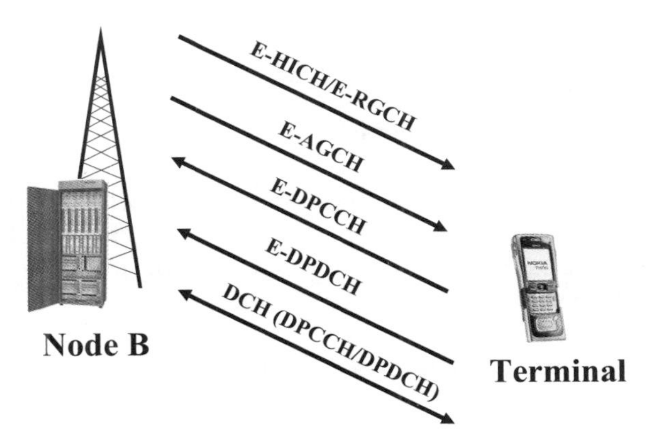


**Figure 5.2** Channels needed for HSUPA operation when downlink is on the Release 99 DCH.

than is necessary for good quality reception. Thus, the usage of higher order modulation can sometimes provide higher data rates with no extra transmitted power for HSDPA. In the uplink the situation is different and data rates high enough to require all the UE transmitted power even relatively close to the Node B are available with BPSK modulation and multicode transmission.

### 5.2.2 Fast L1 HARQ for HSUPA

The basic principle behind HARQ for HSUPA is the same as that for HSDPA. After each transmitted TTI the Node B indicates to the transmitting UE whether the packet was received correctly or not. In the event of incorrect reception the UE will retransmit the packet. The Node B tries to recover the packet by combining the energy of the retransmission with previous transmissions until the packet is received correctly or the maximum number of retransmissions is reached. The HSUPA HARQ may either use Chase combining where each retransmission is an exact copy of the initial transmission, or incremental redundancy where retransmissions contain additional redundancy bits for the initially transmitted bits.

The main differences between HARQ with HSUPA and HARQ with HSDPA are that HSUPA HARQ is fully synchronous and, with incremental redundancy, even transmitted redundancy versions can be predetermined; it also operates in soft handover. HARQ for HSUPA is discussed in more detail later in this chapter.

### 5.2.3 Scheduling for HSUPA

In Release 5, HSDPA moved downlink scheduling from the RNC to the Node B in order to be able to make scheduling decisions with minimum latency as close to the radio interface as possible. HSUPA scheduling does the same thing for the uplink and moves the scheduling to the Node B, but the similarities between HSDPA and HSUPA scheduling end there. With HSDPA all the cell power can be directed to a single user for a short period of time, and in this way can reach very high peak data rates for that particular UE, but simultaneously leaving all the others with a zero data rate. In the next time instant the Node B resources are used to serve some other UE and so on. Obviously, with HSUPA this is not possible, because when HSDPA is a one-to-many type of scheduling HSUPA is a many-to-one scheduling. The uplink transmission power resources of a cell are distributed evenly to the users or, to put it simply, each UE has its own transmitter and can only transmit data from that particular UE. So, obviously, in the uplink the cell's transmission power resources just cannot be given to a single UE at one time and to another UE at some other time, but users have their own transmitter power resource which clearly cannot be shared. This fact alone leads to a need to have a great level of parallelism in uplink scheduling and, thus, the dedicated channel approach was seen as the only feasible one with HSUPA, contrary to the shared channel approach of HSDPA.

As stated above, when considering HSUPA scheduling one should forget about HSDPA scheduling and, instead, think of very fast DCH scheduling. The shared resource of the uplink is the uplink noise rise, or the total received power seen in the Node B receiver. Typically, one UE – even when transmitting with full power – is unable

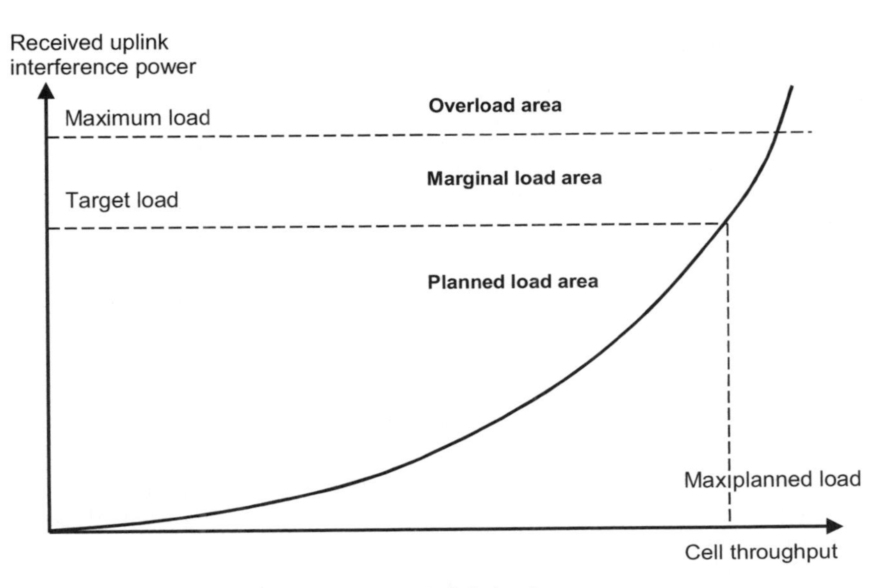


**Figure 5.3** Uplink load curve.

to consume that resource alone completely, but it is very beneficial for the scheduler to know at each time instant how much of that resource each UE will consume and to try to maintain the interference level experienced close to the maximum.

If we assume that all the active UEs want to transmit all the time with as high a data rate as possible, then the scheduler has very little to do in the uplink. Its task is simply to admit new users and while doing so downgrade the allocations of old users. This is, of course, very different from the HSDPA scheduler, but actually very much what the scheduler of a Release 99 system does. As it is not really possible for the RNC to schedule dynamically according to UEs' needs and chances to transmit, it simply provisions all the users with a certain maximum data rate that they are allowed to use whenever they can. If utilization of the maximum is very low or completely non-existent, if uplink loading is rising too high, or if new users are admitted to the cell, then the allocation is lowered.

Figure 5.3 shows the well-known load curve of the uplink where the relationship of cell throughput to uplink interference power is shown. Above a certain point the cell becomes unstable as all the users experience so much interference from the other users that they must increase their transmission power to be heard and start generating higher interference to the cell, affecting again other users and so on. This is known as a party effect, where in the end it does not matter how loud you shout, you still will not be heard. Obviously, this situation is highly undesirable in the WCDMA system and the uplink scheduler's primary task is to ensure that overload does not occur. The secondary, but for the end user just as important, task is to try and use as much of the uplink capacity as possible without running the risk of the cell becoming overloaded.

With HSUPA the scheduler is much closer to the radio interface, has more instantaneous information about the uplink interference situation, and can control uplink data rates in a rapid manner.

## 5.3 E-DCH transport channel and physical channels

### *5.3.1 Introduction*

As said in previous chapters, HSUPA is a new uplink transport channel, E-DCH, which supports enhanced features to those of the uplink transport channels of Release 99. Uplink transport channel processing for E-DCH is similar to the processing of the uplink DCH with two exceptions. There can be only one E-DCH transport channel in the UE as there may be multiple parallel DCHs that are multiplexed together to a single coded composite transport channel (CCTrCH) of DCH type. Nevertheless, the MAC layer can multiplex multiple parallel services to the single E-DCH. The other significant difference is HARQ support for the E-DCH which is provided in the transport channel processing chain and is, of course, something totally new.

After transport channel processing, the E-DCH maps to one or multiple parallel new dedicated physical data channels – E-DPDCHs – for physical layer transmission. This is completely parallel to uplink DCH processing chain and physical channels, so both E-DCH and DCH can coexist in the same UE with the restriction that the maximum DCH data rate is 64 kbps when the E-DCH is configured.

Using E-DPDCH transmissions a simultaneous and parallel control channel is sent a separate code channel – E-DPCCH. This E-DPCCH transmits all the necessary information about the E-DPDCH that is needed in order to know how to receive the data channel.

In the downlink three new physical channels were introduced to provide HARQ feedback and facilitate uplink scheduling. The E-DCH HARQ indicator channel (E-HICH) sends the HARQ acknowledgement information back to the UE. The E-RGCH provides relative step-up/down scheduling commands and the E-AGCH provides an absolute scheduling value for the UE. All the new physical channels are described in detail in [2] and covered in the following sections.

### *5.3.2 E-DCH transport channel processing*

Transport channel processing is the functionality that transforms the transport blocks delivered by the MAC layer to bits transmitted on physical channels. Figure 5.4 shows the overview of DCH and E-DCH transport channel processing from the MAC layer to the physical channels. Transport channel processing for both the DCH and E-DCH are described in detail in [3].

A single E-DCH transport channel processing chain always gets one transport block to process for transmission in one TTI, because – for the DCH – a set of transport blocks for each configured DCH will be delivered to the processing chain. In Figure 5.5 the differences between the elements of transport channel processing chains for the E-DCH and DCH are illustrated:

- *CRC attachment* for the E-DCH always attaches a 24-bit CRC to the transport block received from the MAC layer. In comparison, the CRC length for the DCH is configurable and can be 0, 8, 12, 16, or 24 bits.

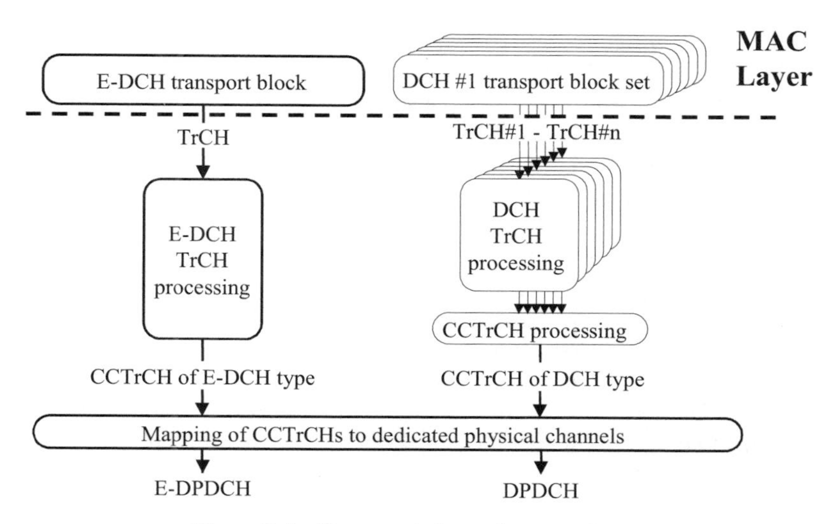


**Figure 5.4** Transport channel processing.

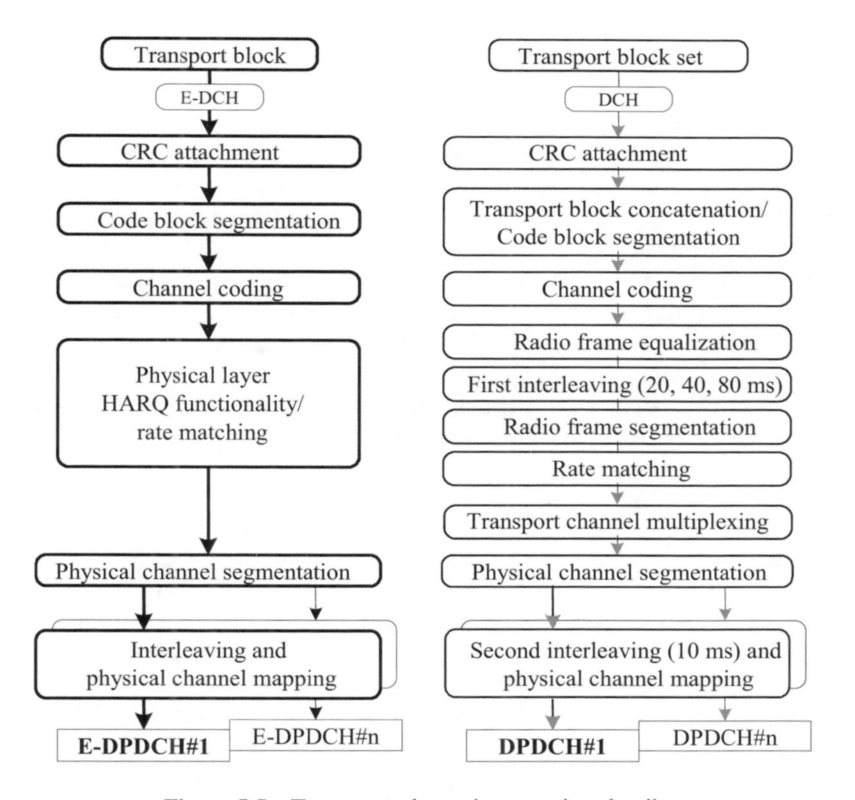


**Figure 5.5** Transport channel processing details.

- *Code block segmentation* for the E-DCH splits its input into equal size code blocks so that the length of the block does not exceed 5114 bits. For the DCH the same block first concatenates the transport block set to a single block of data before splitting. Also the size of the maximum code block with the DCH depends on the coding in use (5114 for turbo-coding and 504 for convolutional coding).
- *Channel coding* for the E-DCH is always turbo-coding with a code rate of 1/3. DCH channel coding may be either convolutional coding with code rates 1/2 or 1/3 or turbo-coding with a code rate of 1/3.
- *Physical layer HARQ funtionality/rate matching* for the E-DCH matches the channel codes output bits to the available physical channel bits and produces the different redundancy versions needed for incremental redundancy HARQ.
- *Physical channel segmentation* for the E-DCH distributes the channel bits among the multiple E-DPDCHs if more than one E-DPDCH is needed. The functionality is also the same in the corresponding block in the DCH processing chain.
- *Interleaving and physical channel mapping* for the E-DCH, as well as for the DCH, interleaves the bits in the radio frame and maps the bits to be transmitted to their final positions in the physical channel.

### *5.3.3 E-DCH dedicated physical data channel*

The E-DPDCH is a new uplink physical channel used for transmitting bits – as a result of E-DCH transport channel processing – from the mobile to the base station. It is a new channel that exists parallel to all the uplink dedicated channels of the 3GPP Release 5 (DPDCH and DPCCH used for uplink data transmission and HS-DPCCH used for HSDPA feedback delivery). Thus, with the introduction of HSUPA there may be up to five different types of dedicated channels transmitted simultaneously in the uplink.

The E-DPDCH has a very similar structure to the DPDCH of Release 99 with a few exceptions. They both support orthogonal variable spreading factors (OVSFs) to adjust the number of channel bits to the amount of data actually being transmitted. They both could go beyond the data rate that one physical data channel can support by transmitting multiple channels in parallel [4]. They both use BPSK modulation and follow the same fast power control loop.

The most notable properties of the E-DPDCH that the DPDCH does not have are the ones that make HSUPA what it is: the E-DPDCH supports fast physical layer level HARQ and fast Node B based scheduling. However, these are not really properties of the physical data channel as such, but the HARQ is visible in the transport channel processing chain and the scheduling is visible in the MAC layer. The differences in the actual physical data channels are a bit more subtle.

The biggest difference for E-DPDCH is the support of a spreading factor (SF) of 2, which allows delivering twice as many channel bits per code than the minimum spreading factor of 4 that the DPDCH supports. However, this is not the whole truth as the DPDCH could support six parallel SF4 codes as the E-DPDCH supports simultaneous transmission of two SF2 and two SF4 codes which both lead to the maximum physical layer bit rate of 5.76 Mbps.

The steps for supporting the data rates is different. Both DPDCH and E-DPDCH support spreading factors 256, 128, 64, 32, 16, 8, and 4 and respective physical channel

**Table 5.2** Physical channel bit rate steps for DPDCH and E-DPDCH.

| *Channel bit rates* | *DPDCH* | *E-DPDCH* |
|---|---|---|
| 15–960 kbps | SF256–SF4 | SF256–SF4 |
| 1.92 Mbps | 2 × SF4[1] | 2 × SF4 |
| 2880 Mbps | 3 × SF4[1] | — |
| 3840 Mbps | 4 × SF4[1] | 2 × SF2 |
| 4800 Mbps | 5 × SF4[1] | — |
| 5760 Mbps | 6 × SF4[1] | 2 × SF4 + 2 × SF2 |

[1] Multicode not supported in practice with DPDCH.

bit rates of 15, 20, 60, 120, 240, 480, and 960 kpbs with a single OVSF code transmission. If the 960-kbps channel bit rate provided by spreading factor 4 is insufficient to transmit all the data from transport channel processing, then both move to using two SF4 codes in parallel and achieve a 1920-kpbs physical channel bit rate. Things only go differently for the two types of dedicated data channels beyond this point. After two parallel SF4 codes the DPDCH has steps for three, four, five, and six parallel SF4 codes up to a 5.76-Mbps channel bit rate. The E-DPDCH has fewer steps and uses a spreading factor of 2. After two SF4 codes, the E-DPDCH goes directly to two SF2 codes. These steps are shown in Table 5.2. Usage of two SF2 codes provides some gains over using three or four parallel SF4 codes for the transmitter's peak to average power ratio and enables implementing a more power-efficient amplifier for UEs supporting very high data rate transmission.

The most notable physical layer difference between the two data channels is the new TTI length of 2 ms supported by the E-DPDCH. This is achieved by maintaining the 10-ms radio frame structure already familiar with the DPDCH, but when a 2-ms TTI is in use the 10-ms radio frame is divided into five independent sub-frames. The differences between the DPDCH and E-DPDCH are listed in Table 5.3. Table 5.4 shows the physical channel data rates for different spreading factors for both the DPDCH and E-DPDCH.

**Table 5.3** DPDCH and E-DPDCH comparison table.

| *Feature* | *DPDCH* | *E-DPDCH* |
|---|---|---|
| Maximum SF | 256 | 256 |
| Minimum channel data rate | 15 kbps | 15 kbps |
| Minimum SF | 4 | 2 |
| Maximum channel data rate | 960 kbps | 1920 kbps |
| Fast power control | Yes | Yes |
| Modulation | BPSK | BPSK |
| Soft handover | Yes | Yes |
| TTI length [ms] | 80, 40, 20, 10 | 10, 2 |
| Maximum no. of parallel codes | 6 × SF4[1] | 2 × SF2 + 2 × SF4 |

[1] Practical maximum for DPDCH is 1 × SF4.

**Table 5.4** Data rates of a single DPDCH and E-DPDCH code.

| *Spreading factor* | *Bit rate* (kbps) | *Bits/slot* | *Bits/radio frame* | *Bits/sub-frame*[2] |
|---|---|---|---|---|
| 256 | 15 | 10 | 150 | 30 |
| 128 | 30 | 20 | 300 | 60 |
| 64 | 60 | 40 | 600 | 120 |
| 32 | 120 | 80 | 1 200 | 240 |
| 16 | 240 | 160 | 2 400 | 480 |
| 8 | 480 | 320 | 4 800 | 960 |
| 4 | 960 | 640 | 9 600 | 1920 |
| 2[1] | 1920 | 1280 | 19 200 | 3840 |

[1] SF2 only supported by E-DPDCH.
[2] Concept of three-slot sub-frame applies to E-DPDCH only.

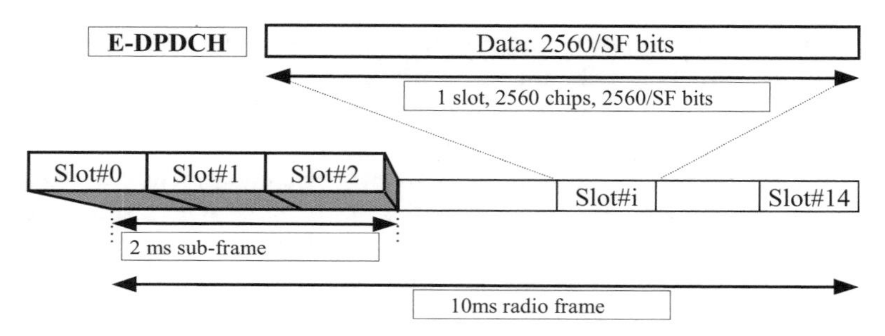


**Figure 5.6** E-DPDCH frame structure.

When a 10-ms TTI is in use all the 15 slots of the E-DPDCH radio frame are used to deliver the transport block processed by the E-DCH transport channel processing chain. In the case of a 2-ms TTI, each 2-ms sub-frame delivers one E-DCH transport block. Figure 5.6 depicts the E-DPDCH frame structure.

The E-DPDCH is not a standalone channel, but requires simultaneous transmission of the DPCCH. The DPCCH pilot bits are needed for channel estimation and signal to interference ratio (SIR) estimation purposes in the receiver, and the power control bit delivered by the DPCCH is needed for downlink power control. In addition to this, another control channel is needed in parallel with the E-DPDCH to deliver the information the receiver needs in order to know what format of E-DPDCH transmission to expect. This new control channel is named the E-DPCCH; the E-DPDCH cannot exist without a simultaneous E-DPCCH transmission.

### *5.3.4 E-DCH dedicated physical control channel*

The E-DPCCH is a new uplink physical channel used for transmitting out-of-band information about E-DPDCH transmission from the mobile to the base station. The E-DPCCH – like the E-DPDCH – is a new channel that exists parallel to all

**Table 5.5** Slot format of E-DPCCH.

| *Spreading factor* | *Bit rate* (kbps) | *Bits/slot* | *Bits/radio frame* | *Bits/sub-frame* |
|---|---|---|---|---|
| 256 | 15 | 10 | 150 | 30 |

uplink dedicated channels of 3GPP Release 5, and always accompanies E-DPDCH transmission.

To some extent the E-DPCCH does the same for E-DPDCH transmission as the DPCCH does for DPDCH transmission – that is, the control channel delivers the information needed to decode the corresponding data channel transmission. The principal difference between the two is that – in addition to information about the DPDCH – the DPCCH also provides common information related to, for example, channel estimation and power control, while the E-DPDCH only contains information about the E-DPDCH.

The E-DPCCH has only one possible slot format (as shown in Table 5.5), which uses a spreading factor of 256 and is capable of delivering 30 channel bits in a 2-ms sub-frame. It is designed to deliver 10 bits of information for each E-DPDCH TTI transmitted. The E-DPCCH uses the same (30, 10) second-order Reed–Muller coding as used for transport format combination indicator (TFCI) coding in the DPCCH. This means that the 10 information bits result in 30 bits to be transmitted in the physical channel. This number of bits can be carried by the E-DPCCH in 2 ms. If the TTI length of the E-DPDCH is 10 ms, then the 30-bit E-DPCCH sub-frame is repeated five times allowing reduced power level. With this procedure the same E-DPCCH structure can be employed regardless of the TTI used for E-DPDCH transmission. The E-DPCCH frame structure is illustrated in Figure 5.7 and the coding and mapping of E-DPCCH bits are described in detail in [3].

The 10 information bits on the E-DPCCH consist of three different segments:

- *E-TFCI*, the E-DCH transport format combination indicator of 7 bits indicating the transport format being transmitted simultaneously on E-DPDCHs. In essence, the E-TFCI tells the receiver the transport block size coded on the E-DPDCH. From this information the receiver can derive how many E-DPDCHs are transmitted in parallel and what spreading factor is used.

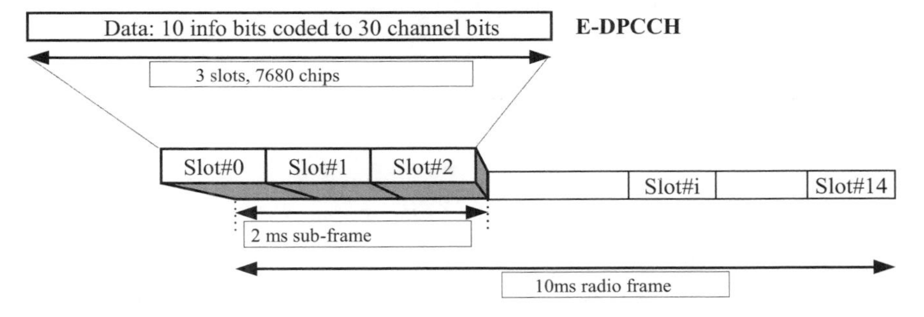


**Figure 5.7** E-DPCCH frame structure.

- *RSN*, the retransmission sequence number of 2 bits informing the HARQ sequence number of the transport block currently being sent on E-DPDCHs. The initial transmission of a transport block is sent with RSN = 0, the first with RSN = 1, the second with RSN = 2, and all subsequent transmissions with RSN = 3.
- *Happy bit* – as inferred from the name – is 1 bit only. It indicates whether the UE is content with the current data rate (or relative power allowed to be used for E-DPDCHs) or whether it could use higher power allocation.

### *5.3.5 E-DCH HARQ indicator channel*

The E-HICH is a new downlink physical channel used for transmitting positive and negative acknowledgements for uplink packet transmission. If the Node B received the transmitted E-DPDCH TTI correctly it will respond with a positive acknowledgement (ACK) and if it received the TTI incorrectly it will respond with a negative acknowledgement (NACK).

E-HICH information is BPSK-modulated with on/off keying and the modulation depends on which cell is transmitting the E-HICH. If the E-HICH is coming from the radio link set contained in the serving E-DCH radio link (transmitted from the base station that has the serving E-DCH cell), then both ACKs and NACKs are transmitted. The E-HICHs transmitted by Node Bs that do not contain the serving E-DCH cell only transmit ACKs. If such a cell does not receive the E-DPDCH TTI correctly, then it does nothing. The UE will continue retransmitting until at least one cell responds with an ACK.

The purpose of this arrangement is to save downlink transmission power. The assumption behind the different modulations is that those Node Bs that do not have the serving E-DCH cell are typically the ones that do not have the best connection to the UE and are more likely not to receive the E-DPDCH TTI correctly and have a significantly larger portion of NACKs than ACKs to be transmitted. In this way only the ACKs actually consume downlink capacity. As for the serving E-DCH radio link set the assumption is that typically more ACKs than NACKs are transmitted. When both ACK and NACK actually result in BPSK bit transmission (+1 and −1, respectively) the peak power required to transmit a reliable ACK is smaller when the receiver needs to separate + 1 from −1 than would be the case if it needed to separate +1 from 0 (as no transmission). ACK/NACK mappings to actual E-HICH transmissions from different types of cells are listed in Table 5.6.

**Table 5.6** ACK/NACK mapping to E-HICH.

| *E-DCH TTI reception* | *Logical response* | *Transmission on E-HICH* | |
|---|---|---|---|
| | | *Cells in the same RLS with the serving HSUPA cell* | *Other cells* |
| TTI received correctly | ACK | +1 | +1 |
| TTI received incorrectly | NACK | −1 | DTX |
| TTI not detected | — | DTX | DTX |

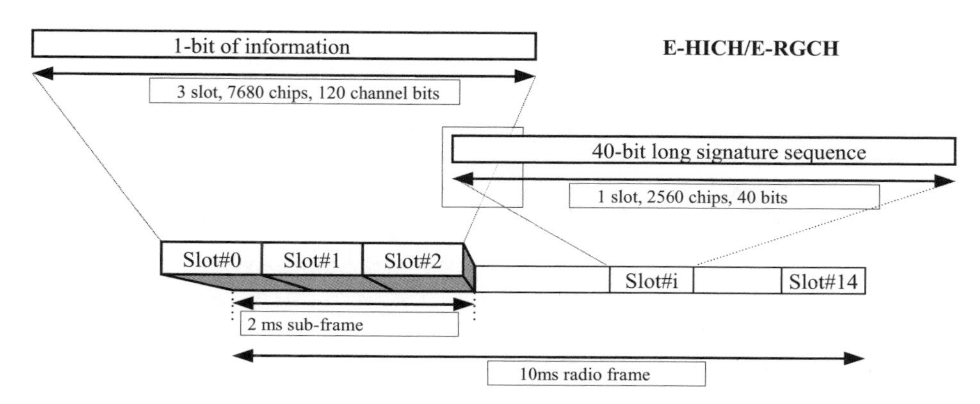


**Figure 5.8** E-HICH/E-RGCH frame structure.

All the cells in the same Node B are assumed to receive uplink E-DPDCH transmission in cooperation and, thus, even if there are multiple cells in the Node B participating in a softer handover the TTI reception either succeeds or fails only once, not separately in all the cells. Due to this all E-HICHs transmitted from the Node B containing the serving E-DCH cell transmit both ACKs and NACKs, effectively enabling the UE to combine the radio links for more reliable ACK/NACK detection.

E-HICH and E-RGCH channel structures are exactly the same, and they are illustrated in Figure 5.8. Each delivers 1 bit of information in three slots. In the case of a 10-ms TTI the three slots are repeated four times resulting in an 8-ms-long message. The exception is the E-RGCH transmitted from cells not belonging to the serving E-DCH radio link set. That channel always – regardless of the E-DCH TTI – transmits a 10-ms-long message (i.e., the three slots are always repeated five times).

The E-HICH/E-RGCH basic building block is a 40-bit-long orthogonal sequence which allows the orthogonal multiplexing of 40 bits in one slot on a single spreading factor 128-code channel. The same E-HICH/E-RGCH bit is repeated three times over three slots, but uses a different signature in each of the three slots following a deterministic code hopping pattern. This is because different signature pairs have different isolations in a real radio environment and, thus, the effect is averaged this way. Signature sequences and hopping patterns are defined in [2].

E-HICHs and E-RGCHs utilize 40-bit-long orthogonal sequences for multiplexing multiple E-HICHs and E-RGCHs (40 in total) to a single downlink code channel of spreading factor 128.

One cell can use multiple channelization codes to exceed the limit of 40 signatures (e.g., 20 E-HICHs and 20 E-RGCHs in a code) with the constraint that the E-HICH and E-RGCH intended for the same UE must be transmitted with the same channelization code. Figure 5.9 illustrates the Node B operation for combining up to 40 signatures on a single downlink channelization code.

### *5.3.6 E-DCH relative grant channel*

The E-RGCH is a new downlink physical channel used for transmitting single step-up/down scheduling commands that affect the relative transmission power the UE is allowed

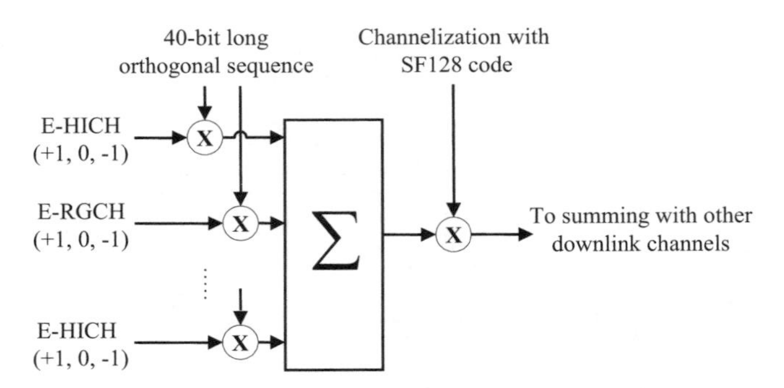


**Figure 5.9** Multiplexing up to 40 E-HICHs/E-RGCHs on a single code channel.

to use for data channel transmission (E-DPDCH), thus effectively adjusting the uplink data rate up/down.

Exactly like the E-HICH, E-RGCH information is BPSK-modulated with on/off keying and the allowed transmission depends on which cell is transmitting the E-RGCH. The cells belonging to the serving E-DCH radio link set of a UE by definition transmit the same E-RGCH content and, thus, enable the UE to soft-combine these channels. The cells not belonging to the E-DCH serving radio link set may only transmit down (otherwise there is no transmission) and thus only the serving cell and the other cells belonging to the same E-DCH radio link set can increase the UE's allowed maximum relative transmission power of the data channel. The up/down/hold mapping to E-RGCH is listed in Table 5.7.

The one difference between the E-RGCH and E-HICH is in soft-combining. All E-HICHs transmitted by the same radio link set (i.e., radio links transmitted from the same Node B and containing the same power control command) must carry the same content and thus are soft-combinable. E-RGCHs that are transmitted from the serving E-DCH radio link set must carry the same content and are soft-combinable. The purpose of the serving E-DCH radio link set is solely to reveal whether the E-RGCHs can be soft-combined or not. The purpose of this arrangement is to leave the network the freedom to either get a soft-combining gain for the E-RGCHs transmitted from whichever Node B also has a serving E-DCH radio link, or to have only a UE-specific E-RGCH from the serving E-DCH cell and use a common E-RGCH for all other UEs.

**Table 5.7** Message mapping to E-RGCH.

| *Scheduler decision* | *Transmitted message* | *Transmission on E-RGCH* | |
|---|---|---|---|
| | | *Cells in the serving E-DCH RLS* | *Other cells* |
| Increase UE's allocation | UP | +1 | Not allowed |
| Decrease UE's allocation | DOWN | −1 | −1 |
| Keep the current allocation | HOLD | DTX | DTX |

As mentioned above, E-RGCH frame structure and signature generation is exactly the same as for the E-HICH (Figures 5.8 and 5.9). Differentiation is in the higher layers – that is, the network informs the UE by means of RRC signalling that signature $x$ on channelization code $k$ is the E-HICH and signature $y$ on the same channelization code is the E-RGCH.

### *5.3.7 E-DCH absolute grant channel*

The E-AGCH is a new downlink physical channel used for transmitting an absolute value of the Node B scheduler's decision that lets the UE know the relative transmission power it is allowed to use for data channel transmission (E-DPDCH), thus effectively telling the UE the maximum transmission data rate it may use.

The E-AGCH delivers 5 bits to the UE for the absolute grant value, indicating the exact power level the E-DPDCH may use in relation to the DPCCH. In addition, the E-AGCH carries a 1-bit indication for the absolute grant scope. With this bit the Node B scheduler can allow/disallow UE transmission in a particular HARQ process. This bit is only applicable for 2-ms TTI E-DCH operation. In addition to this the E-AGCH uses a primary and a secondary UE-id for identifying the intended receiver and delivering one additional bit of information. The E-AGCH coding chain is illustrated in Figure 5.10 and the detailed specification for each step can be found in [3]:

- *Absolute grant value* is a 5-bit integer number ranging from 0 to 31 that has a specific mapping [3] to the E-DPDCH/DPCCH power ratio the UE may use.
- *Absolute grant scope* can be used to activate/de-activate a particular HARQ process (identified by the E-AGCH timing) or all HARQ processes. The absolute grant scope can only be used with a 2-ms E-DCH TTI.

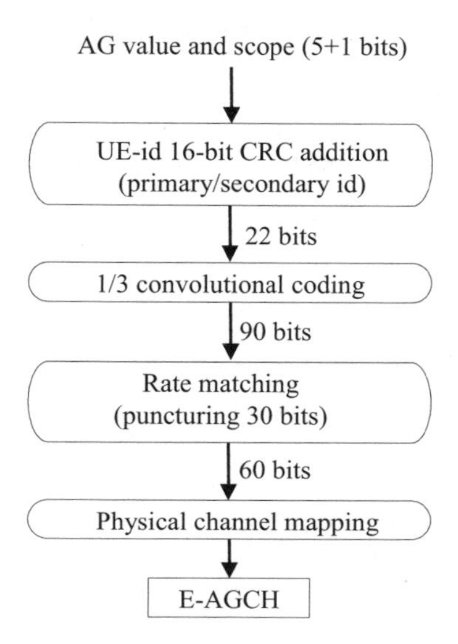


**Figure 5.10** E-AGCH coding chain.

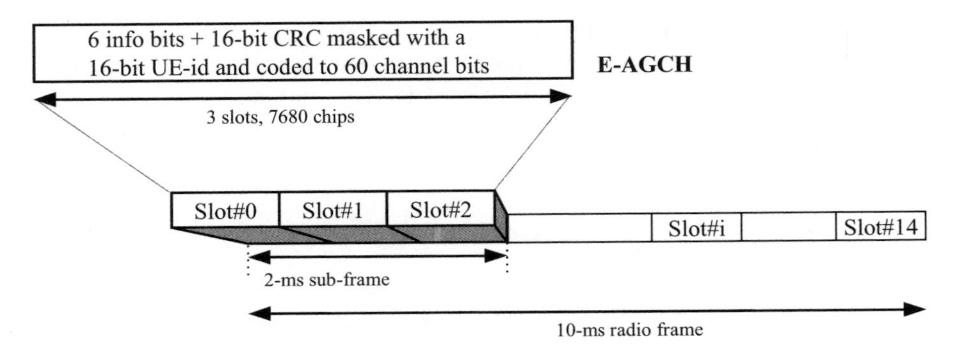


**Figure 5.11** E-AGCH frame structure.

- *Primary/Secondary UE-id* or primary/secondary E-DCH radio network temporary identity (E-RNTI) is used to mask the CRC of the E-AGCH. Each UE may have up to two UE-ids which it checks from each E-AGCH and if it detects one or the other as matching the transmission it knows that the E-AGCH transmission was destined for it.

The structure of an E-AGCH is very similar to an HS-SCCH for HSDPA. A 16-bit CRC is calculated over the 6 information bits and masked with either a primary or a secondary UE-id. With these ids the UE knows whether the E-AGCH transmission was meant for it or not. The package is then coded and rate-matched to fit the three-slot-long (2-ms) SF 256 channel. If a 10-ms E-DCH TTI is used the three slots are repeated five times to fill the whole radio frame. Figure 5.11 depicts the E-AGCH frame structure.

### *5.3.8 Motivation and impact of two TTI lengths*

While HSDPA only supports a single TTI (2 ms), with HSUPA there are two TTI lengths – 2 and 10 ms – that can be chosen. The motivation for the 2-ms length was the potential delay benefit while 10 ms was needed for range purposes to ensure cell edge operation.

A potential delay benefit could be obtained if there are not too many retransmissions using a 2-ms TTI, as the delay between retransmissions is shorter compared with the 10-ms case. A problem occurs when approaching an area of low geometry (closer to the cell edge) where signalling using a 2-ms period starts to consume a lot of transmission power, especially at the BTS end. This is illustrated in Figure 5.12. The difference from HSDPA is that now potentially a much larger number of users are expected to be active simultaneously and, thus, aiming to also provide downlink signalling to such a large number of users using a 2-ms period would become impossible.

As can be seen from Chapter 8, with data rates below 2 Mbps there are no major differences from the capacity point of view regardless of the TTI used. When going above 2 Mbps per user, then the block size using 10 ms would get too big and, thus, data rates above 2 Mbps are only provided using a 2-ms TTI. As with macro-cells, practical data rates in the uplink have limitations due to transmission power limitations. This means

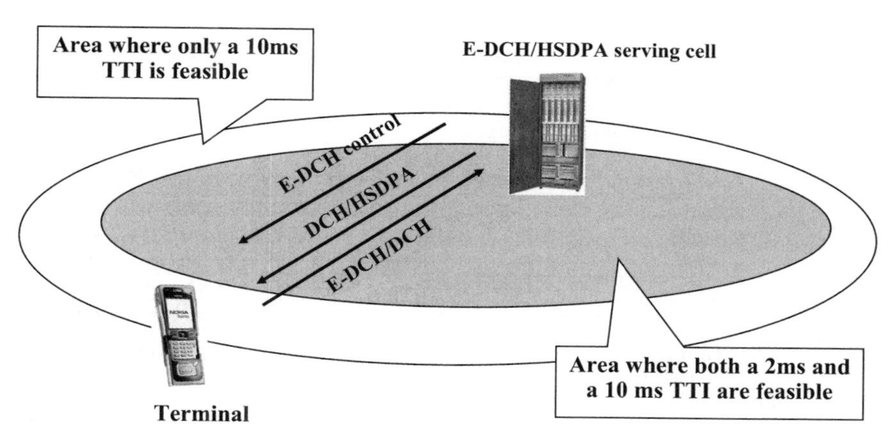


**Figure 5.12** 2-ms and 10-ms TTI applicability in a cell.

the 10-ms TTI is expected as the starting value for system deployment; this has also been reflected in terminal capabilities (where a 2-ms TTI is optional for most categories).

## 5.4 Physical layer procedures

The physical layer operation procedure of HSUPA has fewer elements specified in the physical layer specifications compared with HSDPA. One of the key reasons is that more of the functionality regarding terminal transmission control is located in the MAC specifications [9]. Physical layer procedures cover the handling of physical layer signalling from the Node B and HSUPA operation in compressed mode. Further, the HARQ operation procedure is covered in different parts of the specification in different channel descriptions as well as different timing sections.

### *5.4.1 HARQ*

The fundamental principle underlying HSUPA HARQ operation has a lot of similarities to HSDPA in terms of retransmission combining. Now, as the link direction is different, the soft buffer is maintained by the Node B instead of the terminal. The terminal will keep unacknowledged data in memory, and the MAC layers trigger retransmission if the physical layer provides a NACK as received from the Node Bs in the E-DCH active set. Both soft (Chase) combining and incremental redundancy are available with HSUPA as well (using the principles behind the combining methods covered in Chapter 4). The fundamental difference in operation procedure is the synchronous nature of the HSUPA HARQ process. As a function of the TTI being used – 2 ms or 10 ms – all the remaining timing events are defined, including the number of HARQ processes. With a 10-ms TTI, there are four HARQ processes, and the processing times are fixed within a 2-ms window. The window results from the timing rule set for downlink signalling channel association with uplink data transmissions. Depending on the uplink timing, downlink signalling channels are set with the 2-ms resolution for the possible starting point. HARQ process

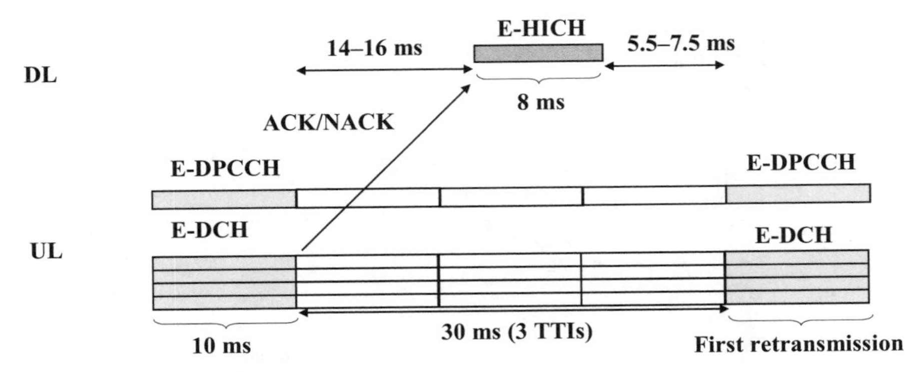


**Figure 5.13** HSUPA HARQ process timing with a 10-ms TTI.

timing with a 10-ms TTI is shown in Figure 5.13, where the resulting four HARQ processes are described.

There is no need to configure the number of HARQ processes, which gives clear implementation guidance to Node B implementation in terms of which processing time needs to be met in all cases. The Node B processing time requirement is thus approximately 14 ms, using the exact definition given in [2] and in [5]. The range up to 16 ms cannot be easily utilized as the resulting time available depends on terminal uplink timing, which is difficult to control in a network with mobility included. The extra delay for a single retransmission is 40 ms (including the waiting/processing/signalling time + the additional TTI for retransmission). With a 2-ms TTI there are respectively eight HARQ processes in use (as shown in Figure 5.14).

The use of a synchronous HARQ process removes the need to know as part of the data stream which HARQ process is being used – as is the case for HSDPA to enable downlink scheduling flexibility. With HSUPA the timing tells exactly which HARQ process is in question, and the only indication needed is whether the data are new or a retransmission. This is needed to avoid buffer corruption – either due to a signalling error (terminal missing the ACK) or, especially in the case of soft handover, due to a Node B not being aware of the possible ACK sent by another Node B in the active set.

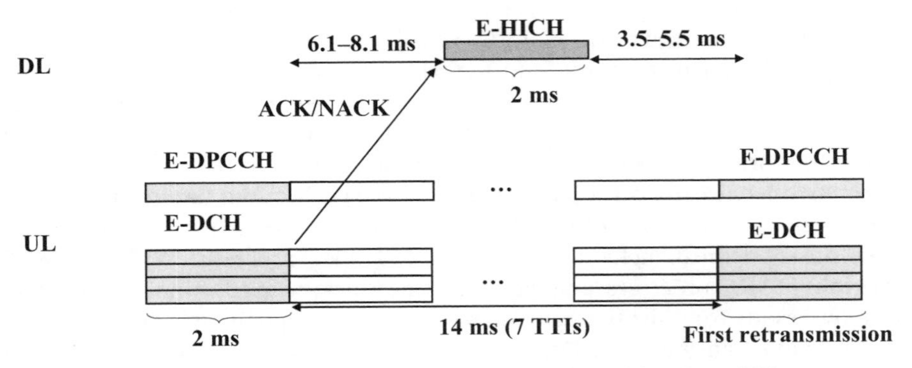


**Figure 5.14** HSUPA HARQ process timing with a 2-ms TTI.

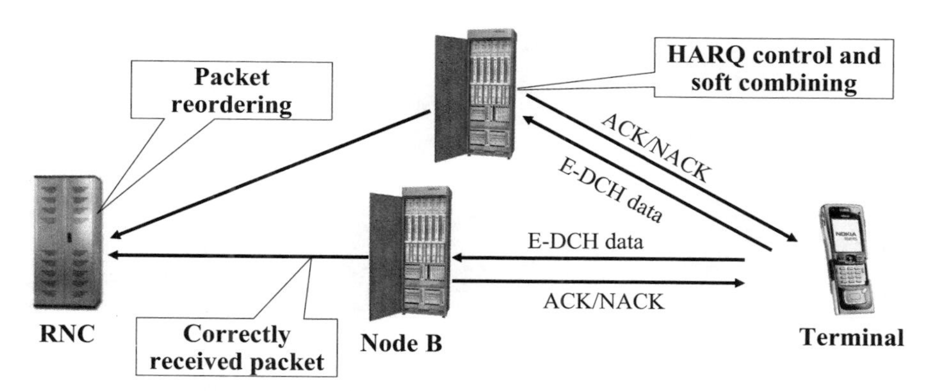


**Figure 5.15** HSUPA ARQ operation in soft handover.

### 5.4.2 *HARQ and soft handover*

Operation with an active set larger than 1 causes additional implications for HARQ operation. While with HSDPA only one Node B was involved in HARQ handling, with HSUPA all the Node Bs in the active set are involved. HARQ operation is done using similar rules to those for uplink power control. If the single Node B part of the active set sends an ACK, then the information given to the MAC layer is that an ACK has been received and the MAC layer will consider the transmission successful and move on to the next packet. The principle underlying HARQ operation in soft handover is shown in Figure 5.15. As the different Node Bs handle the process independently, packet ordering cannot be preserved by them and, thus, has to be adjusted in the RNC. This is also reflected in the protocol architecture – in the form of an additional MAC entity in the RNC.

For NACKs from the non-serving cells mapping is defined as a zero-sequence and, thus, NACKs are not really transmitted from cells other than serving cells. If the serving cell is in softer handover with other cells defined as belonging to the same E-DCH radio link set, then NACKs are transmitted from all of these cells in order to enable combining in the receiver similar to that of the power control commands received in the softer handover case. ACK/NACK mapping is described in detail in Section 5.3.5.

### 5.4.3 *Measurements with HSUPA*

While HSDPA does not directly introduce new terminal physical layer measurements that need to be reported to the network, unless you count channel quality information (CQI) as one, with HSUPA a new physical layer measurement is added that relates to terminal power headroom. UE transmission power headroom (UPH) is defined in [6] as being the ratio of the maximum UE transmission power and the corresponding DPCCH code power. The maximum power is either according to the power level of the terminal power class or a lower level if the maximum power is restricted by the UMTS Terrestrial Radio Access Network (UTRAN). UPH reporting is illustrated in Figure 5.16.

UPH indicates the available power resources – that is, whether or not and by how much the terminal could increase the data rate from the current situation. If the reported

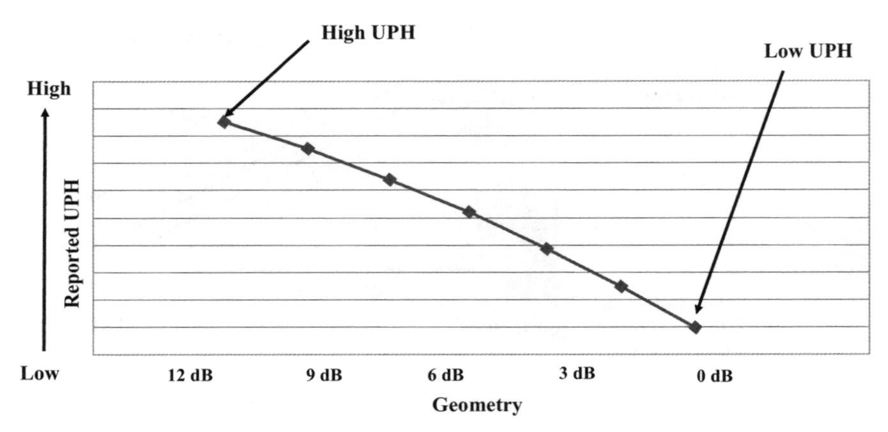


**Figure 5.16** HSUPA UPH measurement.

UPH value is low then the terminal is in a power-limited situation, and if the reported UPH is high then there is room to increase the data rate. The resulting UPH is a combination of the consequent path loss situation to the Node B with the best uplink and a varying uplink channel condition. The measurement does not try to cover HS-DPCCH activity (which is not continuous) or DPCH activity as those would only add inaccuracy in the measurement. Due to the inherent delays and measurement accuracy, UPH measurement cannot be used in the same way as CQI is with HSDPA.

At the Node B end, HSDPA measurements for non-HSDPA power are also extended to cover HSUPA downlink channels, making it 'non-HSDPA/HSUPA' power – this gives an idea of how much of the Node B power resource is used for purposes other than HSDPA/HSUPA. HSDPA/HSUPA Node B transmission power use can be based either on the principle of using all power, or of using power according to RNC allocation. In the latter case the reporting of HSDPA/HSUPA power would only add little value, and in the former case the maximum power is known in any event. Information about non-HSDPA/HSUPA power allows derivation of the other side's transmission power.

At the uplink end, existing Node B measurements can be used for HSUPA purposes. Received total wideband power (RTWP) includes the impact of all transmissions in the network on the particular Node B receiver. Another new feature is the provision of bit rate measurement. This is not a physical layer measurement but is rather determined in the MAC layer (MAC-e). This gives an indication of terminal-specific throughput in the uplink direction.

## 5.5 MAC layer

### *5.5.1 User plane*

The HSUPA MAC layer has more functionality in the specifications than HSDPA. This is due to the fact that scheduling control is carried out by the base station, but the scheduling itself actually controls the uplink data transmission of the E-DCH. The detailed algorithm on how the base station should define what kind of control infor-

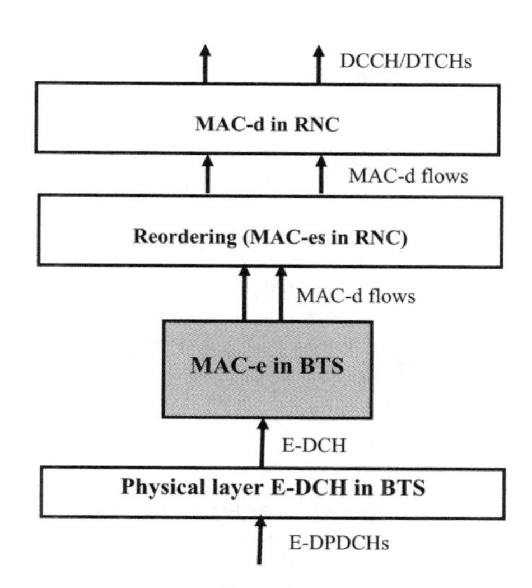


**Figure 5.17** User data flow through MAC layer with HSUPA.

mation to send to the mobile station is not specified, but MAC layer operation for controlling the terminal as well as additional feedback on top of physical layer feedback (the happy bit) has been defined for the UE. HARQ operation – described in Section 5.4.1 – including the soft handover behaviour typical of HARQ also belongs to MAC layer specifications.

As was shown in Chapter 3, the HSUPA MAC layer was divided between the RNC and BTS. MAC-e handles the other tasks, and MAC-es in the RNC guarantees in-sequence delivery to the RLC layer inside the RNC. The flow of user data through the MAC layer with HSUPA is illustrated in Figure 5.17. The logical channels (DCCH and DTCH) are then provided to the RLC layer.

When the logical channel is being mapped to use HSUPA (E-DCH) there is no need for the MAC-d header, thus MAC-d takes no part in the data flow. The MAC-es header has the necessary transmission sequence number (TSN) to enable reordering in the RNC. The MAC-e header has a data description indicator (DDI) which identifies the logical channel as well as information about the MAC-d flow and the MAC-d PDU size. The N parameter in the header reveals the number of consecutive MAC-d PDUs belonging to the same DDI. The MAC-e PDU structure is illustrated in Figure 5.18, indicating how one or multiple MAC-es PDUs are constructed.

### *5.5.2 MAC-e control message – scheduling information*

The scheduling information (SI) in the MAC-e PDU in Figure 5.18 stands for the MAC level SI message the UE provides to the serving HSUPA Node B. The SI consists of four information elements totalling 18 bits:

1. Total E-DCH buffer status (TEBS, 5 bits) reveals the total amount of data in the UE's transmission buffer. This information can be used by the scheduler for deciding the data rate the UE could actually use.

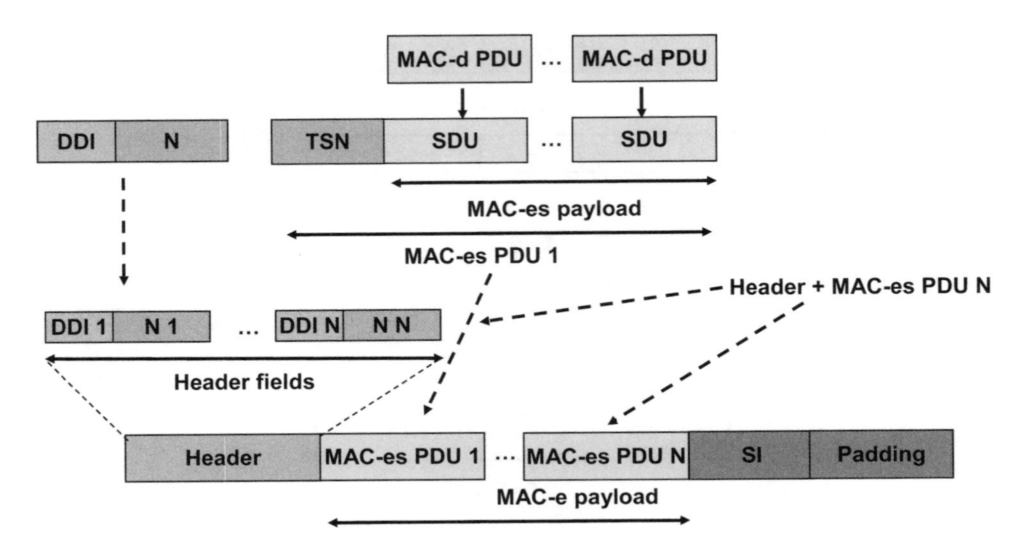


**Figure 5.18** HSUPA MAC-e PDU structure.

2. Highest priority logical channel ID (HLID, 4 bits) indicates the highest priority logical channel that has data in the UE's transmission buffer.
3. Highest priority logical channel buffer status (HLBS, 4 bits) reveals the amount of data in the buffer for the logical channel indicated by the HLID. The HLID and HLBS can be used by the Node B scheduler for deciding which UEs should be served first or served with higher data rates.
4. UPH (5 bits) indicates the Node B of the power ratio of maximum allowed UE transmit power to DPCCH pilot bit transmit power. In essence, the ratio tells the Node B scheduler how much relative power the UE can use for its data transmission. This information can be used by the Node B not to schedule any given UE a higher power allocation than it is actually capable of transmitting due to transmit power limitation.

SI is either transmitted periodically or triggered by data arriving at an empty buffer with the condition that the UE had no permission to transmit at any data rate. The Node B scheduler may use SI in addition to the happy bit in the E-DPCCH and the observed allocation utilization seen for a specific UE.

### *5.5.3 Selection of a transport format for E-DCH*

Scheduler commands take effect in the UE's selection of transport format (number of bits to be transmitted in a TTI) for the E-DCH transport channel. This E-DCH transport format combination (E-TFC) selection process is a part of the MAC layer. As described earlier, the relative grant and absolute grant commands on the E-RGCH and E-AGCH adjust the maximum allowed E-DPDCH to DPCCH power ratio the UE may transmit. This takes effect in the E-TFC selection process in the following way.

Initially, the UE has a list of transport block sizes and the relative powers that a transmission of a specific transport block requires. It also has a MAC-d flow-specific power offset for each assigned MAC-d flow (service). With this the UE can have different retransmission probabilities for same-sized transport blocks depending on which MAC-d flow it is transmitting: for example, a highly delay-critical service could have a higher MAC-d flow-specific power offset and, thus, a higher probability of getting through without requiring a retransmission.

The UE starts preparing to transmit an E-DCH transport block in a TTI by evaluating how much power it can at most use in the transmission of E-DPDCHs. At this stage the DPCCH, E-DPCCH and, potentially, the DPDCH transmission power levels of the TTI to be transmitted are already known. Then, it checks the power offset of the highest priority MAC-d flow that has data in the buffer, as this will be the power offset to be used on top of the transport block size specific power offset when actually transmitting E-DPDCHs. A transport block size with a MAC-d flow-specific power offset selected is called, as already mentioned, an 'E-DCH transport format combination' (E-TFC).

Then, the UE checks the current maximum power offset it may use for E-DPDCH transmission. This is the parameter controlled by the relative grant and absolute grant transmissions. The UE selects the largest transport block that it can fill with data and that fulfills the condition that the power offset of the transport block + the selected MAC-d flow-specific power offset is lower than or equal to the maximum allowed power offset – that is, it selects the highest E-TFC that scheduling has allowed it to use. If transmission of this E-TFC is not blocked due to the maximum transmit power of the UE, the UE processes the transport block and transmits it on the E-DPDCHs with the gain factor selected. If the transmission of that E-TFC is not possible due to UE power limitation, then the UE selects the largest E-TFC it has sufficient power to transmit and goes forward with that. The E-TFC selection procedure is illustrated in Figure 5.19.

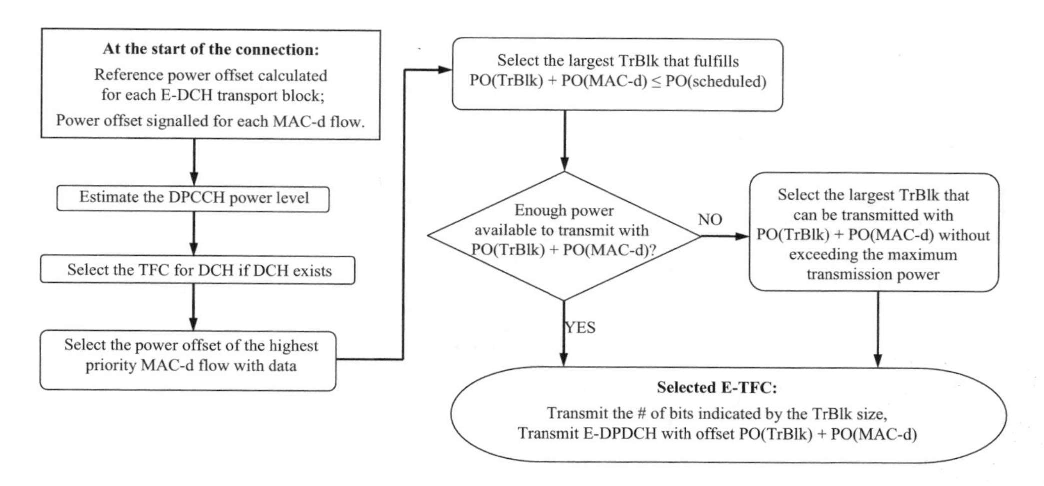


**Figure 5.19** HSUPA E-TFC selection.

### 5.5.4 *E-DCH coexistence with DCH*

As described earlier, transport block processing for the E-DCH and DCH is completely different and is carried over the air interface by completely separate physical channels, too. Apart from one very important fact – that is, the shared power amplifier of the UE – simultaneous transmission of the E-DCH and DCH could be considered to be coming from two different UEs. There is only so much power the UE is capable of transmitting and if the DCH is using some of it there is less for the E-DCH.

The interaction between the DCH and the E-DCH in simultaneous transmission from the same UE is simply so that TFC selection is first done for the DCH and the power used by doing this is, of course, no longer available for the E-TFC selection process. This means that the DCH has an absolute first take right to the UE power resource, or the DCH has absolute priority over the E-DCH. The reason for this is that the E-DCH is designed for uplink packet access and, thus, if there are any circuit-switched services they must be mapped to the DCH. As circuit-switched services are not very tolerant of sudden and frequent data rate changes it makes sense to let, say, a normal AMR voice call to take the power it needs and send it with that on DCH and use whatever is left for packet data transmission on the E-DCH. The power allocation of the TFC and E-TFC selection processes of the UE is illustrated in Figure 5.20. The TFC and E-TFC selection processes are defined in [9]. It should be noted though that the maximum DCH data rate allowed by the specification is 64 kbps when configured parallel to the E-DCH.

All of this has one notable impact on radio bearer mapping. If a DCH is configured parallel to an E-DCH for a single UE, then the signalling radio bearer must also be mapped to the DCH. This is due to the fact that the signalling radio bearer is the most important bearer, and it would not be acceptable to let the power required to transmit a voice frame prevent the transmission of a signalling message – that is, the SRB may only be mapped to the E-DCH if there are no DCHs configured for the UE.

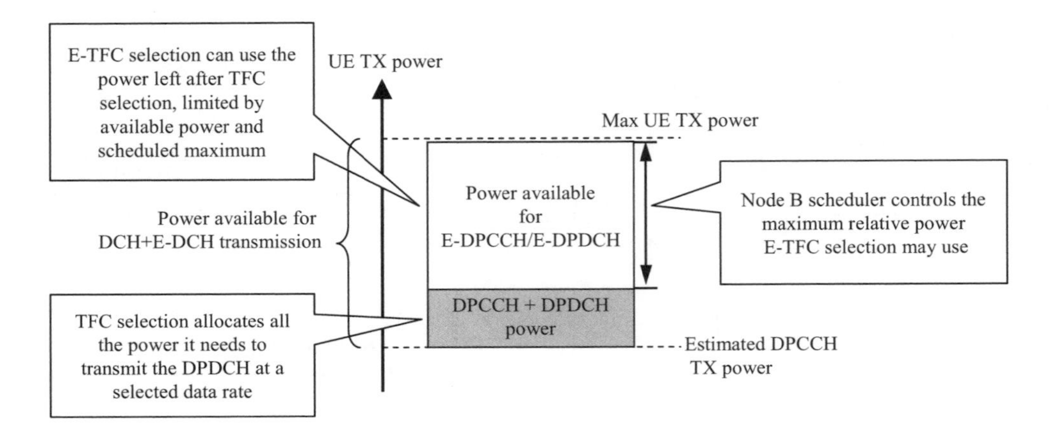


**Figure 5.20** E-DCH and DCH power resource sharing.

### 5.5.5 *MAC-d flow-specific HARQ parameters*

One fundamental difference between uplink DCHs and the transport channel of HSUPA, E-DCH, is that there can be multiple DCHs but there can only be one E-DCH configured for the UE. The possibility of configuring multiple DCHs simultaneously allows for quality of service (QoS) differentiation of the different transport channels in the Release 99 uplink. As there is only one E-DCH, QoS differentiation has to be done in a different way.

Different services can be mapped to different MAC-d flows and each MAC-d flow can have specific attributes:

- A maximum number of HARQ retransmissions before the UE drops the packet.
- A MAC-d flow-specific power offset that is added on top of the transport block specific power offset. A larger power offset means a lower probability of needing a retransmission and, thus, lower latency.

Thus, the two parameters impact on HARQ operation and, for example, delay-tolerant background services can have a lower power offset and higher retransmission probability while streaming services can have a lower maximum retransmission count as they tolerate a few lost packets before there is an impact on the service.

The network can configure which MAC-d flows can be multiplexed together and the highest priority MAC-d flow being transmitted dictates the QoS parametrization of a transmission.

### 5.5.6 *HSUPA scheduling*

HSUPA scheduling works by adjusting the limitations of the UE's E-TFC selection process. This process and its interaction with the TFC selection of Release 99 DCH channels is described in more detail in Section 5.5.3. The basic parameter that Node B scheduling adjusts is the maximum gain factor the UE may use for E-DPDCH transmission. The gain factor means the transmitted E-DPDCH power level in relation to the DPCCH power level. As fast power control keeps the received DPCCH power level fairly constant, the scheduling actually adjusts the maximum allowed received E-DPDCH power level of a given UE, or the share of the uplink noise rise a given UE may use.

HSUPA scheduling is facilitated using three physical channels, E-AGCH and E-RGCH in the downlink and a happy bit carrying E-DPCCH in the uplink. In addition to this the E-DPDCH delivers an SI message in the MAC-e header for additional details for the Node B scheduler. SI can be used in the Node B scheduler in addition to the happy bit received on the E-DPCCH and the observed power allocation usage a given UE is using. It should be noted, though, that SI is only available when it is actually triggered for transmission either as a result of increased data levels in the buffer or as a result of timer elapse, and when the scheduling HSUPA cell receives the transmitted packet correctly. The channels used for HSUPA scheduling as well as the information carried by them are shown in Figure 5.21.

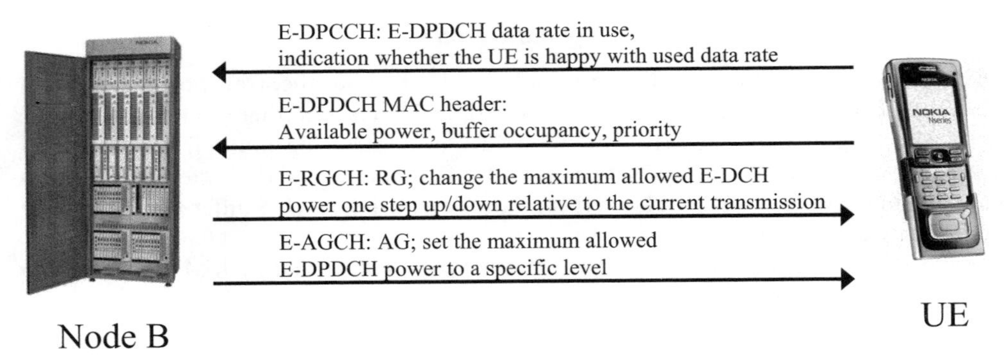


**Figure 5.21** HSUPA – scheduling-related information exchanged over the air.

The MAC layer defines the rules for setting the happy bit transmitted on the E-DPCCH. The basic information the happy bit delivers is whether or not the UE can empty its buffers with the current allocation in N TTIs, where N is a parameter signalled by the network. If the UE estimates that all the data in its E-DCH transmission buffer can be transmitted in N TTIs without increasing its current maximum, then the UE indicates that it is happy. Otherwise, it informs the Node B that it is unhappy.

Simply based on the single bit on the E-DPCCH and knowledge of whether the UE is actually transmitting with the maximum allocation or below it ,the Node B scheduler can make decisions on upgrading, downgrading or keeping a given UE's allocation unchanged. If, say, the UE is transmitting below its current maximum for multiple TTIs in a row, it makes sense to downgrade its allocation and perhaps upgrade some other UE indicating that it is unhappy.

Thus, the Node B receives information in the uplink of the UE's current situation (happy bit, SI, currently used data rate) and adjusts the maximum allowed relative transmit power of a UE's E-DPDCHs in the uplink by issuing relative grant (up/down/hold) and absolute grant messages (use relative power X) in the downlink. E-TFC selection translates the maximum allowed relative power to a transport format (i.e., the number of bits to be transmitted in a TTI). Figure 5.22 illustrates the change in maximum E-DPDCH/DPCCH power ratio and the UE utilizing the higher allowed relative power by transmitting with a higher data rate.

### *5.5.7 HSUPA scheduling in soft handover*

With HSUPA, uplink soft handover impacts the scheduling operation. While HSDPA sends data from one base station only, with HSUPA all the base stations in the E-DCH active set receive the transmission from the terminal. Thus, all the base stations are impacted by the transmission in terms of noise rise, as seen by the base station receiver. Even if there are multiple base stations receiving the data (as also discussed in Section 5.7), there is only one base station that acts as the serving E-DCH cell. The serving E-DCH cell uses all the available scheduling methods – that is, both relative and

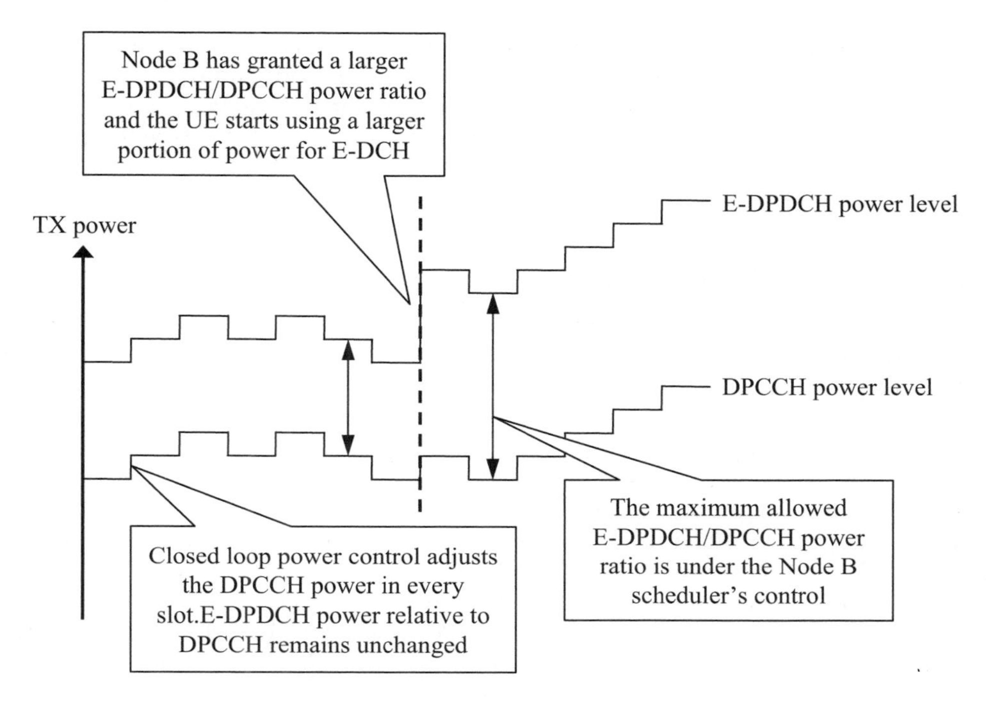


**Figure 5.22** Node B scheduler controls the maximum allowed E-DPDCH-to-DPCCH power ratio.

absolute grants. Other base stations that are part of the active set use only the relative grant and only send either 'hold' or 'down' commands (as shown in Figure 5.23).

The scheduling operation of a cell that is not part of the E-DCH active set should be seen as part of the overload control mechanism for the system. As sending individual down commands could be resource-consuming, the system may configure several

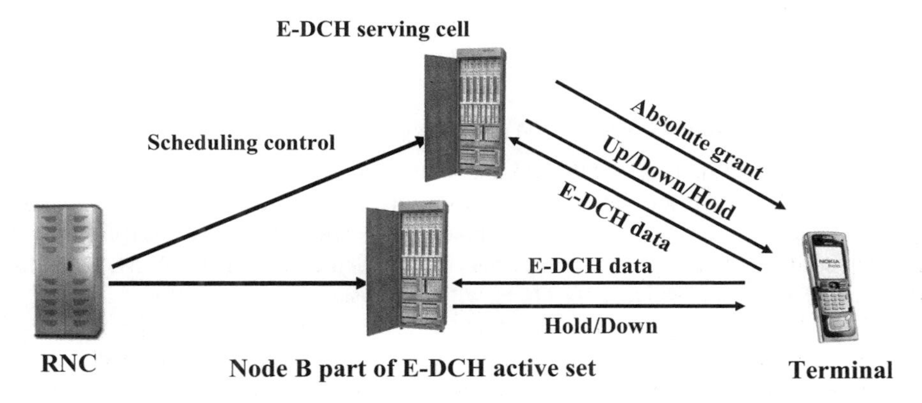


**Figure 5.23** HSUPA scheduling in soft handover.

terminals to listen for the same sequence from the non-serving HSUPA cells. This enables a smaller signalling overhead and rapid reaction for the overload condition. The terminal may only increase the data rate when there is an up command from the serving HSUPA cell and no down commands from any of the other cells in the E-DCH active set.

### *5.5.8 Advanced HSUPA scheduling*

The standard allows some advanced mechanisms for the network to control the UE's E-TFC selection process in cooperation with uplink HARQ.

Both the radio network controller (RNC) with radio resource control (RRC) signalling and the Node B with E-AGCH signalling can turn off specific HARQ processes. This feature is only applicable for a 2-ms TTI as the basic reason behind process control is to allow a reduction in the minimum data rate of a 2-ms TTI from the relatively high minimum peak rate and to have better granularity for the UE's transmitted rates. A 10-ms TTI does not suffer from this problem as the minimum data rate and, thus, the minimum data rate step is 1/5 that of a 2-ms TTI. At least one radio link control (RLC) payload data unit (PDU) must be transmitted in a TTI and, thus – with a typical RLC PDU size of 320 bits – the minimum RLC layer data rate is 160 kbps with a 2-ms TTI and 32 kbps with a 10-ms TTI. By allowing, say, only one HARQ process to transmit in a 2-ms TTI the nominal average RLC layer data rate is reduced to 20 kbps even though the instantaneous data rate remains the same.

An UE can be allocated two different UE-ids (E-RNTIs) that it needs to listen to on the E-AGCH. The UE always follows the absolute grants transmitted using the primary UE-id and can be commanded to follow the absolute grants transmitted using the secondary UE-id as well. The motivation behind this mechanism is to use the primary UE-id as a common UE-id for multiple UEs and schedule a group of UEs with a data rate they can start using whenever data arrive at the UE buffer – that is, the common UE-id could be used to control the data rate used for initial access when the UE moves from no transmission to transmission and the dedicated UE-id can be used to control the active UEs.

### *5.5.9 Non-scheduled transmissions*

In addition to scheduled data, HSUPA also supports something called 'non-scheduled transmission'. This allows the RNC to configure a specific MAC-d flow (a specific service) to have a guaranteed physical layer data rate effectively disabling Node B scheduler control of this particular service. A non-scheduled HSUPA MAC-d flow is best characterized as a Release 99 DCH with a physical layer HARQ.

If a 2-ms TTI is used, the non-scheduled MAC-d flow can be restricted to specific HARQ processes only. In this way the minimum data rate allocation for such a MAC-d flow can be reduced from the relatively high minimum data rate applicable with a 2-ms TTI.

## 5.6 Iub parameters

Similar to HSDPA operation, HSUPA also needs a large number of parameters to be aligned between the terminal and the Node B. These are provided (as shown in Figure 5.24) to the Node B from the RNC, based on vendor-specific algorithms in the RNC and on the terminal's capability. These parameters can be broken down into the following categories:

- Parameters for Node B resource allocation, to indicate, say, which codes are available for the E-HICH/E-AGCH/E-RGCH.
- Scheduler parameters, to control scheduler behaviour, such as scheduling priority indicator, maximum data rate, maximum RTWP value, or guaranteed bit rate.
- Terminal-specific parameters, such as the terminal's capability, maximum bit rate, puncturing limit, which code word on the HICH/E-AGCH/E-RGCH terminal is listening to, or the TTI to be used. A 10-ms TTI could be used for all terminals, but with a 2-ms TTI there will be a need to check the terminal's capability and return the reconfiguration to a 10-ms TTI when closer to the cell edge.

Respectively, terminal-specific parameters are then communicated by the RRC signalling the terminal to have them aligned between the terminal and the Node B [9, 10]. From the core network point of view, nothing else has changed from Release 99, except that higher data rates are also enabled in the uplink direction compared with the theoretical 2-Mbps or the practical 384-kbps uplink being used by HSDPA terminals.

There are a few parameters which are sent from the Node B towards the RNC that relate to Node B HSUPA capability. The Node B can respond to an audit message, say, from the RNC about HSUPA resource availability status. This is useful if Node B resources depend on the other features being used or to see when the Node B might be overloaded from the HSUPA resource point of view.

For the DCH case, there are some key differences with Iub interface usage. First of all, only the so-called 'silent mode' is used – that is, the Node B will send a received data packet to the RNC only after successfully decoding the packet. Empty frames indicating failed decoding or a zero data rate will not be transmitted as is done with the DCH, with the exception of the serving E-DCH cell when the maximum number of HARQ retransmissions is exceeded. The E-DCH serving cell will indicate a failed HARQ process when

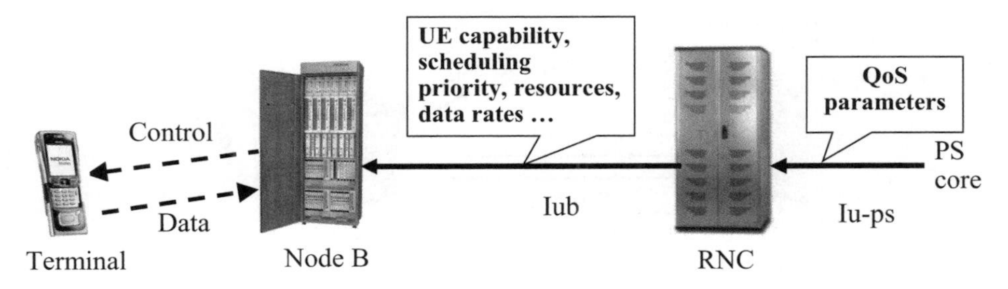


**Figure 5.24** HSUPA parameters over the Iub interface.

the maximum number of retransmissions is reached without success and, thus, the UE will progress to a new packet. If the maximum number of HARQ retransmissions is not exceeded, then the serving Node B assumes that some other BTS has picked up the packet.

In order to allow minimum end-to-end delay as well, the concept behind the TTI was slightly changed in the Iub. With a 2-ms TTI it is possible either to send data every 2 ms or to use a 10-ms TTI and bundle several 2-ms TTIs together. Each MAC-d flow has a separate transport connection on the Iub interface. In case of Iub overload, there are tools the RNC can use to control Node B maximum data rates.

## 5.7 Mobility

### *5.7.1 Soft handover*

HSUPA can be operated in soft handover (as was discussed previously). For HSUPA there can however be different active sets in use – as is the case for the Release 99 DCH. In Release 99 the key motivation was to ensure that uplink power control was handled in such a way that the near–far problem could not occur, thus the active set size was defined to be up to 6. With HSUPA there is no similar motivation for uplink data, thus scheduling and HARQ functionalities do not need to operate with as many base stations – as is the case for the Release 99 DCH. 3GPP specifications require the terminal to handle fewer cells – a maximum of four – for HSUPA operation. An example of a mixed deployment is shown in Figure 5.25.

The possibility of different active sets may typically arise in the following cases. If the active set contains base stations that do not have HSUPA capability, then at least the DPCCH is received by all the base stations in the DCH active set and only HSUPA-capable base stations run the scheduling and HARQ operation (using separate rules, as covered earlier), handling the scheduling commands and HARQ feedback from multiple cells. Also, if the active set – as a result of network conditions – needs to be larger than 4 for power control purposes, then the additional cells act like Release 99 cells and do not

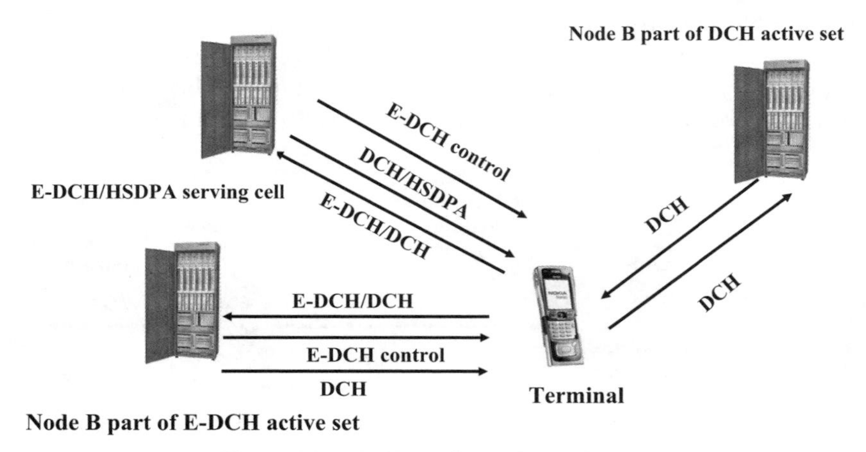


**Figure 5.25** DCH and E-DCH active sets.

take part in HARQ and scheduling operation. This allows the total number of additional signalling channels the terminal has to listen to in soft handover to be limited. Obviously, all the cells in the E-DCH active set must be in the DCH active set as well, but the DCH active set may also contain cells that do not belong to the E-DCH active set if this is seen useful by the network.

The change in serving E-DCH cell can be based on the same criteria as the change in serving HSDPA cell. For this the measurement event – 1D – has already been modified in Release 5 to allow reporting when the best serving cell changes. There is no obvious reason why the serving E-DCH cell would not be the same as the serving HSDPA cell, and this is also required to be the case in the specifications. The network may configure the UE uplink/downlink operation so that HSDPA would not be used in the downlink even if HSUPA is being used in the uplink. From the terminal capability point of view all HSUPA-capable devices are required to support HSDPA as well [8]. A new measurement event 1J can be used for active set management.

### 5.7.2 *Compressed mode*

Compressed mode handling using HSUPA depends on the TTI length. With a 2-ms TTI the solution is simple, as the uplink E-DPDCH is not transmitted at all if compressed mode overlaps partly or fully with the TTI, the solution being the same as with HSDPA. Sub-channel timing is maintained and the retransmission is then postponed to the next available TTI of the same HARQ sub-channel, preventing any sliding of HARQ process timing. With a 10-ms TTI the solution has to be different, as otherwise a frequent compressed mode patter could have suspended the data flow too critically, even fully on some HARQ processes. In the 10-ms TTI case, terminal behaviour depends on such cases as follows:

- If the initial transmission experiences a compressed mode transmit gap (TX GAP), the retransmission will transmit the same number of slots in the TTI as shown in case A in Figure 5.26. Only the same number of slots as the initial transmission will be utilized and the terminal will use discontinuous transmission (DTX) for the E-DPDCH for the extra slots (during which control channels are still transmitted).

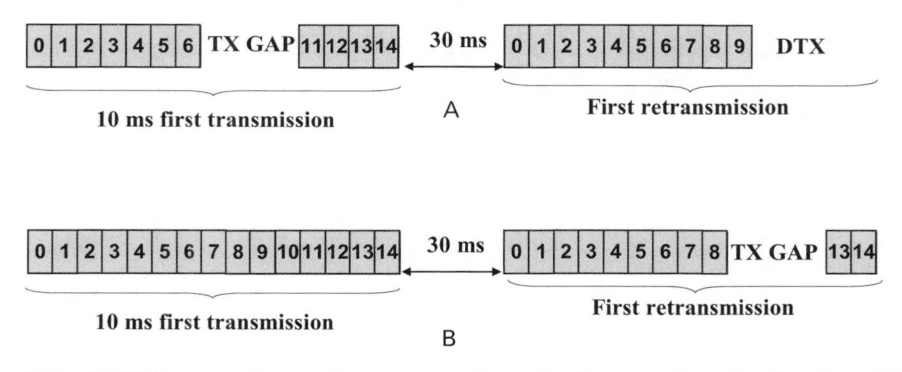


**Figure 5.26** HSUPA operation with compressed mode: in case A only the first TX is compressed; in case B just the retransmission is compressed.

- If a TX GAP occurs for the retransmission, then – as shown in case B in Figure 5.26 – only as many slots as available will be used. The rate matching functionality will assume for spreading factor selection and rate matching the same number of slots as in the initial transmission, but will only transmit as many slots as actually available in the frame [3].

The operation is simplified with the definition that the retransmission will always assume the same number of slots for rate matching and spreading factor selection. From a practical operation point of view, sending slots with reasonable energy per bit is also preferable to sending at very high data rates (which would be subject to terminal capabilities as well) for just a few slots. With downlink HSUPA signalling channels (E-RGCH/E-AGCH/E-HICH), despite the terminal trying to grab most of the energy it is allowed to ignore those slots that overlap with a downlink TX GAP.

Similarly to HSDPA, the use of compressed mode on E-DPDCH can be avoided. There is a possibility to reconfigure to the DCH when performing inter-system or inter-frequency measurements. From the network point of view, configuring from a 10-ms to a 2-ms TTI – to avoid dealing with compressed mode – is not typically a sensible thing to do as inter-system and inter-frequency measurements are often needed at the point when the terminal is closer to the coverage limit and, thus, is having its transmission power limited. As discussed in Section 5.3.8, the link budget with signalling using a 2-ms TTI is worse and, thus, does not suggest using 2 ms for typical cases of compressed mode.

## 5.8 UE capabilities and data rates

The same approach was chosen for HSUPA terminal capabilities as with HSDPA. The terminal informs the network of one from six possible terminal categories instead of signalling individual capabilities. The key differences between different categories are related to the terminal's multicode capability and to the support of a 2-ms TTI. All terminal categories support the 10-ms TTI. The categories are shown in Table 5.8.

**Table 5.8** HSUPA terminal categories.

| *Category* | *Maximum number of E-DPDCHs and smallest spreading factor* | *Supported TTIs* | *Maximum data rate with a 10-ms TTI*[1] | *Maximum data rate with a 2-ms TTI*[2] |
|---|---|---|---|---|
| 1 | 1 × SF4 | 10 ms | 0.72 Mbps | N/A |
| 2 | 2 × SF4 | 2 and 10 ms | 1.45 Mbps | 1.45 Mbps |
| 3 | 2 × SF4 | 10 ms | 1.45 Mbps | N/A |
| 4 | 2 × SF2 | 2 and 10 ms | 2 Mbps | 2.91 Mbps |
| 5 | 2 × SF2 | 10 ms | 2 Mbps | N/A |
| 6 | 2 × SF2 + 2 × SF4 | 2 and 10 ms | 2 Mbps | 5.76 Mbps |

[1] Physical layer data rates.

Table 5.9 RLC layer data rates with 320-bit PDU size.

| *Category* | *Maximum data rate with a 10-ms TTI*[1] | *Maximum data rate with a 2-ms TTI*[1] |
|---|---|---|
| 1 | 0.672 Mbps | N/A |
| 2 | 1.376 Mbps | 1.280 Mbps |
| 3 | 1.376 Mbps | N/A |
| 4 | 1.888 Mbps | 2.720 Mbps |
| 5 | 1.888 Mbps | N/A |
| 6 | 1.888 Mbps | 5.440 Mbps |

[1] Data rates offered by the RLC layer assuming 320-bit PDU size and typical MAC header.

As shown in Table 5.9, all categories except category 1 can carry out multicode transmission. With category 6 the notation in Table 5.8 means that two E-DPDCHs are transmitted with a spreading factor of 4 and two with a spreading factor of 2. Additionally, [7] defines the minimum total terminal RLC/MAC buffer size for combined HSDPA and HSUPA operation. The values range from 50 kbytes (HSPDA category 12 and HSUPA category 1) to 400 kbytes (HSDPA category 10 and HSUPA category 6).

The maximum uplink DCH capability when configured simultaneously with HSUPA is 64 kbps for all the UE categories.

There is a similar impact on bit rates above the physical layer due to the overhead on the MAC and RLC layers, as is the case with HSDPA. Thus, data rates above the RLC layer are reduced compared with maximum physical layer data rates. The resulting RLC layers for different categories are shown in Table 5.9. In a few cases, a 2-ms TTI suffers a bit more as individual TTI has smaller payload and relative overhead is slightly higher.

## 5.9 References and list of related 3GPP specifications

[1] H. Holma and A. Toskala (eds) (2004), *WCDMA for UMTS* (3rd edn), John Wiley & Sons.

[2] 3GPP, Technical Specification Group RAN, Physical channels and mapping of transport channels onto physical channels (FDD), 3GPP TS 25.211 version 6.7.0, Release 6, available at *www.3gpp.org*

[3] 3GPP, Technical Specification Group RAN, Multiplexing and channel coding (FDD), 3GPP TS 25.212 version 6.7.0, Release 6, available at *www.3gpp.org*

[4] 3GPP, Technical Specification Group RAN, Spreading and modulation (FDD), 3GPP TS 25.213 version 6.4.0, Release 6, available at *www.3gpp.org*

[5] 3GPP, Technical Specification Group RAN, Physical layer procedures (FDD), 3GPP TS 25.214 version 6.7.1, Release 6, available at *www.3gpp.org*

[6] 3GPP, Technical Specification Group RAN, Physical layer – measurements (FDD), 3GPP TS 25.215 version 6.4.0, Release 6, available at *www.3gpp.org*

[7] 3GPP, Technical Specification Group RAN, UE radio access capabilities definition, 3GPP TS 25.306 version 6.7.0, Release 6, available at *www.3gpp.org*

[8] 3GPP, Technical Specification Group RAN, FDD enhanced uplink; Overall description; Stage 2, 3GPP TS 25.309 version 6.4.0, Release 6, available at *www.3gpp.org*

[9] 3GPP, Technical Specification Group RAN, Medium Access Control (MAC) protocol specification, 3GPP TS 25.321 version 6.7.0, Release 6, available at *www.3gpp.org*
[10] 3GPP, Technical Specification Group RAN, Radio Resource Control (RRC) protocol specification, 3GPP TS 25.331 version 6.8.0, Release 6, available at *www.3gpp.org*
[11] 3GPP, Technical Specification Group RAN, Iub, 3GPP TS 25.433 version 6.8.0, Release 6, available at *www.3gpp.org*

# 6

# Radio resource management

Harri Holma, Troels Kolding, Klaus Pedersen, and Jeroen Wigard

The radio resource management (RRM) algorithms are responsible for mapping physical layer enhancements introduced by both high-speed downlink packet access (HSDPA) and high-speed uplink packet access (HSUPA) to a capacity gain while providing attractive end user performance and system stability. This chapter presents the RRM algorithms for HSDPA in Section 6.1 and for HSUPA in Section 6.2. Both radio network controller (RNC) based algorithms and Node B based algorithms are described. The resulting performance of the algorithms is presented in Chapters 7 and 8.

## 6.1 HSDPA radio resource management

Figure 6.1 shows a schematic overview of the most essential HSDPA RRM algorithms at the RNC and the Node B. At the RNC, the new HSDPA algorithms include HSDPA resource allocation, admission control, and mobility management. In this context, HSDPA resource allocation refers to the functionality that allocates power and channelization codes to the Node-B for HSDPA transmission in each cell. HSDPA admission control is different from the Release 99 dedicated channel (DCH) admission control algorithms, since HSDPA relies on a shared channel concept. Mobility management for HSDPA is also a new functionality, since data are only transmitted from one cell to the user equipment (UE) at a time, and effective Node B buffer management is needed during handovers due to the distributed architecture. The new HSDPA RRM algorithms at the RNC are further described in Section 6.1.1. At the Node B, a new high-speed downlink shared channel (HS-DSCH) link adaptation functionality is needed to adjust the HS-DSCH bit rate every transmission time interval (TTI), depending on the user's reception quality. High-speed shared control channel (HS-SCCH) power control is needed to minimize the power overhead while guaranteeing reliable reception. Finally, the high-speed medium access control (MAC-hs) packet scheduler in the Node B controls how often admitted HSDPA users are served on the HS-DSCH. A well-designed MAC-hs packet scheduler is able to maximize cell capacity while ensuring

*HSDPA/HSUPA for UMTS* Edited by Harri Holma and Antti Toskala

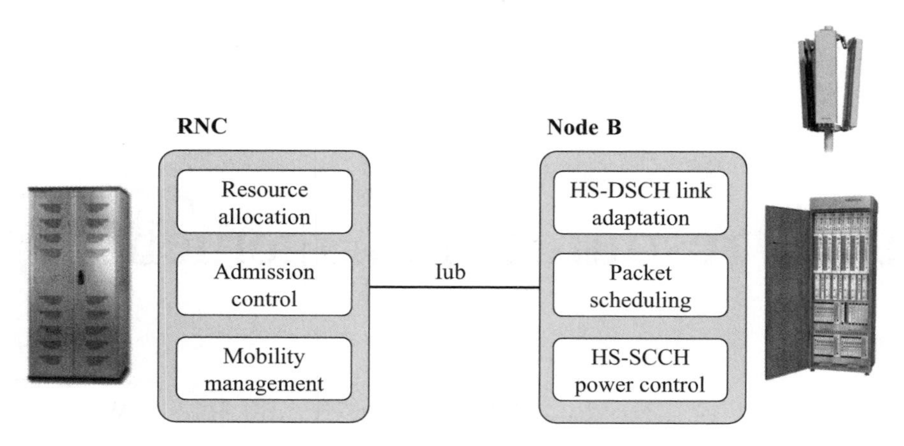


**Figure 6.1** Overview of the most relevant HSDPA RRM algorithms.

an attractive end user experience. The new HSDPA RRM algorithms at the Node B are discussed in Section 6.1.2. Note that the Third Generation Partnership Project (3GPP) only defines the interfaces and minimum UE performance requirements. Hence, network equipment manufacturers can individually design their Node B and RNC RRM algorithms according to market demands.

### *6.1.1 RNC algorithms*

#### 6.1.1.1 Resource allocation

Before the Node B can start transmitting data on the HS-DSCH, the controlling RNC needs to allocate channelization codes and power for the transmission of HSDPA. As a minimum, one HS-SCCH code with a spreading factor (SF) of 128 and one HS-PDSCH code with SF16 should be allocated to the Node B. The RNC and Node B signal to one another using the Node B Application Part (NBAP) protocol specified by 3GPP in [1]. The resources are allocated by sending an 'NBAP: Physical shared channel reconfiguration request' message from the controlling RNC to the Node B (as shown in Figure 6.2). Hence, the allocation of channelization codes for HSDPA transmission only requires signalling between the RNC and the Node B. In general, it is advantageous to allocate as many high-speed physical downlink shared channel (HS-PDSCH) codes to the Node B as possible, since the spectral efficiency of the HS-DSCH is thereby improved. On the other hand, channelization codes reserved for HS-PDSCH transmission cannot be simultaneously used for transmission of Release 99 channels, so allocation of many HS-PDSCH codes might eventually result in call blocking of Release 99 users. Fortunately, if channelization code congestion is detected, the controlling RNC can quickly

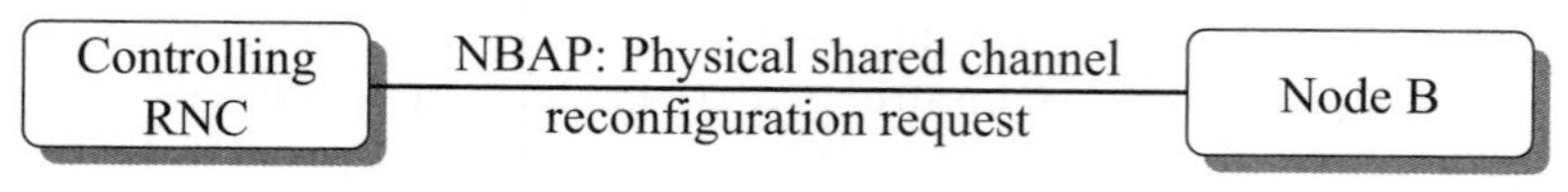


**Figure 6.2** Signalling for HSDPA resource allocation.

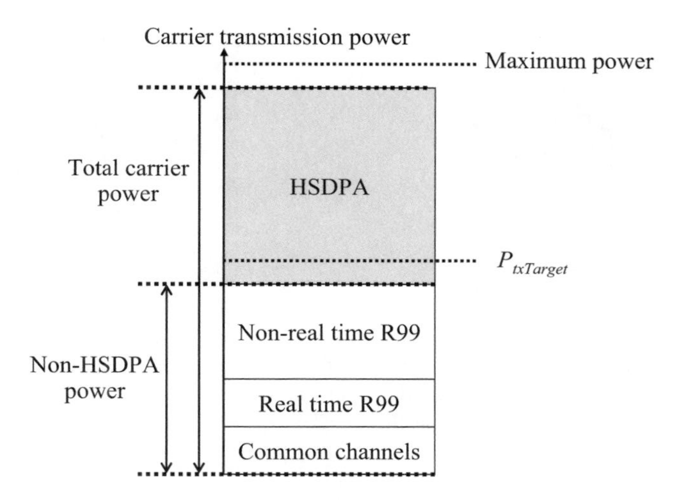


**Figure 6.3** Illustration of downlink power budget.

release some of the allocated HS-PDSCH codes to prevent blocking of Release 99 voice or video connections.

HS-DSCH transmission to multiple users in parallel during a single TTI requires multiple HS-SCCH codes and multiple HS-PDSCH codes. Code multiplexing is typically found useful for scenarios where a Node B has more HS-PDSCH codes allocated than what is supported by HSDPA mobiles; the Node B may support 10–15 HS-PDSCH codes while the HSDPA terminal supports only 5 HS-PDSCH codes. The algorithm for allocation of HS-SCCH codes to the Node B can therefore be derived as a function of allocated HS-PDSCH codes and HSDPA UE categories in the cell. The performance results are presented in Chapter 7 as a function of the number of channelization codes reserved for HSDPA.

In most cases, the most scarce downlink transmission resource is power. Figure 6.3 shows the downlink power budget for a cell with transmission on both HSDPA and Release 99 channels. The power budget consists of the power needed for common channels such as the P-CPICH, power for Release 99 DCH transmissions, and power for HSDPA transmission. As discussed in [2], the power for real time DCHs is managed by RNC admission control, while non-real time DCHs are controlled by the RNC packet scheduler. The power for non-real time DCH is characterized as controllable power since it can be adjusted via bit rate modifications, while the power for common channels and real time DCH is considered to be non-controllable. An example power allocation case is illustrated in Figure 6.3. Assuming a power-based RRM paradigm, the RNC RRM algorithm aims at keeping the total power for all Release 99 channels below *PtxTarget* [2]. In order to allow implementation of such schemes with HSDPA as well, the Node B can be configured to report average measurements of non-HSDPA power per carrier (as illustrated in Figure 6.3). Based on these measurements, the RNC is able to conduct admission control and packet scheduling for Release 99 channels in cells with simultaneous HSDPA transmission (see, e.g., the studies in [6] and [8]).

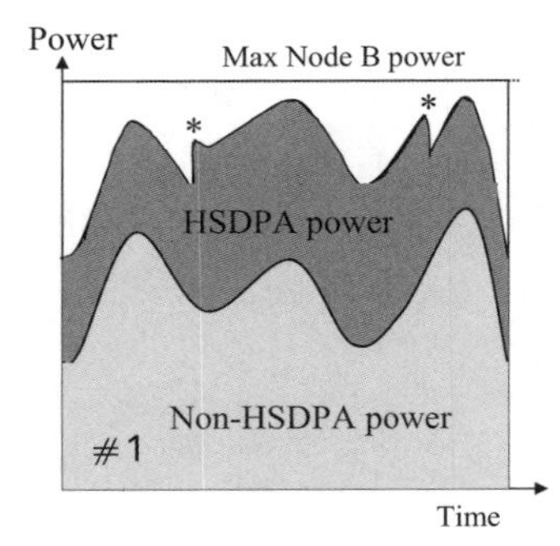


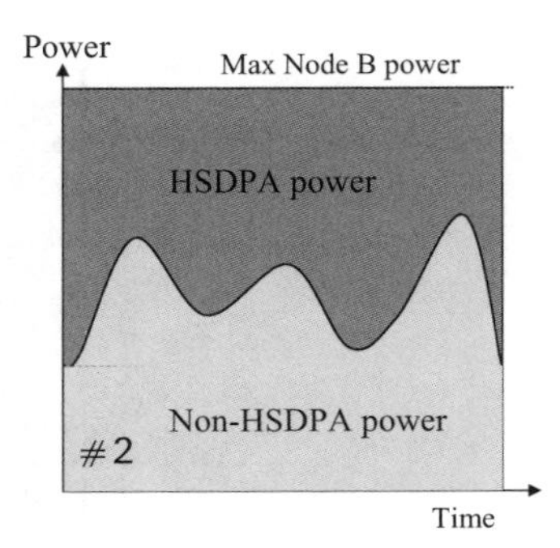


**Figure 6.4** HSDPA power allocation principles. Carrier transmission: Option #1 is explicit HSDPA power allocation from the RNC, while Option #2 is fast Node B based HSDPA power allocation. * Power adjustment by the RNC.

There are two main options for allocating HSDPA transmission power to each Node B cell:

- Option #1 – the controlling RNC allocates a fixed amount of HSDPA transmission power per cell. The Node B can afterwards use this power for transmission of HS-SCCH(s) and HS-PDSCH(s). The controlling RNC may update HSDPA transmission power allocation any time later.
- Option #2 – if the controlling RNC does not explicitly allocate HSDPA transmission power to the Node B, the Node B is allowed to use any unused power in the cell for HSDPA transmission. This means that the Node B can adjust HSDPA transmission power, so it equals the maximum transmit power minus the power used for transmission of non-HSDPA channels.

The behaviour of Option #1 and Option #2 is illustrated in Figure 6.4. Note that non-HSDPA power is time-variant due to (i) the fast power control of DCHs, (ii) new incoming real time DCH calls in the cell, (iii) termination of DCH calls, and (iv) bit rate modifications of packet calls on the DCH. Using Option #2, the total available carrier transmission power can be better utilized, since the Node B can quickly adjust HSDPA transmission power based on short-term measurements of the current power used by all non-HSDPA channels. Option #2 is therefore considered to be more attractive than Option #1. This is especially true in coverage-limited scenarios where an increase in total carrier transmission power maps directly to increased cell capacity. However, in capacity-limited scenarios there is no cell capacity gain from further increasing Node B transmit power for all cells in the network.

Regardless of whether Option #1 or Option #2 is used for HSDPA power allocation, the RNC is still in control of overall power sharing between HSDPA and other channels. If the RNC allows a power increase in the non-HSDPA channels by, for instance, increasing *PtxTarget*, then less power will become available for HSDPA transmission. The state-of-the-art solution does therefore call for a dynamic algorithm at the RNC that can adjust power sharing between HSDPA and non-HSDPA channels based on the quality of service (QoS) attributes for ongoing calls on these two channel types. Cell capacity performance results for HSDPA and the DCH are presented in Chapter 7 for different power allocations between those two channel types.

#### 6.1.1.2 QoS parametrization

QoS for Release 99 DCHs is conducted as a function of the user's traffic class (TC), traffic handling priority (THP), allocation retention priority (ARP), and potentially also other Universal Mobile Telecommunications System (UMTS) bearer attributes (as discussed in [2], [3]). These QoS parameters from the Iu interface are not available in the Node B for MAC-hs packet scheduling. New QoS parameters have been defined for the Iub interface between the RNC and the Node B. HSDPA QoS parameters in the Iub are:

- guaranteed bit rate (GBR);
- scheduling priority indicator (SPI); and
- discard timer (DT).

Figure 6.5 illustrates 3GPP QoS parameters and their interfaces. 3GPP neither defines how parameter mapping is designed in the RNC, nor how the QoS parameters are used by the MAC-hs packet scheduler.

Scheduling priority indicator (SPI) takes values in the range [0, 1, ..., 15], where a high number indicates high priority and vice versa. The DT specifies the maximum time that a packet is allowed to be buffered in the Node B's MAC-hs before it should be discarded. For the conversational and streaming traffic class, the HSDPA GBR parameter can be set according to the bit rate requirement specified in the UMTS bearer attribute for this traffic class. A high SPI could be allocated to video streaming or other real time services, while general Internet access applications could be assigned a low SPI value. 3GPP specifications also allow implementation of advanced options where the SPI is adjusted dynamically during a packet call. As will be discussed next, the value of the GBR and SPI for new HSDPA users who are requesting access can also be used in the HSDPA admission control decision.

#### 6.1.1.3 Admission control

HSDPA admission control is the functionality that determines whether new users with HSDPA terminals can be granted access to the cell and whether they will be served using HSDPA or the DCH. The admission control decision is taken by the RNC. In case

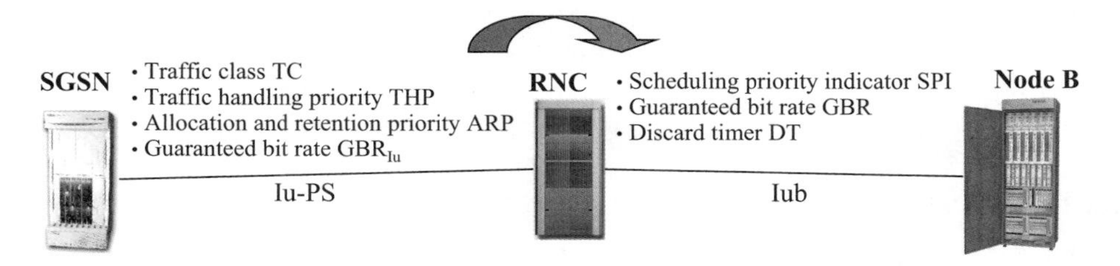


**Figure 6.5** 3GPP QoS parameters in the Iu-PS and Iub interfaces.

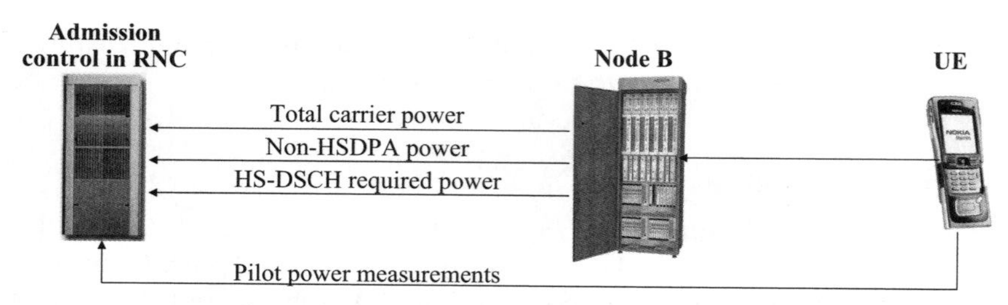


**Figure 6.6** Sketch of measurements and parameters applicable for HSPDA admission control [5].

of circuit-switched services, like AMR speech or video, it is obvious to stick with the DCH. For packet-switched services the algorithm in the RNC needs to consider the QoS parameters provided by the core network as well as the general resource situation in the network. If only best effort traffic with no strict QoS requirements are transmitted on HSDPA, then the admission control algorithm can be made fairly simple by only checking the availability of RNC and Node B hardware resources to serve a new HSDPA user. If more demanding services with stricter QoS requirements are considered for HSDPA, then a more advanced admission control algorithm is needed to ensure that the QoS requirements for existing HSDPA users in the cell as well as the requirements of the new user can be fulfilled after the potential admission. Examples of quality-based HSDPA access control algorithms are studied in [5]–[7], where the QoS attributes of new HSDPA users are taken into account in admission control decisions. Hence, using this type of algorithm, high-priority users tend to experience a lower blocking probability than low-priority users.

Figure 6.6 shows an example of the measurements and parameters that are available for HSDPA admission control in the RNC: the Node B reports the total average carrier transmit power and the non-HSDPA transmit power. Given these two measurements, the RNC can compute the amount of available HSDPA transmit power in the cell. The Node B also reports the HS-DSCH power needed to serve all the existing HSDPA users in the cell with their guaranteed bit rates. Finally, the new user requesting HSDPA access sends a pilot common pilot channel (CPICH) $E_c/N_0$ measurement report to the RNC. The latter measurement can be used by the RNC to estimate the HS-DSCH signal quality of the user. Given these measurements – together with the user's QoS attributes – the RNC can estimate whether there is HSDPA capacity available to grant the new user access without violating the QoS requirements for the existing users in the cell [5]. As discussed in [4]–[5], such an HSDPA admission control algorithm efficiently supports high-quality streaming and VoIP services on HSDPA. Finally, note that the non-HSDPA power measurement from the Node B can also be used for conventional power-based admission control of the Release 99 channels that coexists on the same carrier [8].

#### 6.1.1.4 Mobility management

HSDPA does not use soft handover, since HS-DSCH and HS-SCCH transmission take place only from a single cell, called the 'serving HS-DSCH cell'. The RNC determines the serving HS-DSCH cell for the active HSDPA UE. The serving HS-DSCH cell is one of the cells in the UE's active set. A synchronized change in the serving HS-DSCH cell is supported between the UTRAN and the UE. This feature allows benefitting from the HSDPA service with full coverage and with full mobility [7]. The serving HS-DSCH cell may be changed without updating the user's active set for Release 99 dedicated channels or in combination with establishment, release, or reconfiguration of the DCHs. An HS-DSCH serving cell change is typically based on measurement reports from the UE. 3GPP Release 5 includes a new measurement procedure to inform the RNC of the best serving HS-DSCH cell.

In the following subsections, we briefly discuss the new UE measurement event for supporting mobility for HSDPA users, as well as outline the procedures for intra and inter Node B HS-DSCH to HS-DSCH handover. Finally, we address handover from HS-DSCH to the Release 99 DCH. Inter-frequency handovers with compressed mode measurements are also supported for HSDPA users but are not considered in this section [2].

#### *6.1.1.4.1 Measurement event for best serving HS-DSCH cell*

The serving RNC determines what cells are in the active set for transmission of DCHs. The serving RNC makes the handover decisions directly based on the CPICH measurement reports from the UE. A measurement event '1d' has been defined for HSDPA – that is, a change in best serving HS-DSCH cell [9]. This measurement reports the CPICH $E_c/N_0$ of the best cell, and is triggered when the best cell changes (as illustrated in Figure 6.7). It is possible to configure this measurement event so that all cells in the user's candidate set are taken into account, or to restrict the measurement event so that only the current cells in the user's active set for DCHs are considered. Usage of a hysteresis margin to avoid fast change in the serving HS-DSCH cell is also possible for this measurement event, as well as specification of a cell-individual offset to favour certain cells – for instance, to extend their HSDPA coverage area.

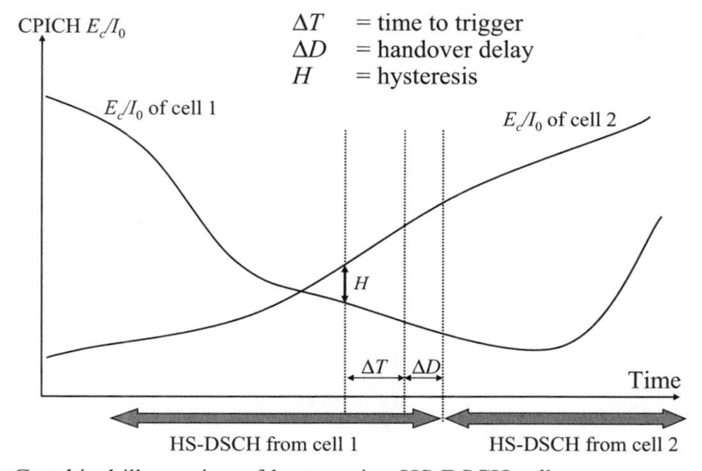


**Figure 6.7** Graphical illustration of best serving HS-DSCH cell measurement from the user.

Although serving HS-DSCH cell changes are typically triggered by downlink UE measurements, they may also be triggered by specific uplink Node B measurements. Uplink Node B measurements can be used to ensure that the data connection is not lost due to poor uplink coverage to the serving cell. The high-speed dedicated physical control channel (HS-DPCCH) must be received by the serving cell since it carries channel quality information (CQI) and ACK/NACK messages. The HS-DPCCH cannot use macro-diversity and, therefore, higher power level and repetition is typically used on HS-DPCCH in soft handover to improve signalling reliability. If the quality of the uplink connection to the serving cell becomes poor, an HS-DSCH cell change may be necessary to maintain reliable uplink signalling. The standardized Node B measurement of *SIRerror* is an example of an uplink measurement that potentially can be used for triggering serving HS-DSCH cell changes. The *SIRerror* is a measurement of the difference between the actual uplink signal to interference ratio (SIR) on the dedicated physical channel (DPCH) and the SIR target used by closed loop power control. Hence, if the *SIRerror* gets too high, then it indicates relatively poor uplink signalling quality in that particular cell.

#### *6.1.1.4.2 Inter Node B HS-DSCH to HS-DSCH handover*

HSDPA supports mobility both between sectors of the same Node B, and between two different Node Bs. Inter Node B handover is illustrated in Figure 6.8, where the UE is about to change serving HS-DSCH cell from the source cell to the target cell.

The handover procedure and the delay for the inter Node B case is illustrated in Figure 6.9. The delay analysis assumes that the signalling radio bearer (SRB) is mapped to the HS-DSCH and to the enhanced uplink dedicated channel (E-DCH) for the uplink with a 10-ms TTI. First, the UE sends a measurement report on the SRB when the trigger

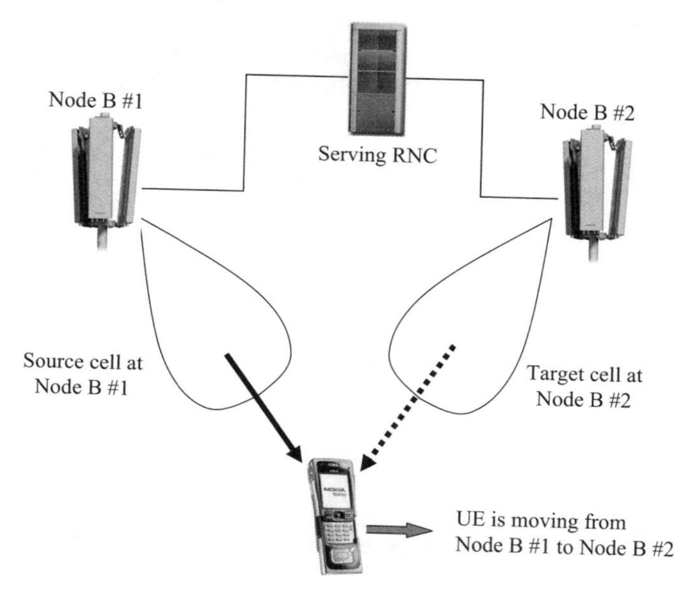


**Figure 6.8** Inter Node B HS-DSCH to HS-DSCH handover.

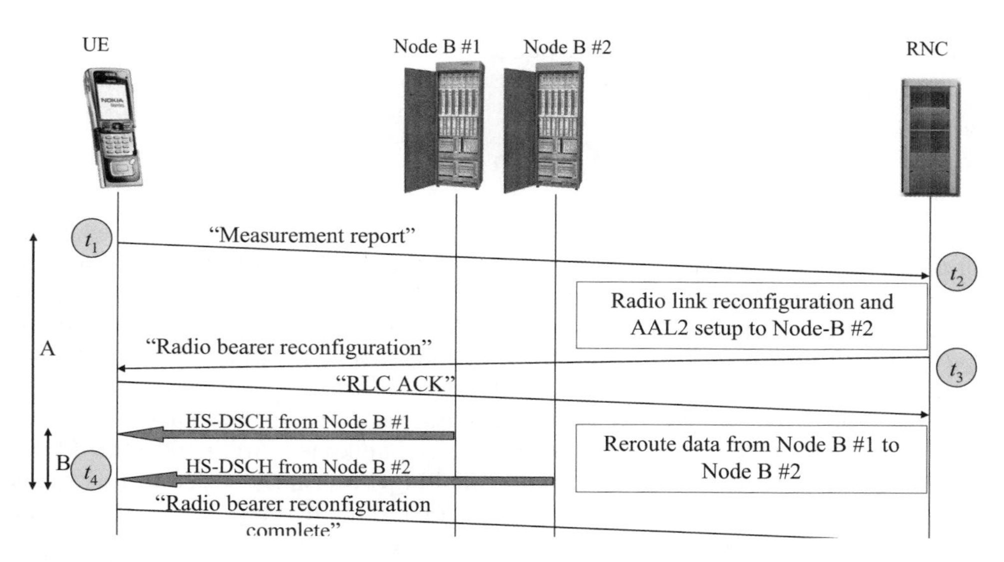


**Figure 6.9** Inter Node B HS-DSCH to HS-DSCH handover procedure and timing. A = procedural delay; B = break in data reception.

for measurement event 1d is fulfilled. The transmission starts at time $t_1$ and the RNC receives the message at $t_2$. The serving RNC next reserves the base station resources and Iub resources for the target Node B. Resource reservation can be made faster using pre-configurations if the resources have been booked beforehand. Once the resources are ready at time $t_3$, the RNC sends a radio bearer reconfiguration message to the UE, which is still receiving data from the source Node B. When the UE has decoded the reconfiguration message and the activation time has expired at time $t_4$, the UE will move reception from the source cell to the target cell. The UE starts to listen to the HS-SCCH from the new target cell. It also measures the channel quality of the new cell and sends CQI reports accordingly to the new cell. The MAC-hs for the user in the source cell is reset at the time of the cell change and the buffered payload data units (PDUs) are deleted. At the same time, the flow control unit in the MAC-hs in the target cell starts to request PDUs from the serving RNC, so that it can begin transmitting data on the HS-DSCH to the user. It is also possible for the RNC to send duplicate transmissions of the packet to both Node Bs during the cell change. When the RNC receives the 'Reconfiguration Complete' message from the UE, it can release the resources from the source cell.

The transmission gap, denoted as time $B$ in Figure 6.9, is negligible since the UE is making the cell change synchronously with the network switching the transmission from the source cell to the target cell. That allows seamless mobility for a low-delay real time service like Voice over IP (VoIP) as well.

The procedural delay $A$ is defined from time $t_1$ – when the UE sends a measurement report – until time $t_4$ – when the UE receives data from the new cell. This delay is relevant in the event that channel conditions and fading change very rapidly. Assuming a low RLC retransmission probability, the delay is 200–250 ms. The delay $t_3$–$t_2$ for network resource reservation depends on the usage of pre-configuration and on radio network configuration. The approximate delay budget equals:

- $t_2$–$t_1$ = 50 ms;
- $t_3$–$t_2$ = 50–100 ms;
- $t_4$–$t_3$ = 100 ms;
- total = 200–250 ms.

Prior to the serving HS-DSCH cell change, there might be several PDUs buffered in the source cell's MAC-hs for the user – that is, both PDUs that have never been transmitted to the user and pending PDUs in the hybrid-ARQ (HARQ) manager that are either awaiting ACK/NACK on the uplink HS-DPCCH or PDUs that are waiting to be retransmitted to the user. Those buffered PDUs in the source cell are deleted and can be recovered by radio link control (RLC) retransmissions if acknowledged mode RLC is used. When the RLC protocol realizes that the original PDUs to the source cell are not acknowledged, it will initiate retransmissions, which implies forwarding the same PDUs to the new target cell. In order to reduce potential PDU transmission delays during this recovery phase, the RLC protocol at the UE can be configured to send an RLC status report to the RNC immediately after the serving HS-DSCH cell has been changed. This implies that the RLC protocol in the RNC can immediately start to forward the PDUs that were deleted in the source cell prior to the HS-DSCH cell change.

There are applications that do not include any higher layer retransmission mechanisms – such as applications running over the User Datagram Protocol (UDP) and using RLC transparent or unacknowledged mode. Such applications running on RLC transparent or unacknowledged mode are typically low-delay applications, like VoIP, and they only use very short buffering in the Node B. Therefore, the number of PDUs deleted can be small or even 0. 3GPP specifications also allow duplicating the PDUs from the RNC to both Node Bs during the cell change to make sure there are no packet losses.

#### *6.1.1.4.3 Intra Node B HS-DSCH to HS-DSCH handover*

Intra Node B HS-DSCH to HS-DSCH handover between two sectors of the same Node B is also supported (as illustrated in Figure 6.10). The handover procedure is similar to that of inter Node B, except for the forwarding of buffered packets and for uplink reception of the HS-DPCCH.

Assuming that the Node B supports MAC-hs preservation, all the PDUs for the user are moved from the MAC-hs in the source cell to the MAC-hs in the target cell during the HS-DSCH handover. This means that the status of the HARQ manager is also preserved without triggering any RLC retransmission during intra Node B HS-DSCH to HS-DSCH handover.

The uplink DPCH uses softer handover during intra Node B HS-DSCH to HS-DSCH handover. Under such conditions the uplink HS-DPCCH may also be regarded as being in a two-way softer handover, so Rake fingers for demodulation of the HS-DPCCH are allocated to both cells in the user's active set. This implies that uplink coverage of the HS-DPCCH is improved for users in softer handover.

#### *6.1.1.4.4 HS-DSCH to DCH handover*

Handover from an HS-DSCH to a DCH may potentially be needed for HSDPA users that are moving from a cell with HSDPA to a cell without HSDPA (as illustrated in

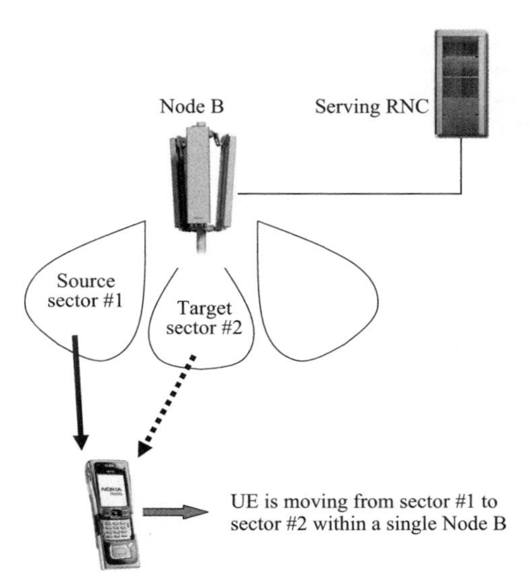


**Figure 6.10** Intra Node B HS-DSCH to HS-DSCH handover between sectors.

Figure 6.11). Once the serving RNC decides to inititiate such a handover, a synchronized radio link Reconfiguration Prepare message is sent to the Node Bs involved, as well as an RRC physical channel reconfiguration message to the user. Similarly, for inter Node B HS-DSCH to HS-DSCH handover, the HS-DSCH to DCH handover results in a reset of the PDUs in the MAC-hs in the source cell, which subsequently requires recovery via higher layer retransmissions – such as RLC retransmissions.

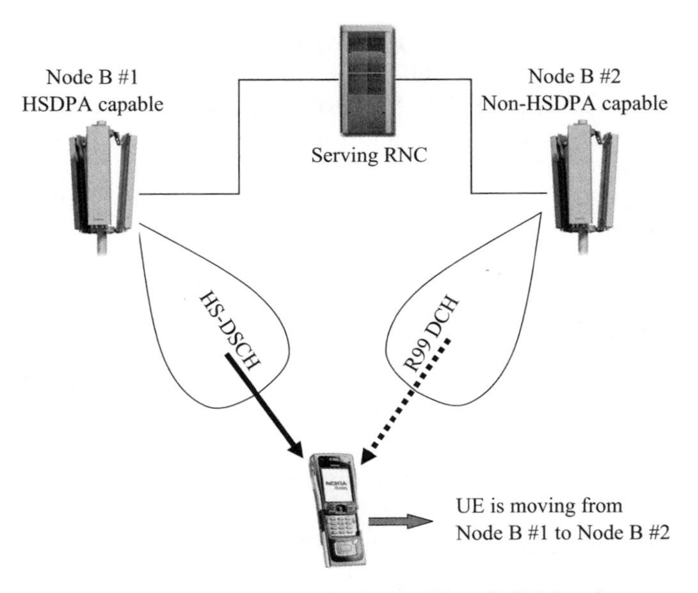


**Figure 6.11** Example of HS-DSCH to DCH handover.

**Table 6.1** Summary of HSDPA handover types and their characteristics.

| | *Intra Node B HS-DSCH to HS-DSCH* | *Inter Node B HS-DSCH to HS-DSCH* | *HS-DSCH to DCH* |
|---|---|---|---|
| Handover measurement | Typically by UE, but possibly also by Node-B | | |
| Handover decision | By serving RNC | | |
| Packet retransmissions | Packets forwarded from source MAC-hs to target MAC-hs | Packets not forwarded. RLC retransmissions used from SRNC | RLC retransmissions used from SRNC |
| Packet losses | No | No, when RLC acknowledged mode is used, or when duplicate packets are sent on RLC unacknowledged mode | No, when RLC acknowledged mode is used |
| Uplink HS-DPCCH | HS-DPCCH can use softer handover | HS-DPCCH received by one cell | |

Release 5 specifications also support implementation of handover from the DCH to the HS-DSCH. This handover type may, for instance, be used if a user is moving from a non-HSDPA capable cell to an HSDPA-capable cell.

##### *6.1.1.4.5 Summary of intra-frequency HSDPA handover characteristics*
Table 6.1 presents a summary of the different handover modes and their characteristics. Various performance results for HSDPA over the Release 99 DCH are presented in [7] for users with different active set sizes, and for different HSDPA handover strategies.

### *6.1.2 Node B algorithms*

#### 6.1.2.1 HS-DSCH link adaptation techniques

The HS-DSCH link adaptation algorithm at the Node B adjusts the transmit bit rate on the HS-DSCH every TTI when a user is scheduled for transmission. Ideally, the HS-DSCH transmit bit rate should be adjusted as a function of the per-TTI HS-DSCH signal-to-interference plus noise ratio (SINR) experienced at the user end. The general principle underlying HS-DSCH link adaptation is illustrated in Figure 6.12.

Various sources contribute to variance of HS-DSCH SINR even though the HS-DSCH transmit power is assumed to be constant. The different factors are illustrated in Figure 6.13. The total transmit power from the serving HS-DSCH cell is time-variant due to the transmission of power-controlled DCHs, the downlink radio channel is time-variant if the user is moving, and, finally, the other-cell interference experienced at the

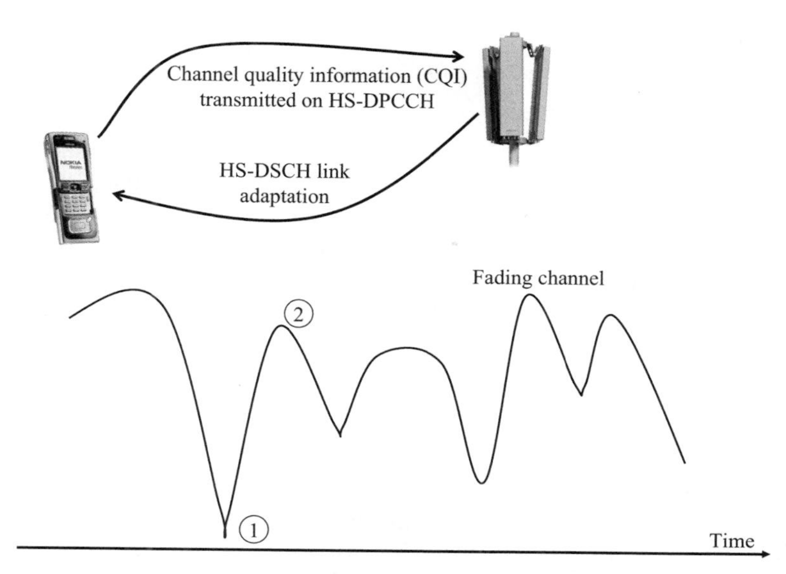


**Figure 6.12** HS-DSCH link adaptation principle: ① = the UE reports low-quality channel information and the Node B allocates a low bit rate; ② = the UE reports high-quality channel information and the Node B allocates a high bit rate.

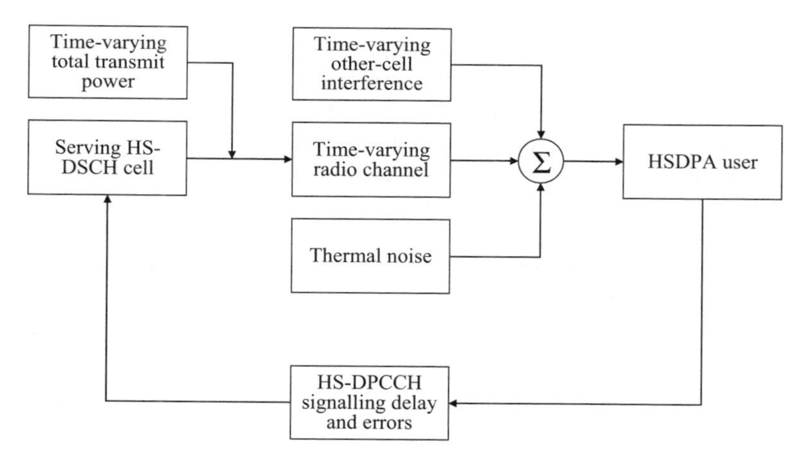


**Figure 6.13** Block diagram showing the received signal at the HSDPA user end and reporting of the CQI to the serving HS-DSCH cell.

user end is also time-variant. For the purpose of HS-DSCH link adaptation, the UE periodically sends a CQI to the serving HS-DSCH cell on the uplink HS-DPCCH (as described in Chapter 4). The CQI indicates the maximum transport block size that can be received correctly with at least 90% probability. This information is signalled via a CQI index in the range from 0 to 31, where each step corresponds appoximately to a 1-dB step in HS-DSCH SINR [10].

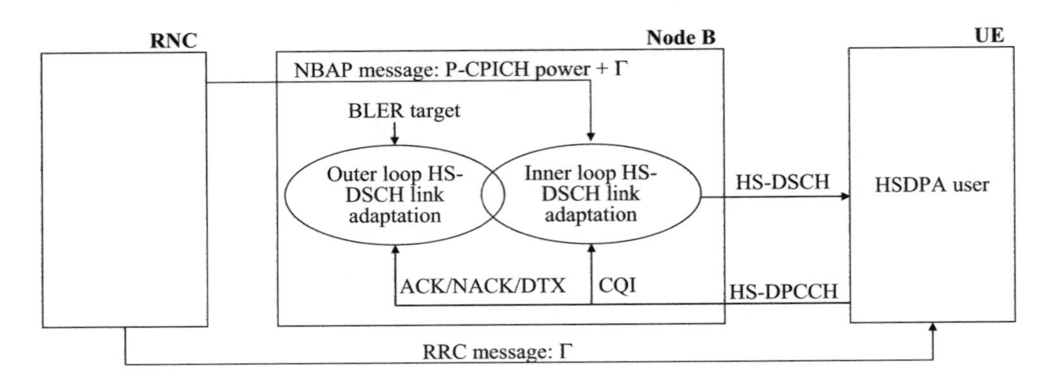


**Figure 6.14** Block diagram for the HS-DSCH link adaptation algorithm at the Node B.

A simple link adaptation algorithm would directly follow the CQI values reported by the UE. However, there may be a need to adjust the UE-reported CQI by adding an offset for the following reasons. The HS-DSCH transmit power from the Node B to the user might be different from the assumed HS-DSCH transmit power by the UE at the time it derived the CQI report. The UE assumes that HSDPA power equals the power of the primary common pilot channel (P-CPICH) plus Γ, where Γ is a power offset parameter signalled to the UE via RRC signalling from the RNC.

The effect of feedback delays on link adaptation performance has previously been addressed in [11], [12]. The conclusion from those studies indicate a need for an outer loop HS-DSCH link adaptation algorithm to further adjust the CQI index received from a user before applying it for adjustment of the HS-DSCH transmission format.

The outer loop algorithm can be based on ACKs/NACKs from past transmissions. The algorithm adjusts the offset values to ascertain the average targetted retransmission probability. Too many retransmissions add an unnecessary delay while too few indicate that the transport block sizes used are not large enough, unnecessarily lowering throughput. Outer loop HS-DSCH link adaptation can be based on the same principle as the Release 99 outer loop power control algorithms that were first published in [13]. Examples of outer loop HS-DSCH link adaptation algorithms are studied in [14], [15]. The HS-DSCH link adaptation algorithm is summarized in the simplified block diagram in Figure 6.14.

### 6.1.2.2 HS-SCCH power control

Reliable reception quality of the HS-SCCH is important since the transport block on the HS-DSCH can only be decoded if the HS-SCCH has first been correctly received. Therefore, sufficient power should be allocated to transmission of the HS-SCCH to ensure reliable reception. On the other hand, it is also desirable to reduce HS-SCCH transmission power to lessen interference levels in the network. Hence, it is generally recommended to have HS-SCCH power controlled every TTI, in which HS-SCCH transmit power is adjusted such that the desired user has a high probability of correctly decoding the channel (see Figure 6.15). A large amount of HS-SCCH power is used for

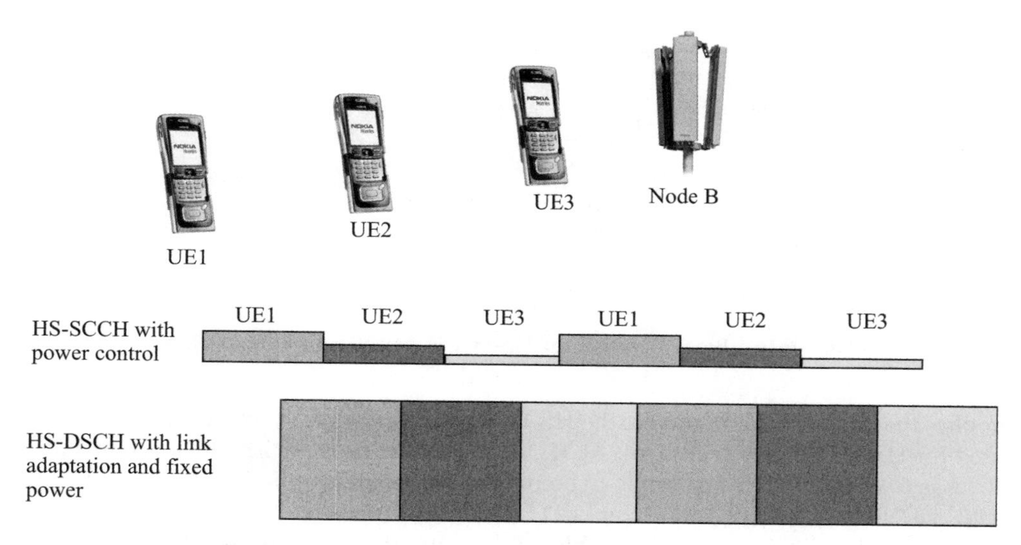


**Figure 6.15** Principle behind HS-SCCH power control.

UE1 at the cell edge while a smaller amount can be used for UE3 close to the base station. The HD-DSCH uses link adaptation rather than fast power control.

The 3GPP specifications do not explicitly specify any power control mechanism for the HS-SCCH. HS-SCCH power control can be based on the following inputs:

1. *Associated DPCCH power control commands* – HS-SCCH transmit power is adjusted relative to the transmit power of the associated downlink DPCCH. This is possible because the DPCCH is subject to closed loop power control, and the power offset between the HS-SCCH and the DPCCH can be set assuming *a priori* knowledge of the relative SINR performance between the two channels.
2. *CQI reports* – HS-SCCH transmit power is adjusted as a function of the CQI report received from the user. This is possible if there is an internal table at the Node B expressing a power offset between each CQI index and the required HS-SCCH power.

Thus, in both cases it is possible to implement a pseudo closed loop power control scheme for the HS-SCCH, relying on either feedback information from the user about the reception quality of the associated DPCH or the HS-DSCH (CQI). Common to both approaches is the Node B need for *a priori* knowledge of a power offset parameter before it can adjust HS-SCCH transmit power as a function of either DPCCH power or CQI. The magnitude of this power offset determines the residual block error probability (BLEP) on the HS-SCCH. It is therefore recommended to also use an outer loop power control algorithm at the Node B which fine-tunes the aforementioned power offset to meet a target BLEP on the HS-SCCH. Once again, we can apply here a similar outer loop algorithm to that for link HS-DSCH adaptation.

The Node B knows whether the UE received the HS-SCCH successfully if it subsequently receives a corresponding ACK or NACK. If the Node B receives neither – that is, the UE had sent a discontinuous transmission (DTX) on the HS-DPCCH – this

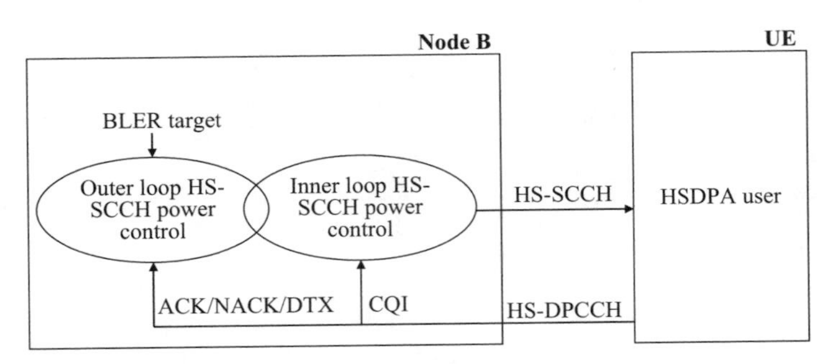


**Figure 6.16** Block diagram for HS-SCCH power control algorithm.

implies that the UE did not detect the HS-SCCH transmission. This information can be used for HS-SCCH power control. ACK/NACK reporting is further enhanced in 3GPP Release 6 where the UE first sends a specific pre-amble message for an ACK/NACK to help the Node B distinguish it from a DTX.

Figure 6.16 shows a simplified block diagram that summarizes the HS-SCCH power control algorithm at the Node B. The HS-SCCH power control algorithm is connection-specific. The design, performance, and power control of the HS-SCCH has been further studied in [16], [17]. HS-SCCH performance results are furthermore presented in Chapter 7.

#### 6.1.2.3 Packet scheduler

This section presents the functionality of the packet scheduler and introduces the different types of scheduling algorithms and explains the trade-offs in terms of user bit rate and cell capacity.

##### *6.1.2.3.1 Basic theory*

The basic problem a packet scheduler has is how to share the available resources to the pool of users eligible to receive data. A commonly used formulation to describe this challenge is the one proposed by Kelly [19]. Kelly uses the concept of a utility function, $U_n(r_n)$, where $n$ denotes a particular HSDPA user and $r_n$ is the average throughput for the $n$th user. Simply put, the utility function is a measure of the 'happiness or satisfaction' gained from being scheduled. The best scheduling solution is the one that maximizes the sum of utility functions for all the users at any given time. The sum of the utility functions is called the 'objective function'.

Assuming that a proper utility function, $U_n(r_n)$, has been defined, another challenge lies in the time-variant behaviour of the cellular system. Each user's channel capacity, as well as the total cell capacity, varies over time. It was shown in [20] that the best that one can do is to use a gradient search algorithm, which aims at further maximizing the objective function for each scheduling decision. Hence, the system should schedule the HSDPA user in the next TTI that fulfills [20]:

$$n^* = \arg\max_n\{M_n\}, \quad \text{where } M_n = d_n \cdot \frac{\partial U_n(r_n)}{\partial r_n} \tag{6.1}$$

Here, $M_n$ denotes the so-called scheduling metric, $d_n$ is the instantaneous data rate that HSDPA user number $n$ can support in the next TTI. Note that $d_n$ is obtained by consulting the HS-DSCH link adaptation algorithm (as discussed in Section 6.1.2.1). The throughput delivered to HSDPA users in the past can be updated every TTI for all users with a recursive expression (exponential smoothing); that is:

$$r_n = \begin{cases} (1-a)r_{n,old} + ad_n & \text{if user } n \text{ is served} \\ (1-a)r_{n,old} & \text{otherwise} \end{cases}$$

where $r_{n,old}$ is the old value of $r_n$ and $a$ is the forgetting factor. Hence, $a^{-1}$ equals the equivalent averaging period in a number of TTIs for the exponential smoothing filter. Throughput calculation for a user is only done for periods of time when the user has data in the Node B buffer. This is important for stability of QoS-aware packet scheduling methods which would otherwise try to compensate for inactive users with no data to transmit.

#### *6.1.2.3.2 Packet scheduler algorithms*

Different classical packet schedulers from the literature are listed in Table 6.2 according to their utility and scheduling policy function. The round robin (RR) scheduler is a

**Table 6.2** Packet scheduling principles.

| *Scheduler* | *Utility function, $U_n(r_n)$* | *Scheduling metric, $M_n$* |
|---|---|---|
| Round robin (RR) | 1 | 0 |
| Maximum C/I or throughput (max-C/I) | $r_n$ | $d_n$ |
| Proportional fair (PF) [19], [21], [23], [24], [27] | $\log(r_n)$ | $\frac{d_n}{r_n}$ |
| Minimum bit rate scheduling (min-GBR) [20] | $r_n + (1 - \exp[-\beta(r_n - r_{\min})])$ | $d_n[1 + \beta \exp(-\beta(r_n - r_{\min}))]$ |
| Minimum bit rate scheduling with proportional fairness (min-GBR + PF) | $\log(r_n) + (1 - \exp[-\beta(r_n - r_{\min})])$ | $d_n\left[\frac{1}{r_n} + \beta \exp(-\beta(r_n - r_{\min}))\right]$ |
| Maximum delay scheduling (max-Del) [28] | $-\log(\delta_n)\log(r_n)\frac{d_{HOL,n}}{d_{req,n}}$ | $d_n\left[\frac{-\log(\delta_n) \cdot d_{HOL,n}}{r_n \cdot d_{req,n}}\right]$ |

$r_{\min}$ = Minimum bit rate target – for example, the guaranteed bit rate (GBR);
$\beta$ = Constant controlling the aggressiveness of the scheduler (recommended value is $\beta = 0.5$ [20]);
$d_{HOL,n}$ = Head-of-line packet delay;
$d_{req,n}$ = Maximum packet delay requirement;
$\delta_n$ = Violation probability (or aggressive factor) for the algorithm [28].

popular reference scheduler where HSDPA users are scheduled with an equal probability, independent of radio channel conditions.

The maximum carrier-to-interference ratio (max-C/I) scheduler or, more stringently, the maximum throughput scheduler is designed to maximize HSDPA cell throughput. The max-C/I scheduler monopolizes the cell resources for a small subset of users, and there may be a number of users at the cell edge that will never be scheduled. In order to provide a fairer split of resources between users, the so-called 'proportional fair' (PF) packet scheduler is often considered. The PF scheduler provides a trade-off between fairness and achievable HSDPA cell throughput and provides a significant coverage extension. The popular interpretation of this relation is that users are scheduled 'on top of their fades' – for example, when their instantaneous data rate exceeds the average (see Figure 6.17). The denominator in the scheduling metric provides robustness since a user who is getting little scheduling resources will increase its priority over time. The scheduler has been analysed and studied extensively in the literature (see, e.g., [19], [21], [24]). In [27] it is shown that the PF scheduler can be modified to provide the same average throughput to all HSDPA users by a simple modification of the scheduling policy. The admission and load control entities can then adjust the number of allocated users as well as the available HSDPA resources such that the average throughput is achieved at the target service level.

To address the need for more advanced QoS differentiation, [20] introduces a minimum guaranteed bit rate (min-GBR) scheduler where the utility function returns a relatively low value for cases where the user's experienced throughput is below the GBR, while the utility function only increases moderately for throughputs experienced higher than the GBR. By adjusting the value of $\beta$ (see Table 6.2), it is possible to control how aggressive the MAC-hs packet scheduler should be if the bit rate offered to the HSDPA user drops below the GBR. In Table 6.2, a second variant is also included which adds the basic PF scheduling principle. Other possibilities for defined utility functions are presented in, for example, [20]. The last scheduler presented in Table 6.2 aims at fulfilling packet delay requirements by increasing the scheduling priority when the head-of-line

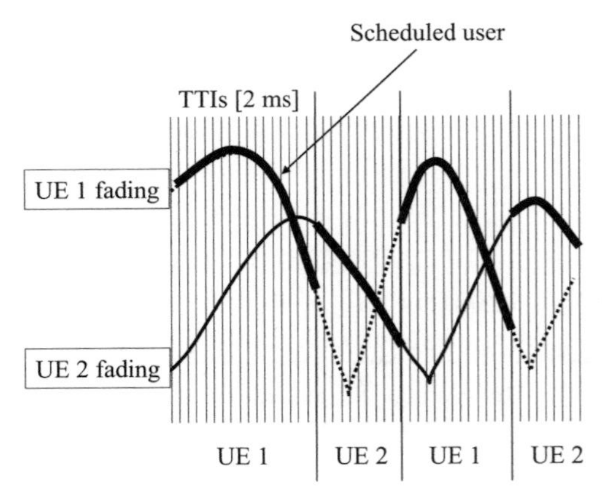


**Figure 6.17** Principle underlying proportional fair scheduling with a 3-TTI delay.

packet delay comes close to the maximum delay requirement [28]. It, too, is based on the PF scheduling principle.

The MAC-hs packet scheduler should also handle the scheduling of pending L1 retransmissions in the HARQ manager. Here there are two basic approaches:

1. Always select users with pending L1 retransmissions that are scheduled with the highest priority in the next TTI. If there are multiple users with pending L1 retransmissions, then one of the scheduling algorithms in Table 6.2 can be used to select which of them should be scheduled.
2. Always select users to be scheduled in the next TTI based on one of the algorithms in Table 6.2. If the user selected for scheduling has pending L1 retransmissions, then they should be sent before initiating new transmissions. Hence, retransmissions are given a high priority within each data flow.

As a general remark, approach #2 is considered to be the most attractive solution from a cell capacity point of view, since this gives the packet scheduler more degrees of freedom to primarily schedule users when they experience good radio conditions – that is, when they benefit from multi-user scheduling diversity gain. On the other hand, approach #1 is more attractive from a packet delay jitter point of view, since pending L1 retransmissions are immediately given a high priority independent of the user's radio channel conditions and other parameters that are potentially part of the scheduling metric. However, in practical scenarios with a BLEP of 10% to 20% on first L1 transmissions, there is only a marginal performance difference between approaches #1 and #2.

Packet schedulers that use knowledge of instantaneous radio channel quality are often associated with the term 'multi-user diversity'. If there is a large number of users in the scheduling candidate set, then typically there ought to be some users in good channel conditions such that a relatively high data rate can be allocated. The results from dynamic network simulations are presented in Chapter 7, where the multi-user diversity gain of using PF scheduling over RR scheduling is reported. In addition, the many aspects of multi-user diversity gain are also addressed in [21]–[23], [25], [26] among others.

#### *6.1.2.3.3 Code-multiplexing*

Code-multiplexing refers to the case where more than one HSDPA user is scheduled in a single TTI in a cell. There are basically two scenarios where the use of code-multiplexing is recommended:

1. Up to 15 HS-PDSCHs may be used in the Node B. However, UEs typically support simultaneous reception of five HS-PDSCHs. Hence, in order to be able to maximize spectral efficiency, code-multiplexing should be used by scheduling three parallel users with five codes each.
2. Code-multiplexing may also be required to optimize the performance if there are many HSDPA users allocated per cell with a low source data rate and strict delay requirements. As an example, VoIP on HSDPA typically requires use of code-multiplexing to achieve good performance (see, e.g., [18]).

However, there are some additional costs associated with the use of code-multiplexing: (i) the overhead from HS-SCCH transmission is increased, since one HS-SCCH is required per code-multiplexed user; and (ii) the effective multi-user diversity order decreases as more than one user is scheduled in every TTI. Hence, code-multiplexing should only be used if one of the above conditions is fulfilled. If code-multiplexing for $N$ users is used, then the packet scheduler first selects those $N$ users with the highest priority. The simplest way of dividing the power and code resources between simultaneous users is by applying an equal code and equal power strategy where all parallel users get the same amount of HS-DSCH power and number of codes.

#### *6.1.2.3.4 Scheduling of control-plane signalling on HS-DSCH*

For the fractional DPCH (F-DPCH) introduced in 3GPP Release 6, the signalling radio bearer for layer 3 signalling from the RNC to the UE will be transmitted on the HS-DSCH. This means that the MAC-hs scheduler in the Node B should be designed to handle joint scheduling of user-plane and control-plane data on the shared HS-DSCH (as illustrated in Figure 6.18). Control-plane signalling consists of RRC messaging and core network signalling. One of the benefits of transmitting those messages on the HS-DSCH, instead of using a standard associated DCH, is the potential reduction in signalling delays due to the higher data rate on the HS-DSCH. As some RRC messages are considered to be delay-sensitive, it is important that the MAC-hs packet scheduler transmits RRC messages shortly after they arrive at the Node-B. This is especially important for RRC messages during serving HS-DSCH cell changes. This can be implemented by assigning a high SPI value to control-plane data flows on HSDPA, so the Node B knows that these flows should be given high scheduling priority. Hence, whenever a new PDU arrives at the Node B with an SPI indicating that it is an RRC message, it will be scheduled in the next TTI. As the RRC message size is typically limited to a few hundred bits, such messages can be transmitted using a single HS-PDSCH code. The MAC-hs scheduler should therefore first consult the link adaptation function and compute the power needed for transmission on the RRC message, whereafter the remaining power and HS-PDSCH codes are used for normal scheduling of user-plane traffic according to the algorithms described in the previous sections. Using this approach, it is possible to facilitate joint efficient scheduling of control-plane and user-plane traffic on the HS-DSCH by using QoS differentiation according to priority settings, as well as Node-B code-multiplexing.

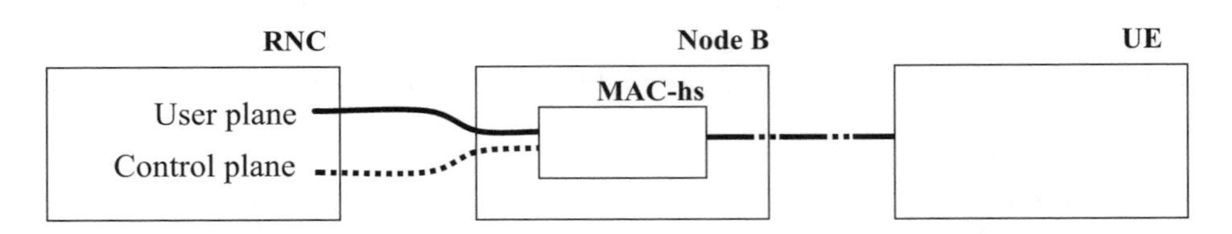


**Figure 6.18** MAC-hs scheduling of both control plane and user plane traffic on HS-DSCH.

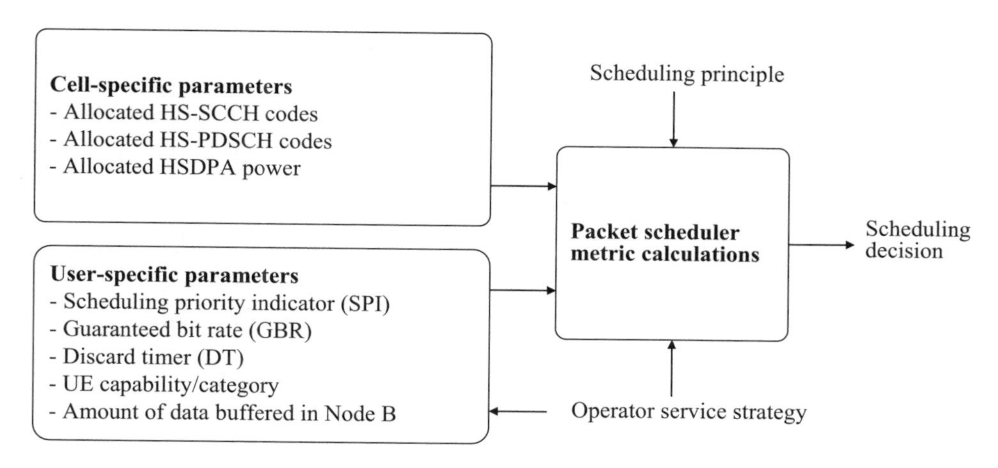


**Figure 6.19** Scheduling principle with input parameters and influences on the overall chosen scheduling strategy.

#### *6.1.2.3.5 Practical scheduling with 3GPP parameters*

The previous section listed the different scheduling algorithms and some conditions for their successful operation. Here, some of the practical aspects of scheduling are briefly addressed. The aforementioned scheduling metrics implicitly indicate that all users have the same utility function and, thus, the same scheduling metric formulation. However, in a network the operator may wish to distinguish between different users and services – for example, reprioritizing the scheduling according to priorities. As mentioned in Section 6.1.1.2, 3GPP provides various parameters to control the QoS more accurately. Figure 6.19 gives an indication of the possible inputs provided within the 3GPP framework. It is by no means complete, but does indicate some of the key parameters. The final scheduling metric is thus based on the basic principles presented earlier modified in a way that suits the operator's service and subscription strategy.

## 6.2 HSUPA radio resource management

RRM for HSUPA consists of functions located in the RNC, Node B, and UE. Figure 6.20 shows the different RRM functions. The RNC is responsible for allocating resources to HSUPA, for admission control and for handover control. The RNC is also in control of the Release 99 DCH channels and can thus control the balance between the Release 99 DCH and HSUPA. The Node B shares the resources among the different HSUPA UEs. The UE is responsible for selecting the transport block based on the available transmit power and on the available data in the buffer.

In the following subsections, the different functional blocks in the RNC and Node B are explained. The naming convention relates to the block diagram in Figure 6.20.

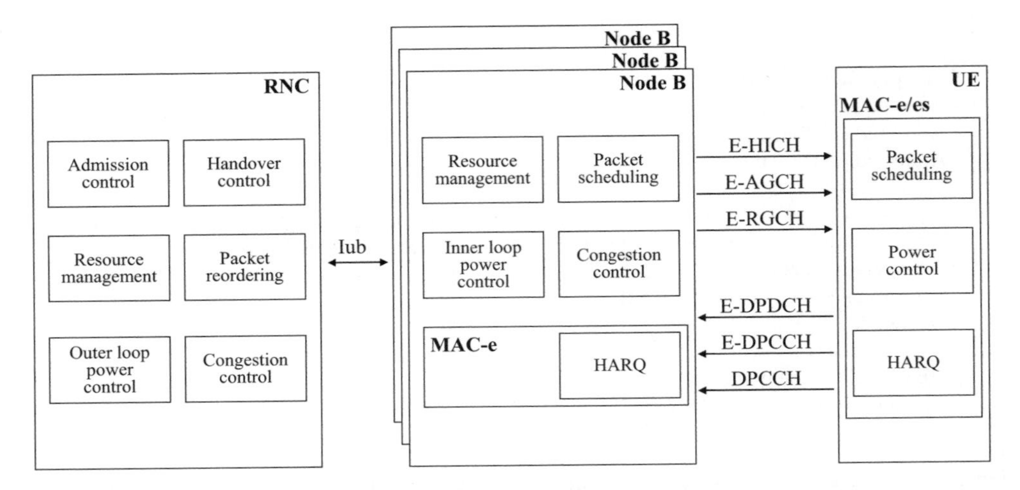


**Figure 6.20** Overview of the different functional RRM blocks for HSUPA in the RNC, Node B, and UE. Physical layer channels are described in Chapter 5.

## *6.2.1 RNC algorithms*

This section considers resource allocation, QoS parameterization, admission control, and mobility management. The other RNC functionalities – that is, reordering in the MAC-es layer and outer loop power control – are described in Chapter 5.

### 6.2.1.1 Resource allocation

The RNC sets the target value for the maximum received wideband power (noise rise) for the Node B. The received power consists of thermal noise, inter-cell interference, intra-cell interference from DCH connections and intra-cell interference from E-DCH (HSUPA) connections. The DCH connections are controlled by the RNC either in admission control or in packet scheduling. E-DCH connections are controlled by the HSUPA packet scheduler in Node B. The HSUPA scheduler can allocate the power for E-DCH users that is not used by DCH connections and is still below the maximum wideband power level. Resource allocation control in the uplink is illustrated in Figure 6.21.

The HSUPA scheduler has instantaneous information about the uplink interference situation because the scheduler is located in the Node B. The scheduler also has a faster means of controlling interference from active UEs than an RNC-based scheduler. Figure 6.22 shows uplink interference as a function of cell throughput. The RNC can set a higher target value for interference levels when HSUPA is used since the interference variations are smaller than in WCDMA. Higher interference levels enable higher cell throughput.

The serving RNC can also send a congestion indication to the Node B. This is a UE-specific parameter indicating congestion in the transport network and can take one of

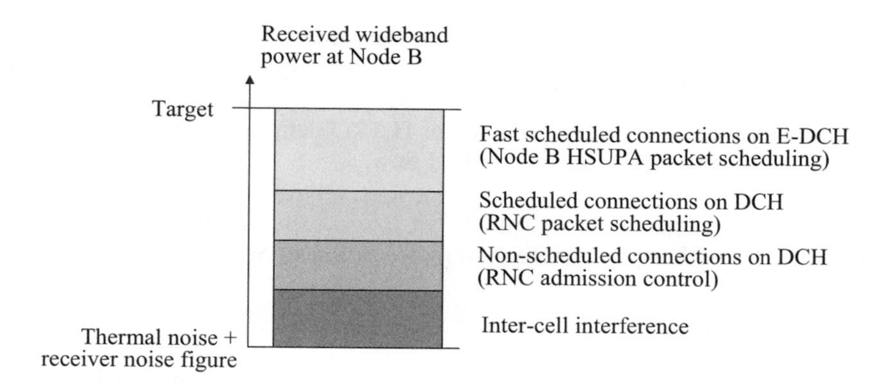


**Figure 6.21** Resource allocation control with HSUPA.

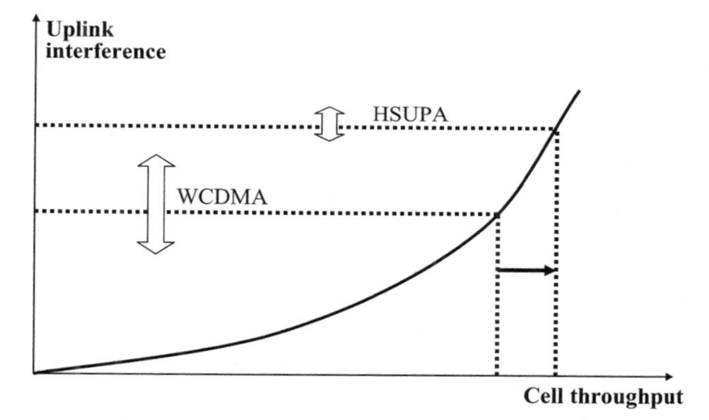


**Figure 6.22** Uplink load curve and the impact of fast scheduling.

three values:

1. No congestion.
2. Delay build-up: meaning the delay of packets in the transport network is increasing.
3. Lost packets: meaning that a certain number of packets in the transport network are lost.

Upon reception of a congestion indicator for a certain UE, the Node B can lower the bit rate of at least this particular user, such that the congestion situation is solved.

### 6.2.1.2 QoS parameterization

The RNC gives a number of QoS parameters to the Node B, which can use the parameters in packet scheduling:

- The scheduling priority indicator indicates the relative priority of different MAC-d flows by allocating 16 different values. The value 15 indicates the highest priority and 0 the lowest.

- The MAC-es guaranteed bit rate indicates the guaranteed number of bits per second to be delivered over the air interface under normal operating conditions for which the Node B shall provide sufficient uplink resources.
- The maximum number of transmissions for HARQ defines the maximum number of L1 HARQ transmissions for each MAC-d flow.

The mapping of QoS parameters from the Iu interface to the Iub interface has been discussed together with HSDPA QoS parameterization in Section 6.1.1.2.

#### 6.2.1.3 Admission control

Admission control decides whether or not to admit a new user to HSUPA. The following information is available for this decision:

- Number of active HSUPA users – the RNC may want to limit the maximum number of active HSUPA users. Network hardware dimensioning may also limit the number of active users.
- Uplink interference levels – the interference level can be obtained from the measured received total wideband power (RTWP). If the interference level is too high compared with the pre-defined target value and if the data rates of the existing users cannot be downgraded, the new user may need to be blocked.
- Scheduling priority indicator – this indicates the priority of a new call, which can be compared with the SPI of existing calls. If the new call has a high priority and existing calls have a low priority, admission control can choose to admit the new call, leading potentially to a degradation in quality for existing calls.
- Guaranteed bit rate – admission control needs to take the GBR into account by estimating whether there are enough resources for a new GBR call, while it also needs to ensure the GBR of existing users in the network.
- Provided bit rate on the E-DCH – the Node B reports the provided bit rate on the E-DCH for each priority class. This can be compared with target bit rates for the different SPI classes, when making the admission control decision for a certain user.
- Provided bit rate on the DCH – the RNC is aware of the provided bit rates on the DCH. This can be compared with a target bit rate for DCH users, when making the decision whether or not to admit a new E-DCH user.
- Downlink limitations – when a new HSUPA user is admitted, it also requires HSDPA in the downlink direction. If there are no resources for HSDPA, then the user needs to be blocked even if there are resources in the uplink.

#### 6.2.1.4 Mobility management

Handover control in the RNC decides, first, which cells are in the active set and, second, which cell is the serving HSUPA cell. The first is very similar to wideband code division multiple access (WCDMA) Release 99 handover control, except for the fact that the maximum active set size equals 4 for HSUPA while Release 99 UEs must support up to 6 cells in the active set. More information on this can be found in [2]. The serving cell algorithm decides which cell is in control of the HSUPA user. The serving cell for

HSUPA can be different from the serving cell in HSDPA [29], but typically the serving cell for HSDPA and for HSUPA are the same and the serving cell change would take place at the same time.

### *6.2.2 Node B algorithms*

In the Node B the main functions related to HSUPA are packet scheduling and HARQ. HARQ is described in detail in Chapter 6, whereas packet scheduling is dealt with in the next subsection.

#### 6.2.2.1 Packet scheduling

Two different scheduling modes are defined for HSUPA: Node B scheduling mode with L1/MAC control signalling in the uplink and downlink, and RNC-controlled non-scheduled mode. The RNC-controlled approach can be used for GBR bearers – such as for VoIP. RNC-controlled non-scheduled mode is similar to WCDMA DCH allocation, but uses fast L1 retransmissions. This section considers Node B based packet scheduling.

HSUPA has two main advantages over WCDMA Release 99: L1 HARQ and Node B based packet scheduling. The first gives a benefit in terms of spectral efficiency, since one is able to operate at a higher BLEP without increasing the delay. This is analysed in Chapter 8. The second gain comes from faster packet scheduling, which enables operation at higher load factors and higher cell throughputs.

The Node B scheduling environment can be seen in Figure 6.23. The figure shows that the packet scheduler is connected to several MAC-e's. Each HSUPA UE has its own MAC-e entity in the Node B. The most important function of the MAC-e is to take care of the reception and acknowledgement side of the HARQ process.

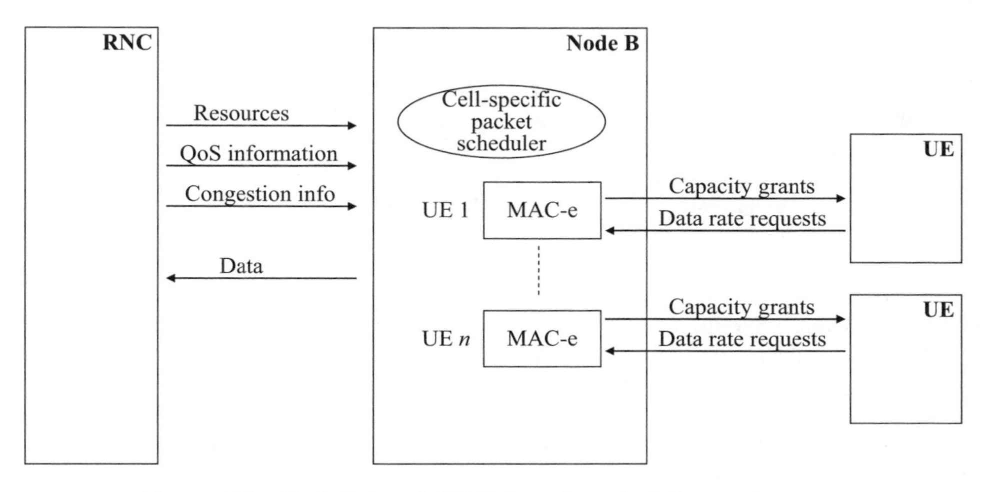


**Figure 6.23** Node B based HSUPA packet scheduling environment.

The Node B can upgrade UE capacity allocation based on happy bit or on UE buffer status information. The UE also reports the available transmission power to indicate whether it can still support higher uplink data from the power point of view. The Node B can downgrade UE capacity allocation if allocated capacity grants are not fully utilized and channel utilization is low.

The Node B can give two different types of capacity grants to the UE when it wants to change its allocation: absolute grants, which give an absolute power ratio between the E-DPDCH and DPCCH to the UE, and relative grants, which can be either *up*, *down* or *keep*. When the UE receives an *up* or a *down* command it will adjust the allocation to one step above or below the selected allocation during the last TTI of the relevant HARQ process. Relative grants are sent in the E-RGCH channel, whereas absolute grants are sent on the E-AGCH channel. Absolute grants can only sent by the serving cell while relative grants in the *down* direction can also be sent by non-serving cells in order to solve overload.

## 6.3 References

[1] 3GPP Technical Specification, TS 25.433, UTRAN Iub Interface Node-B Application Part (NBAP) Signalling.

[2] H. Holma and A. Toskala (eds) (2004), *WCDMA for UMTS* (3rd edn), John Wiley & Sons, Chichester, UK.

[3] S. Dixit, Y. Guo, and Z. Antoniou (2001), Resource management and quality of service in third-generation wireless networks, *IEEE Communications Magazine*, 125–133, February.

[4] P. Hosein (2003), A class-based admission control algorithm for shared wireless channels supporting QoS services, in *Proceedings of the Fifth IFIP TC6 International Conference on Mobile and Wireless Communications Networks, Singapore, October.*

[5] K. I. Pedersen (2005), Quality based HSDPA access algorithms, in *IEEE Proc. Vehicular Technology Conference, Fall, September.*

[6] A. Furuskar, S. Parkvall, M. Persson, and M. Samuelsson (2002), Radio resource management principles and performance for WCDMA release five, *Proc. Radio Vetenskab och Kommunikation (RVK02), Stockholm, June 10–13.*

[7] K. I. Pedersen, A. Toskala, and P. E. Mogensen (2004), Mobility management and capacity analysis for high speed downlink packet access in WCDMA, *IEEE Proc. Vehicular Technology Conference, September.*

[8] K. Hiltunen, M. Lundevall, and S. Magnusson (2004), Performance of link admission control in a WCDMA system with HS-DSCH and mixed services, in *Proc. Personal Indoor and Mobile Radio Communications.*

[9] 3GPP Technical Specification, TS 25.133, Requirements for Support of Radio Resource Management, v. 6.7.0.

[10] 3GPP Technical Specification, TS 25.214, Physical Layer Procedures (FDD), version 6.4.0.

[11] H. Vishwanathan (1999), Capacity of Markov channels with receiver CSI and delayed feedback, *IEEE Trans. on Information Theory*, **45**(2), 761–771, March.

[12] A. Boariu (2003), Effect of delay commands on adaptive modulation scheme in fading channels', *IEEE Proc. Vehicular Technology Conference (VTC), October.*

[13] A. Sampath, P. S. Kumar, and J. M. Holtzman (1997), On setting reverse link target SIR in a CDMA system, in *IEEE Proc. VTC-97, May*, pp. 929–933.

[14] M. Nakamura, Y. Awad, and S. Vadgama (2002), Adaptive control of link adaptation for high speed downlink packet access (HSDPA) in W-CDMA, in *Proc. Wireless Personal Multimedia Communications*, pp. 382–386.

[15] D. W. Paranchych and M. Yavuz (2002), A method for outer loop rate control in high data rate wireless networks, *IEEE Proc. VTC*, pp. 1701–1705.

[16] A. Das, F. Khan, A. Sampath, and H. Su (2002), Design and performance of downlink shared control channel for HSDPA, *Proc. Personal Indoor and Mobile Radio Communications, September*, pp. 1088–1091.

[17] A. Ghosh, R. Ratasuk, C. Frank, R. Love, K. Stewart, and E. Buckley (2003), Control channel design for HSDPA for 3GPP WCDMA Rel-5, *IEEE Proc. Vehicular Technology Conference-2003 Spring, May*.

[18] W. Bang, K. I. Pedersen, T. E. Kolding, and P. E. Mogensen (2005), Performance of VoIP on HSDPA, *IEEE Proc. VTC, Stockholm, June*.

[19] F. Kelly (1997), Charging and rate control for elastic traffic, *European Trans. on Telecommunications*, **8**, 33–37.

[20] P. A. Hosein (2002), QoS control for WCDMA high speed packet data, *IEEE Proc. Vehicular Technology Conference*.

[21] J. M. Holtzman (2000), CDMA forward link water filling power control, *IEEE Proc. Vehicular Technology Conference, May*, pp. 1663–1667.

[22] T. E. Kolding, F. Frederiksen, and P. E. Mogensen (2002), Performance aspects of WCDMA systems with high speed downlink packet access (HSDPA), *IEEE Proc. Vehicular Technology Conference, Fall*.

[23] T. E. Kolding (2003), Link and system performance aspects of proportional fair packet scheduling in WCDMA/HSDPA, *IEEE Proc. Vehicular Technology Conference, September*, pp. 1717–1722.

[24] J. M. Holtzman (2000), Asymptotic analysis of proportional fair algorithm, *IEEE Proc. Personal Indoor and Mobile Radio Communications, September*, pp. F33–F37.

[25] A. Jalali, R. Padovani, and R. Pankaj (2000), Data throughput of CDMA-HDR a high efficiency–high data rate personal communication wireless system, *IEEE Proc. Vehicular Technology Conference, May*, pp. 1854–1858.

[26] R. C. Elliott and W. A. Krzymieh (2002), Scheduling algorithms for the cdma2000 packet data evolution, *IEEE Proc. Vehicular Technology Conference, September*.

[27] G. Barriac and J. Holtzman (2002), Introducing delay sensitivity into the proportional fair algorithm for CDMA downlink scheduling, *IEEE Proc. International Symposium on Spread Spectrum Techniques and Applications, September*, pp. 652–656.

[28] M. Andrews, K. Kumaran, K. Ramanan, A. Stolyar, and P. Whiting (2001), Providing quality of service over a shared wireless link, *IEEE Communications Magazine*, **39**(2), 150–154, February.

[29] 3GPP, Technical Specification Group RAN, FDD Enhanced Uplink; Overall Description; Stage 2, 3GPP TR 25.309 version 6.4.0, Release 6, October 2005, available at *www.3gpp.org*

# 7

# HSDPA bit rates, capacity and coverage

Frank Frederiksen, Harri Holma, Troels Kolding, and Klaus Pedersen

This chapter presents high-speed downlink packet access (HSDPA) performance results from the end user perspective and from the system capacity point of view. First, the link performance of the high-speed downlink shared channel (HS-DSCH) transport channel and the high-speed shared control channel (HS-SCCH) and high-speed dedicated physical control channel (HS-DPCCH) control channels are analysed. Link level results are used as an input to system level studies. System simulations provide the cell capacity results for dedicated HSDPA carriers and for the cases where wideband code division multiple access (WCDMA) and HSDPA coexist on the same carrier. The chapter also presents results for HSDPA Iub transmission efficiency. As the end user performance is affected by both data rate and network latency, these two factors are addressed in the chapter. We furthermore provide real life HSDPA measurement results. First, HSDPA throughput estimates based on field measurements in the existing WCDMA networks are presented. Second, HSDPA laboratory and field measurements are presented and compared with the simulation results. Finally, a number of solutions are presented to boost baseline HSDPA performance further using advanced antenna concepts and receiver algorithms.

## 7.1 General performance factors

It is important to recognize the complexity of the system and the importance of system assumptions in the performance assessment of advanced wireless systems. The performance of a concept like HSDPA depends significantly on deployment and service parameters and absolute performance varies significantly from scenario to scenario. Some of the most essential conditions for HSDPA performance relate to:

*HSDPA/HSUPA for UMTS* Edited by Harri Holma and Antti Toskala

- *Network algorithms* – HSDPA-specific algorithms such as the MAC-hs packet scheduler, HS-DSCH link adaptation, HS-SCCH power control and HSDPA transmission resource allocation.
- *Deployment scenario* – the interference levels in the cell defined by factors such as propagation loss, RF dominance areas, mobility patterns, and multipath propagation.
- *UE receiver performance and capability* – peak data rate, number of high-speed physical downlink shared channel (HS-PDSCH) codes, number of transmit and receive antennas, and baseband receiver algorithms.
- *Traffic and quality of service (QoS)* – mixture of Release 99 dedicated channel (DCH) and HSDPA suitable traffic, number of active users, and their corresponding QoS requirements.

We will attempt to adhere closely to the basic assumptions agreed in the Third Generation Partnership Project (3GPP). During the discussion of results, it will be indicated which factors in the above play the most important role.

### 7.1.1 Essential performance metrics

The fundamental features of HSDPA link adaptation functionality is adaptive modulation and coding (AMC), multicode transmission (multiple HS-PDSCHs), and fast L1 hybrid automatic repeat request (HARQ). With these different domains for adaptation, the notation for HSDPA performance evaluation is slightly different from that traditionally used for Release 99 DCHs. WCDMA Release 99 typically uses $E_b/N_0$ to denote the received-energy-per-user-bit-to-noise ratio. The $E_b/N_0$ corresponds uniquely to a certain block error rate (BLER) for a given data rate where the only adaptation parameter is the spreading gain. However, the $E_b/N_0$ metric is not an attractive measure for HSDPA because the bit rate on the HS-DSCH is varied every transmission time interval (TTI) using different modulation schemes, effective code rates, and a number of HS-PDSCH codes. We therefore define the average HS-DSCH signal-to-interference-plus-noise ratio (SINR) as the narrowband SINR ratio after de-spreading the HS-PDSCH. Referring to the notation in Figure 7.1, we can express the average HS-DSCH SINR for a single-antenna Rake receiver as:

$$SINR = SF_{16} \frac{P_{HS\text{-}DSCH}}{(1-\alpha) \cdot P_{own} + P_{other} + P_{noise}} \tag{7.1}$$

where $SF_{16}$ is the HS-PDSCH spreading factor of 16, $P_{HS\text{-}DSCH}$ is the received power of the HS-DSCH summing over all active HS-PDSCH codes, $P_{own}$ is the received own-cell interference, $\alpha$ is the downlink orthogonality factor [2]–[4], $P_{other}$ is the received other-cell interference, and $P_{noise}$ is the received noise power. In some 3GPP documents, own-cell interference is denoted $I_{or}$ while other-cell interference is expressed as $I_{oc}$. Note from (7.1) that the HS-DSCH SINR is independent of the number of HS-PDSCH codes used, the modulation scheme, and the effective code rate. The HS-DSCH SINR metric is an essential measure for HSDPA link budget planning and network dimensioning. Another commonly used and related parameter is the wideband ratio of

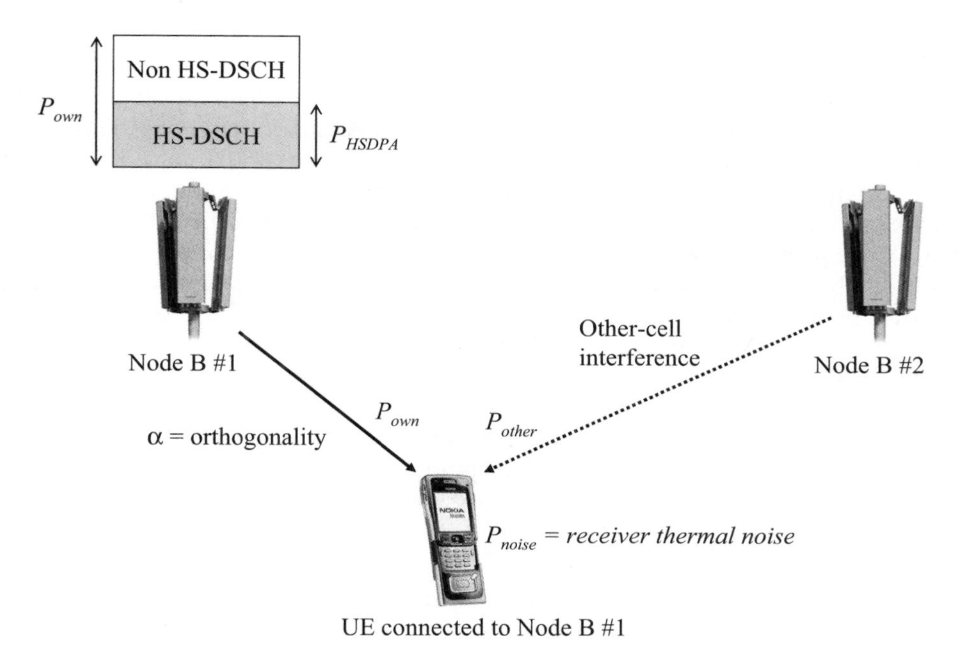


**Figure 7.1** HSDPA performance factors.

own-cell to other-cell plus noise interference at the user, geometry factor, defined as:

$$G = \frac{P_{own}}{P_{other} + P_{noise}} \tag{7.2}$$

When discussing details about performance, the following measures are also considered:

- *Instantaneous required HS-DSCH SINR* – this measure is the per-TTI required SINR on the HS-DSCH to obtain a certain BLER target for a given number of HS-PDSCH codes and for the modulation and coding scheme.
- *Average HS-DSCH SINR* – this is the HS-DSCH SINR experienced by a user averaged over fast fading.

## 7.2 Single-user performance

In this section, the single-user performance aspects are discussed. HS-DSCH link performance is first analysed and results from link adaptation are presented. The performance of the control channels, HS-SCCH and HS-DPCCH, is illustrated. Next, single-user throughput results are presented and data rate coverage is analysed. The section concludes with simulations for 3GPP performance test cases.

### 7.2.1 Basic modulation and coding performance

HSDPA provides adaptive modulation and coding to adjust the data rate to the available channel quality. To maximize coverage and robustness, quadrature phase shift keying (QPSK) is used together with relatively low-rate turbo-coding and a single HS-PDSCH code. The effective code rate (ECR) can be lowered to $1/7 = 137/960$ meaning that 137 bits are carried using 960 channel bits per TTI per HS-PDSCH code. ECR values less than 1/3 are achieved using repetition coding [1]. When channel conditions are better, power can be divided among several HS-PDSCHs to ensure the most effective SINR operating point for each code. For very high SINR conditions, 16-quadrature amplitude modulation (16QAM) with only a little channel coding is employed. To illustrate the dynamic range of the system, consider the example modulation and coding sets in Table 7.1 that are called 'transmission format and resource combinations' (TFRCs). The values used for TFRCs denote the signalling values of the transport block size that is sent over the HS-SCCH. For the exact definition see [4].

The performance of the above-mentioned TFRCs has been simulated using the following assumptions:

- A standard single antenna Rake receiver at the UE.
- The user equipment (UE) supports 5, 10, 15 HS-PDSCH codes and 16QAM.
- A Max-Log-MAP turbo-decoder using eight iterations.
- Primary common pilot channel (P-CPICH) based channel estimation is used.
- Other-cell interference is modelled using additive white Gaussian noise (AWGN).
- The default UE velocity is 3 km/h.

The link between the first transmission block error probability (BLEP) and the available HS-DSCH SINR is shown in Figure 7.2 for a flat Rayleigh fading channel at a slow mobile speed. Each curve relates to the TFRCs of Table 7.1 with curves to the right

**Table 7.1** Example transmission format and resource combinations (TFRCs).

| *TFRC* | *Modulation* | *Effective code rate* (ECR) | *Instantaneous data rate with 1 HS-PDSCH code* [kbps] |
|---|---|---|---|
| #00 | QPSK | 0.14 | 68.5 |
| #10 | QPSK | 0.27 | 128.5 |
| #20 | QPSK | 0.39 | 188.5 |
| #30 | QPSK | 0.52 | 248.5 |
| #40 | QPSK | 0.64 | 308.0 |
| #50 | QPSK | 0.77 | 368.5 |
| #00 | 16QAM | 0.32 | 302.5 |
| #10 | 16QAM | 0.38 | 362.0 |
| #20 | 16QAM | 0.45 | 433.0 |
| #30 | 16QAM | 0.54 | 518.0 |
| #40 | 16QAM | 0.65 | 619.5 |
| #50 | 16QAM | 0.77 | 741.5 |

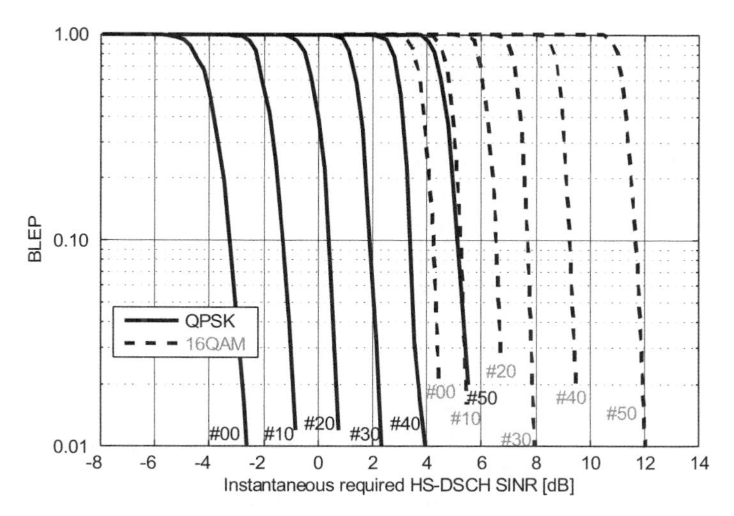


**Figure 7.2** Plot of block error probability (BLEP) as a function of the instantaneous HS-DSCH SINR. The BLEP is shown for the parameters given in Table 7.1. This simulation setup is for a $1 \times 1$ RAKE receiver using an AWGN channel.

generally indicating a higher transmission data rate. There is a crossover point around ECR = 3/4 when it is beneficial to employ 16QAM with strong coding rather than using QPSK with reduced coding.

The results presented in Figure 7.2 are for the single HS-PDSCH case showing the basic modulation and coding performance without link adaptation. Link adaptation selects the modulation and coding scheme so that throughput and delay are optimized for the instantaneous SINR. Figure 7.3 shows the required SINR as a function of the data rate for the first transmission when link adaptation is included.

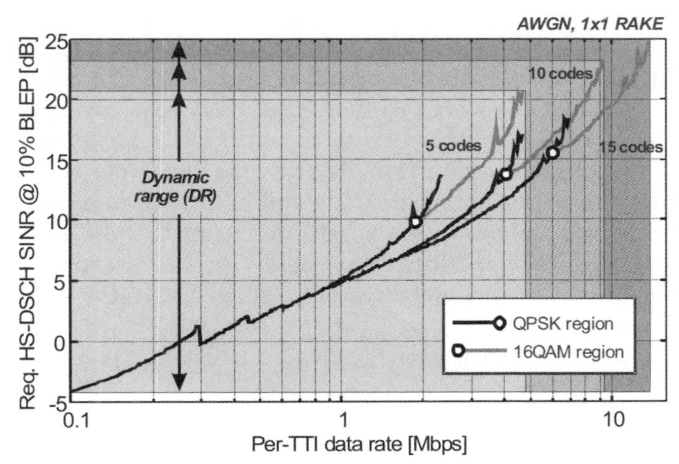


**Figure 7.3** Illustration of the required HS-DSCH SINR to achieve a certain first transmission data rate with the above-mentioned TFRCs.

Note the steps down at certain points in the required SINR in Figure 7.3 when increasing the data rates, especially those between SINR 0 and 5 dB. These steps indicate the points when the number of HS-PDSCH codes is increased from one to two, two to three, etc.

The curves with 5, 10, and 15 codes follow each other up to approx 1 Mbps. At higher data rates 10 and 15 codes require lower SINR than 5 codes. The reason is that 5-code transmission has to reduce the amount of channel coding, which leads to lower spectral efficiency than is the case when employing more HS-PDSCHs. The same is true when comparing the curves for 10 and 15 codes for data rates above 2–3 Mbps.

Figure 7.3 also shows when QPSK or 16QAM modulation is preferred. For a low SINR, QPSK is preferred, while for high SINR conditions, 16QAM is required to provide the higher data rates. For 5-code transmission, 16QAM is preferred when the instantaneous SINR is higher than approximately 10 dB.

The curves indicate that with 5 HS-PDSCH codes and 16QAM, the HS-DSCH dynamic range equals 20 dB – that is, from −3 dB to 17 dB. Increasing the number of HS-PDSCH codes to 15, the total dynamic range is increased to 24 dB. During HSDPA network dimensioning it is important to set the Node B HS-DSCH transmit power such that the HS-DSCH SINR experienced by the different users in the cell is within the HS-DSCH dynamic range. Setting the HS-DSCH transmit power too high will result in excessive interference in the network without achieving higher HSDPA cell throughput if many of the users operate at the high end of the dynamic range. Similarly, if the HS-DSCH transmit power is too low, the highest data rates cannot be achieved in the network.

### 7.2.2 *HS-DSCH performance*

The previous section considered the performance of different modulation and coding schemes without retransmission and without link adaptation. This section presents HS-DSCH performance when link adaptation and retransmission schemes are included.

Figure 7.4 illustrates the HS-DSCH link level performance with 15 codes as a function of the HS-DSCH wideband carrier-to-interference ratio (C/I). The wideband C/I is the received power of HS-DSCH divided by noise and interference without de-spreading. The HS-DSCH data rate is compared with the theoretical Shannon capacity for a 3.84-MHz bandwidth. The Shannon formula gives the maximum error-free data rate that can be transmitted with a specified bandwidth in the presence of noise and interference [5]. There is only an approximate 2-dB difference between the Shannon limit and the simulated HS-DSCH performance, mainly due to decoder limitations and receiver estimation inaccuracies. Simulation results show that HSDPA link performance approaches theoretical limits.

The following assumptions are used in the detailed link adaptation studies later in this chapter:

- The HS-DSCH link adaptation algorithm in Chapter 6 is used, with both an inner loop and an outer loop algorithm.

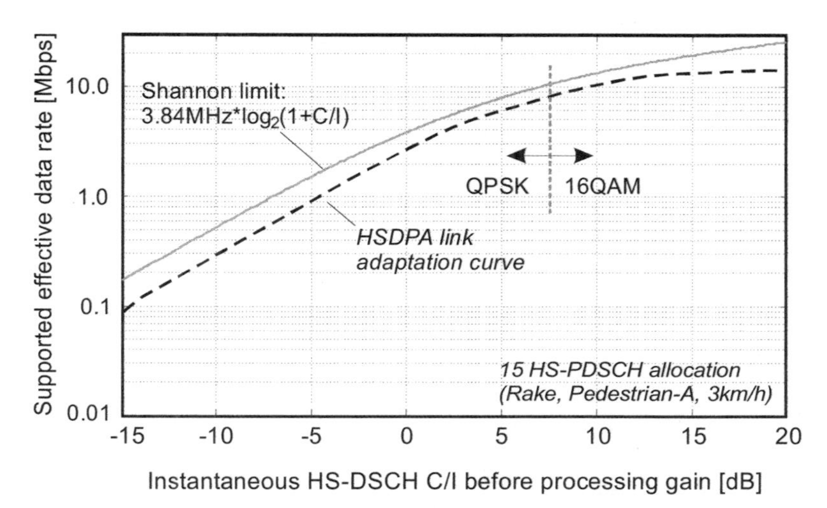


**Figure 7.4** HSDPA data rate compared with the Shannon limit as a function of an average HS-DSCH C/I.

- Incremental redundancy with a maximum of four transmissions. A constellation rearrangement for 16QAM is modelled for retransmissions (the constellation rearrangement is discussed in Chapter 4).
- A log-normally distributed channel quality information (CQI) measurement error at the UE with a standard deviation of 1 dB.
- The delay from CQI measurement until it is reported to the Node B and used for adaptation is assumed to be 6 ms.

The possible inaccuracies in link adaptation are partly compensated by the use of HARQ with soft combining, since the energy from erroneous transmissions helps to improve the probability of correctly decoding retransmissions. The performance aspects of HARQ have been studied in [6]. HSDPA supports both incremental redundancy as well as simple chase combining. Incremental redundancy provides additional gain over chase combining when the effective code rate of the first transmission exceeds 1/3 [1], [6]. The typical link adaptation target level is 10% BLER after the first transmission.

Figure 7.5 shows the single-user average throughput including link adaptation and HARQ as a function of the average HS-DSCH SINR experienced. The results are shown for 5, 10 and 15 codes. An average HS-DSCH SINR of 23 dB is required to achieve the maximum data rate of 3.6 Mbps with 5 HS-PDSCH codes, and over 30 dB for 10 Mbps with 15 codes.

The curves are very similar for different multipath profiles. Note that with a given other-cell-to-own-cell-interference ratio, the SINR is not constant but depends on a number of factors including orthogonality and UE receiver capabilities. These factors must be considered when calculating the value for SINR. The benefit of SINR modelling is that the mapping from SINR to the data rate is fairly constant for the different environments and different UE receiver capabilities.

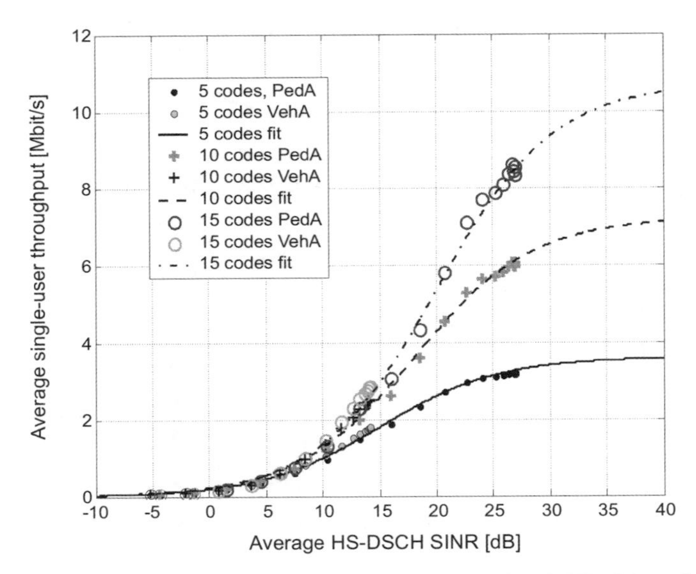


**Figure 7.5** HSDPA data rate as a function of average HS-DSCH SINR. Note that the effect of link adaptation and HARQ is included in these results.

The results in Figure 7.5 can be used for HSDPA link budget planning and HSDPA network dimensioning, since the required HS-DSCH SINR is given for different maximum numbers of HS-PDSCH codes and HS-DSCH throughputs. As an example, the HSDPA single-user throughput is evaluated as a function of the path loss towards the serving HS-DSCH cell by using the results in Figure 7.5 and the expression in Eq. (7.1). Let's first assume that the HSDPA user is subject only to own-cell interference and thermal receiver noise – that is, there is no other-cell interference. Other basic assumptions are summarized in Table 7.2.

Given these assumptions, the HS-DSCH single-user throughput is plotted in Figure 7.6 vs the path loss. At a path loss of 160 dB the throughput is approximately 200–400 kbps depending on HSDPA power allocation. For smaller path loss values,

**Table 7.2** Basic simulation parameters used for the results presented in Figure 7.6.

| *Parameter* | *Settings* |
|---|---|
| Total Node-B Tx power | 12 W |
| Node-B cable loss | 4 dB |
| Node-B antenna gain | 18 dBi |
| UE antenna gain | 0 dBi |
| UE thermal noise power | −101 dBm |
| Allocated HS-PDSCH codes | 5 |
| Allocated HS-PDSCH power | 3 or 7 W |
| P-CPICH Tx power | 2 W |

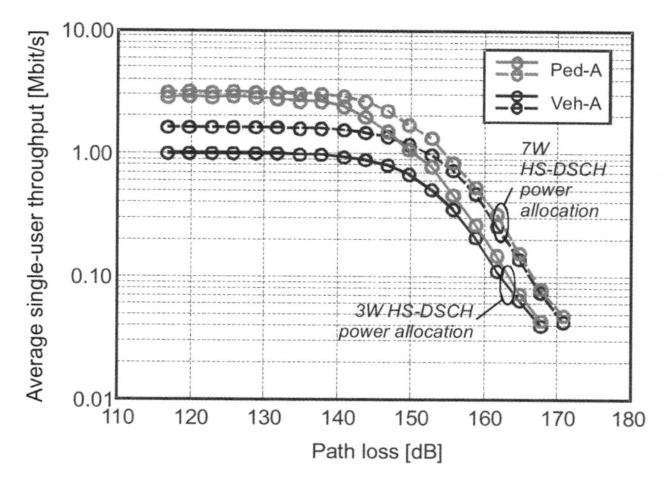


**Figure 7.6** Plot of the average single user throughput as a function of the path loss.

throughput increases. For a path loss smaller than 140 dB, HSDPA throughput is limited by inter-path interference and is, therefore, much lower for the Vehicular A than the Pedestrian A channel. For a 140-dB path loss in the Pedestrian A channel profile, HSDPA throughput equals 3 Mbps, and it can be seen that, even for lower path loss values, throughput does not increase.

Next, we consider HSDPA coverage including the effect of other-cell interference. Assuming that the G-factor at the cell edge is known, the results in Figure 7.5 can be used as input to dimension the required Node B HS-DSCH transmit power to guarantee a minimum HS-DSCH throughput at the cell edge. Rearranging (7.1), we can express the average HS-DSCH SNIR as:

$$SINR = SF_{16} \frac{P_{HS\text{-}DSCH}}{P_{own}} \frac{1}{1 - \alpha + G^{-1}} \tag{7.3}$$

and, hence, we can write:

$$P_{HS\text{-}DSCH} \geq SINR[1 - \alpha + G^{-1}] \frac{P_{own}}{SF_{16}} \tag{7.4}$$

where $P_{HS\text{-}DSCH}$ and $P_{own}$ are the Node B HS-DSCH transmit power and the total carrier transmit power, respectively. Thus, for a typical macro-cell edge with $G = -3$ dB, $\alpha = 0.5$, and $P_{own} = 12$ W, we can calculate the minimum required HS-DSCH transmit power to achieve a certain minimum single-user HS-DSCH data rate at the cell edge. As an example, the HS-DSCH transmit power should equal 5.6 W if the required HS-DSCH throughput at the cell edge needs to be 200 kbps – that is, corresponding to a required HS-DSCH SINR of 5 dB according to Figure 7.5.

In the following we calculate the percentage of the cell area in which an HSDPA single-user can be served with a given minimum data rate on the HS-DSCH. For these calculations, we assume a typical macro-cellular scenario with the following parameter settings:

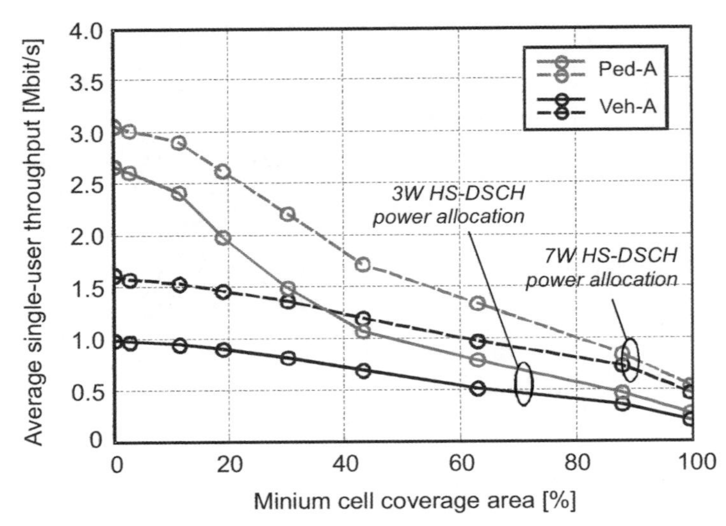


**Figure 7.7** Average single-user throughput as a function of minimum cell coverage area.

- A standard homogeneous three-sector macro-cellular topology with 65° Node B antennas and a site-to-site distance of 2.8 km.
- The COST 231 Hata Okumura path loss model is applied, plus a log-normal distributed shadow-fading component with 8-dB standard deviation.
- All cells in the network are assumed to transmit at 12 W, corresponding to a fairly high air–interface load in all cells.

The results are plotted in Figure 7.7. Using 7 W and five HS-PDSCH codes for HSDPA, an HSDPA single-user can be served with 1.1 Mbps in 50% of the cell area in Vehicular A and with 1.6 Mbps in Pedestrian A. Note here that these results are obtained under the assumption that the remaining power up to 12 W is used by Release 99 channels on the same carrier.

Finally, a commonly used measure for network dimensioning is the pilot $E_c/I_o$. Based on the wideband average P-CPICH $E_c/I_o$ it is possible to estimate the achievable HSDPA single-user throughput. It can be shown that the average HS-DSCH SINR can be expressed as a function of the P-CPICH single $E_c/I_o$ as:

$$SINR = SF_{16} \frac{P_{HSDPA}}{\frac{P_{pilot}}{\rho_{pilot}} - \alpha P_{own}} \tag{7.5}$$

where $P_{HSDPA}$, $P_{own}$, and $P_{pilot}$ are the HSDPA transmit power, the total Node B transmit power, and the P-CPICH transmit power, respectively. The parameter $\rho_{pilot}$ denotes the P-CPICH $E_c/I_o$ when HSDPA power is on. HSDPA single-user throughput as a function of the average HS-DSCH SINR can be seen in Figure 7.5 for cases where 5, 10, and 15 HS-PDSCHs are available.

**Table 7.3** The additional parameters used to illustrate the impact of disabling 16QAM.

| *UE category* | *Multicodes and modulation* |
|---|---|
| Category 6 | 5-code 16QAM max 3.6 Mbps |
| Category 12 | 5-code QPSK max 1.8 Mbps |

### *7.2.3 Impact of QPSK-only UEs in early roll-out*

HSDPA UE Categories 6 and 12 both support five HS-PDSCH codes. The difference is in the support of 16QAM modulation: Category 12 only supports QPSK modulation, while Category 6 also supports 16QAM. The UE capabilities for Categories 6 and 12 are summarized in Table 7.3. Single-user link performance is shown in Figure 7.8 for these UE categories. 16QAM can clearly provide higher data rates with high SINR values especially in the Pedestrian A channel. On the other hand, at low SINR values there is no difference in throughput between 16QAM/QPSK and QPSK-only. For data rates lower than 1 Mbps, performance is practically identical for the two UE categories considered. The reason can be seen from Figure 7.3, where 16QAM is used only for higher SINR values.

### *7.2.4 HS-SCCH performance*

The error probability of HS-SCCH should be low since the HS-DSCH can only be decoded by the UE if the HS-SCCH is first correctly received. Link simulation results illustrating the HS-SCCH BLEP vs the SINR per TTI are shown in Figure 7.9. The

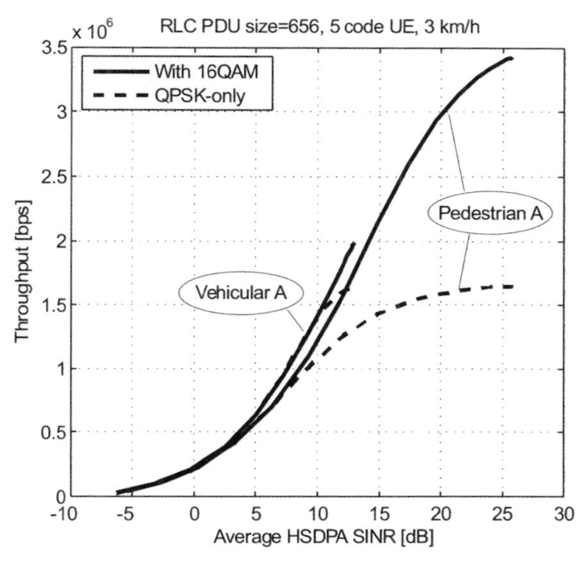


**Figure 7.8** Single-user five-code performance with 16QAM/QPSK and with QPSK-only.

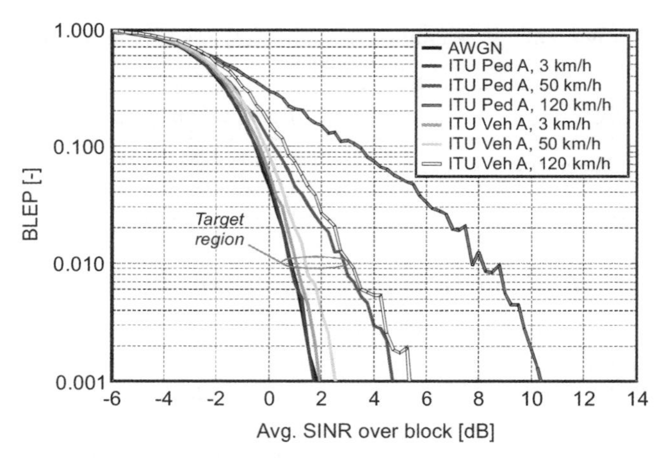


**Figure 7.9** HS-SCCH decoding performance vs average received HS-SCCH SINR.

HS-SCCH SINR is the effective SINR after de-spreading the HS-SCCH including the effect of the orthogonality factor and the processing gain of 128. For UE velocities up to 50 km/h, the required SINR to achieve a BLEP of 1% varies between 1 and 3 dB.

Assuming a target SINR for the HS-SCCH of 1.5 dB, the HS-SCCH overhead is measured in terms of the required Node B HS-SCCH transmit power, assuming HS-SCCH power control every TTI depending on the scheduled user and HS-SCCH SINR it experiences [8]. For details see the HS-SCCH power control discussion in Chapter 6. The results of the simulation are given in Figure 7.10. In the 3GPP macro-cell environment the required HS-SCCH power at the cell edge is approximately

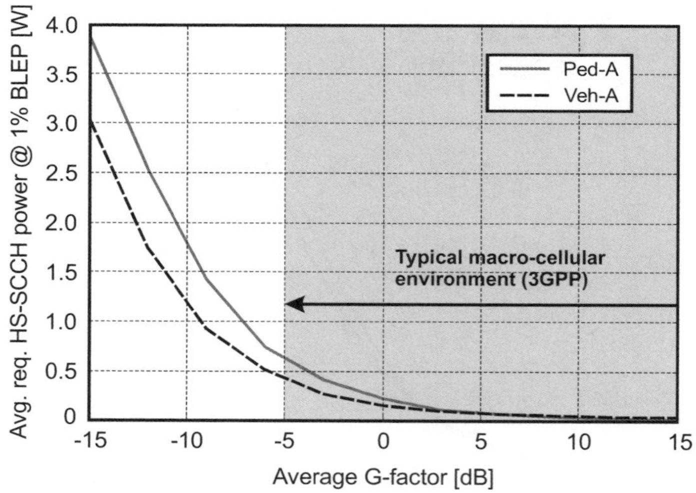


**Figure 7.10** Required HS-SCCH transmission power to achieve good detection performance vs the user's G-factor.

0.5 W and on average is 0.2–0.3 W over the cell area, which is <2% of average Node B transmission power. These simulations assume round robin scheduling, which does not take the current radio channel conditions into account. The impact of more advanced scheduling on the HS-SCCH overhead has been studied in [7]. When scheduling the data – primarily when the user's channel is favourable – a side-benefit is that the HS-SCCH overhead is reduced.

Each code-multiplexed user requires an HS-SCCH. If we code-multiplex three users, three parallel HS-SCCHs are required. Therefore, the optimization of HS-SCCH power control becomes more relevant when code-multiplexing is used.

More detailed design considerations and performance figures for the HS-SCCH can be found in [10].

### *7.2.5 Uplink HS-DPCCH performance*

The uplink HS-DPCCH is power-controlled relative to the uplink DPCCH in every slot interval. The power relation between the DPCCH and the HS-DPCCH is given by the parameters [$\Delta_{ACK}$; $\Delta_{NACK}$; $\Delta_{CQI}$], which are set by the radio network controller (RNC) via radio resource control (RRC) signalling to the UE. The power offset can be up to 6 dB within nine possible quantization steps [14]. The power offsets [$\Delta_{ACK}$; $\Delta_{NACK}$; $\Delta_{CQI}$] must be selected to fulfill the required uplink HS-DPCCH detection probabilities. The normal target values for these offsets are defined in terms of the probability of mis-detecting. Typical values for these are $P_{ACK \rightarrow NACK} < 10^{-2}$ and $P_{NACK \rightarrow ACK} < 10^{-3}$. The probability of decoding a NACK as an ACK has to be very low since that causes missing L1 retransmission and the data must be retransmitted by the RLC layer.

In a previous study [13], the default values for HS-DPCCH power offsets have been extracted that fulfill the HS-DPCCH detection probabilities as specified by 3GPP. It should be noted that the recommended default HS-DPCCH power offsets depend on the service on the uplink-associated DCH. The DPCCH/DPDCH power offset is set to −2.69 dB for the 12-kbps service [32]. The recommended values are [13]: $\Delta_{ACK} = 4$ dB, $\Delta_{NACK} = 4$ dB, and $\Delta_{CQI} = 2$ dB. With an increasing data rate on the DCH, the above offsets generally decrease.

These numbers are valid for HSDPA users with an active set size of 1. The uplink power control in soft handover is based on the best uplink connection – that is, if one of the cells receives a high enough signal level, mobile power can be decreased. The HS-DPCCH, however, is received only by the serving HS-DSCH cell and the received power level of the HS-DPCCH may be too low if that cell instantaneously does not have the best uplink connection. Therefore, a larger HS-DPCCH power offset is required in soft handover. If the maximum power offset of 6 dB is not enough to maintain HS-DPCCH quality, 3GPP specifications allow the use of ACK/NACK repetition as well. For HSDPA users in soft handover, it is recommended to use an HS-DPCCH power offset of 6 dB and an ACK/NACK repetition of 2 resulting in an additional gain of approximately 3 dB. For further information on HS-DPCCH uplink coverage for users in soft handover, the reader is referred to [13].

### 7.2.6 3GPP test methodology

HSDPA performance requirements are defined in [11]. In this text, the focus is on the single link performance of the HS-DSCH with a constant modulation and coding scheme. For the simulations shown here, it has been assumed that the return channel for HSDPA (HS-DPCCH) is error-free when transmitting ACKs/NACKs in response to a HARQ.

In [11], various test channels are defined – denoted as H-Set1 to H-Set6 – each describing different UE capabilities. These channels are called 'fixed reference channels' (FRCs). Here, we will only consider the first test channel (H-Set1). The test channel definitions are shown in Table 7.4. The minimum performance requirements for QPSK and 16QAM are listed in Tables 7.5 and 7.6, respectively. QPSK modulation is tested with own-to-other-cell-interference ratio $I_{or}/I_{oc}$ values of 0 dB and 10 dB. The value of 0 dB corresponds to a location close to the cell edge while 10 dB is located in a clear dominance area of a single cell. 16QAM modulation is tested only with an own-to-other-cell-interference ratio of 10 dB since 16QAM would not be used at a low $I_{or}/I_{oc}$ value. Two HS-DSCH power allocations are considered: 25% of Node B power for HS-DSCH ($E_c/I_{or} = -6$ dB) and 50% power for HS-DSCH ($E_c/I_{or} = -3$ dB).

We compare the link simulations with the 3GPP minimum performance requirements given in Figure 7.11 as a function of HS-DSCH power allocation. ITU Vehicular A multipath profile at 30 km/h is assumed in Figure 7.11. The simulated results are

**Table 7.4** Parameter settings for simulation of the fixed reference channel H-Set1.

| *Parameter* | *Unit* | *Value* | |
|---|---|---|---|
| Modulation | | QPSK | 16QAM |
| Inter-TTI distance | TTI's | 3 TTI | 3 TTI |
| Effective code rate | | 0.67 | 0.61 |
| Number of physical channel codes | Codes | 5 | 4 |

**Table 7.5** Minimum requirement QPSK, fixed reference channel (FRC) H-Set1.

| *Test number* | *Propagation conditions* | *Reference value* | | |
|---|---|---|---|---|
| | | *HS-PDSCH* $E_c/I_{or}$ [dB] | *Data rate at* $I_{or}/I_{oc} = 0$ dB [kbps] | *Data rate at* $I_{or}/I_{oc} = 10$ dB [kbps] |
| 1 | Pedestrian A | −6 | 65 | 309 |
| | 3 km/h | −3 | N/A | 423 |
| 2 | Pedestrian B | −6 | 23 | 181 |
| | 3 km/h | −3 | 138 | 287 |
| 3 | Vehicular A | −6 | 22 | 190 |
| | 30 km/h | −3 | 142 | 295 |
| 4 | Vehicular A | −6 | 13 | 181 |
| | 120 km/h | −3 | 140 | 275 |

**Table 7.6** Minimum requirement 16QAM, fixed reference channel (FRC) H-Set1.

| *Test number* | *Propagation conditions* | *Reference value* | |
|---|---|---|---|
| | | *HS-PDSCH* $E_c/I_{or}$ [dB] | *Data rate at* $I_{or}/I_{oc} = 10\,\text{dB}$ [kbps] |
| 1 | Pedestrian A | −6 | 198 |
| | 3 km/h | −3 | 368 |
| 2 | Pedestrian B | −6 | 34 |
| | 3 km/h | −3 | 219 |
| 3 | Vehicular A | −6 | 47 |
| | 30 km/h | −3 | 214 |
| 4 | Vehicular A | −6 | 28 |
| | 120 km/h | −3 | 167 |

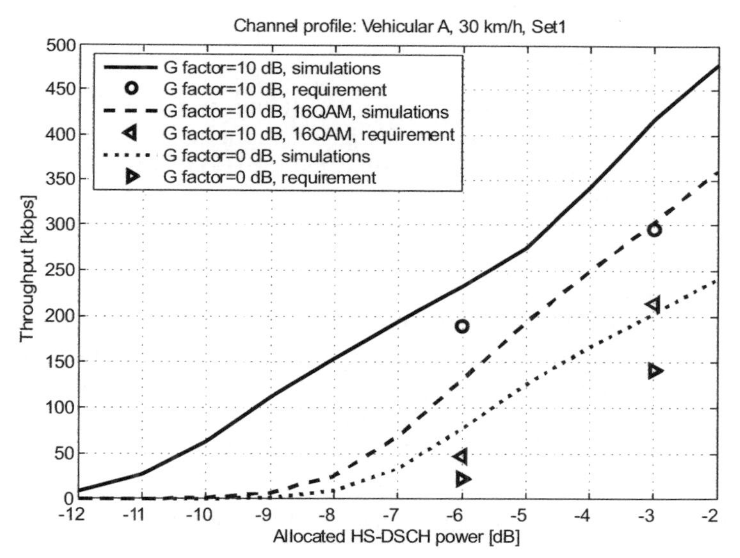


**Figure 7.11** Throughput performance of H-Set1 as a function of allocated HS-DSCH power for the Vehicular A 30 km/h channel profile.

approximately 1–2 dB better than 3GPP performance requirements. The results presented in Figure 7.11 can be viewed as either a margin in terms of throughput or as an implementation margin.

## 7.3 Multiuser system performance

The following subsections contain multiuser simulation results. We provide both cell capacity and end user performance results. Most of the results are obtained for best effort web browsing applications without any strict QoS requirements. Some results for constant bit rate streaming applications are also presented.

### 7.3.1 Simulation methodology

The performance of HSDPA in a multiuser scenario is quantified by conducting advanced dynamic network level simulations with multiple cells and users [8]. The simulator supports scenarios with a mixture of WCDMA Release 99 DCH and HSDPA Release 5 users being served by the same carrier. For each Release 5 HSDPA user, we explicitly simulate reception of HS-DSCH, HS-SCCH, and one associated DCH at 3.4 kbps. The associated DCH is only assumed to carry the signalling radio bearer for Layer 3 signalling. Release 99 DCHs are simulated with inner loop and outer loop power control, soft handover functionality, and power-based admission control. The primary default simulation parameters are summarized in Table 7.7 [8].

### 7.3.2 Multiuser diversity gain

We first study the gain of proportional fair scheduling over round robin, also called 'multiuser diversity gain'. Multiuser diversity gain is reported in Figure 7.12 for the case where 7 W and five HS-PDSCH codes are allocated for HSDPA transmission, while the remaining available resources are used for transmission of Release 99 channels [8]. The average number of HSDPA users per cell equals six for these results. Multiuser diversity gain comes from scheduling users when they experience relatively good SINR, while avoiding scheduling users that experience deep fades. Multiuser diversity gain is only available at moderate UE speeds where the packet scheduler is able to track fast fading radio channel variations. At high UE speeds where the packet scheduler can no longer track radio channel variations by monitoring received CQI reports from UEs, multiuser diversity gain is marginal. The same observation is true for stationary users where the channel is constant. The proportional fair scheduler can provide a minor gain at 0 km/h

**Table 7.7** Summary of default parameter settings for network simulations.

| *Network topology* | *Macro-cellular, three-sector sites, 2.8-km site-to-site distance* |
|---|---|
| Radio propagation | COST231 Okumura–Hata, log-normal shadowing (8 dB std), ITU Vehicular A with 3 km/h. |
| MAC-hs functionalities | Proportional fair packet scheduling, HARQ with chase combining. |
| Traffic model | Closed loop TCP web browsing model, 1500 bytes TCP packet size, 100 kbytes average packet call size, RLC acknowledged mode. |
| Handovers | Synchronized HS-DSCH cell change for HSDPA, standard soft handover for Release 99 with an add window of 2 dB. |
| UE receiver type | Single antenna Rake. |
| Node B power settings | Maximum transmit power per cell 20 W, P-CPICH power 2 W. |

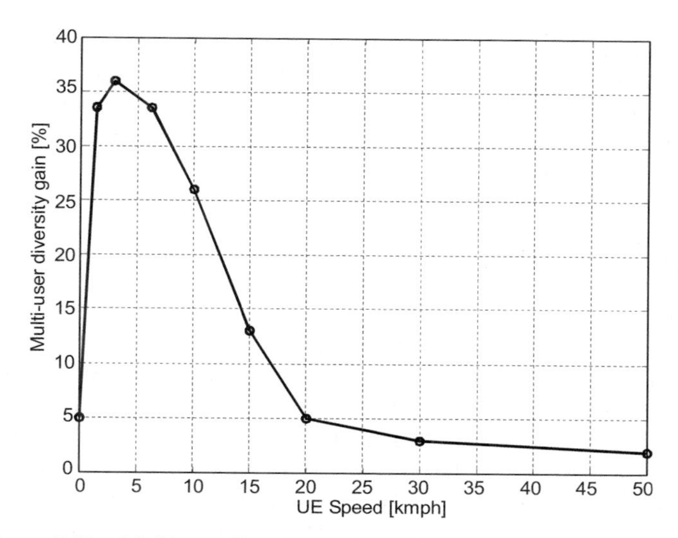


**Figure 7.12** Multiuser diversity gain as a function of the UE speed.

as well due to variations in the interference. The results in Figure 7.12 are obtained for a Vehicular A radio channel.

Multiuser diversity gain also depends on the number of simultaneous active HSDPA users in the cell. The gain of proportional fair over round robin is reported in Figure 7.13 vs the number of HSDPA users, assuming that all users are moving with a velocity of 3 km/h. A clear gain can already be achieved with two parallel users. Multiuser diversity gain only increases marginally as the number of users goes beyond five. It is also observed

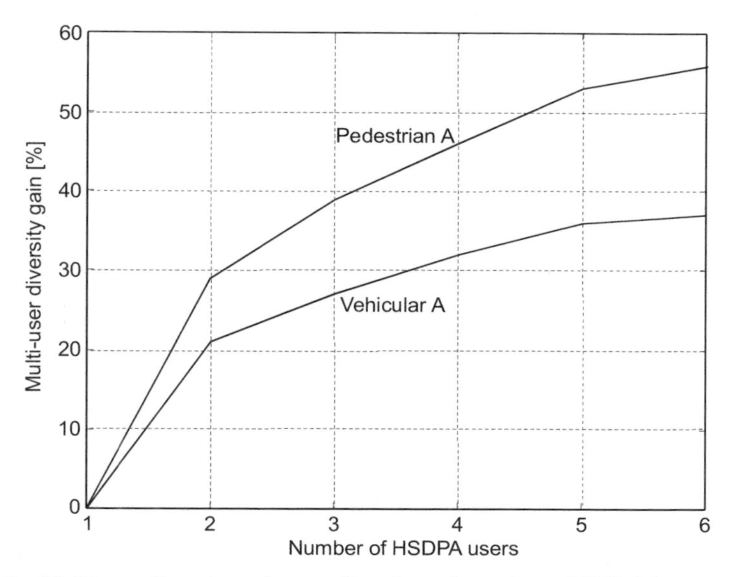


**Figure 7.13** Multiuser diversity gain as a function of number of HSDPA users at 3 km/h.

that multiuser diversity gain is larger in International Telecommunication Union (ITU) Pedestrian A than in ITU Vehicular A. This behaviour is due to the larger fading dynamic range in ITU Pedestrian A. Recall that multiuser diversity gain increases when the fading dynamic range of the radio channel increases. Additional simulation results for ITU Pedestrian A are shown in Section 7.7.3.6.1.

### *7.3.3 HSDPA-only carrier capacity*

Cell throughput results are reported in Figure 7.14 for different allocations of HS-PDSCH codes per cell in an ITU Vehicular A multipath. For five HS-PDSCH codes and no DCH traffic in the cell, achievable cell throughput is found to equal 1.2 Mbps, while it increases to 1.3 Mbps if some of the power is also used for transmission of traffic on the DCH. Hence, with five HS-PDSCH codes per cell, it is better to split the traffic between HSDPA and the DCH rather than allocating all the traffic and power to HSDPA. The practical split between HSDPA and the DCH obviously depends on the available traffic from HSDPA and non-HSDPA UEs. Increasing the number of HS-PDSCH codes from five to ten results in a capacity gain of approximately 50%. The capacity increase is achieved because it is more spectrally efficient to first increase the number of HS-PDSCH codes before starting to increase the effective coding rate and/or modulation order. The results for ten HS-PDSCH codes in an HSDPA-only cell are reported for the case without code-multiplexing assuming that all HSDPA mobiles are able to receive up to ten HS-PDSCH codes and with code-multiplexing of two users per TTI, each capable of receiving five HS-PDSCH codes. It is observed that achievable

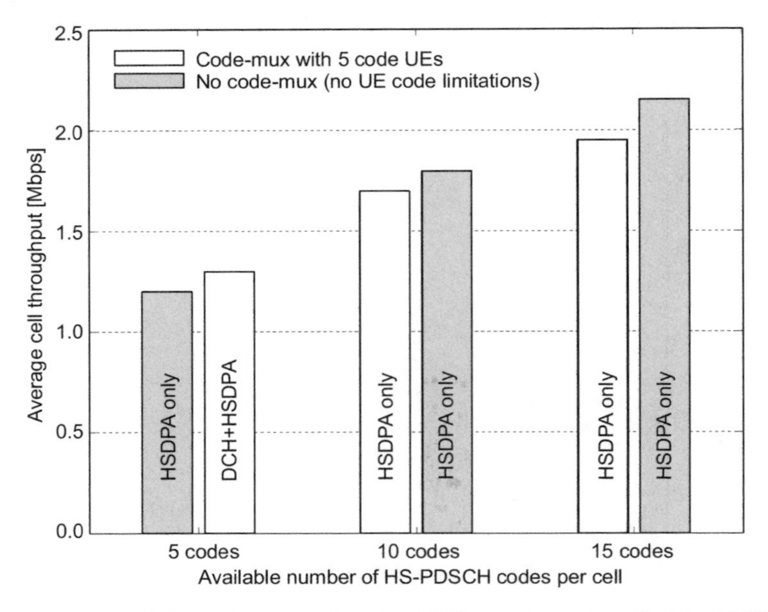


**Figure 7.14** Average cell throughput as a function of allocated number of HS-PDSCH codes and whether code-multiplexing used.

cell throughput is slightly lower in the case of code-multiplexing. This is caused by the higher overhead from having two HS-SCCHs per cell as well as the additional scheduling constraints from having to schedule two users per TTI. The cell capacity with all 15 codes is approximately 2.0 Mbps when code-multiplexing is used.

### *7.3.4 HSDPA capacity with Release 99*

We consider the case where five HS-PDSCH codes and one HS-SCCH code are allocated for HSDPA transmission per cell. This can be considered a typical initial roll-out scenario for HSDPA. The remaining channelization codes are used for the transmission of Release 99 channels. Figure 7.15 shows the average cell throughput on HSDPA and the DCH vs the allocated HSDPA power, as well as the total cell throughput that is the sum of the throughput on HSDPA and the DCH. As expected, HSDPA cell throughput increases the more HSDPA power that is being allocated, while DCH throughput simultaneously decreases. The maximum cell throughput is observed to equal 1.3 Mbps for 7-W HSDPA power allocation. Hence, the introduction of HSDPA with five HS-PDSCH codes brings a capacity gain of 70% over Release 99, which typically offers a capacity of 780 kbps. The capacity gain comes from fast link adaptation and HARQ, and multiuser diversity comes from using proportional fair scheduling. Hence, it is evident that the performance of HSDPA vs DCH depends on how the common transmission resources are shared between these two channel types. However, total cell throughput does not vary significantly with HSDPA transmit power allocated.

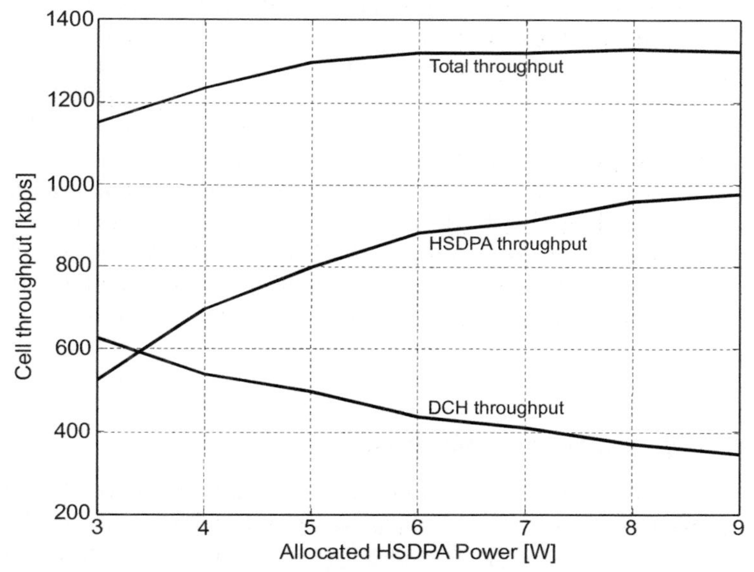


**Figure 7.15** Average cell throughput as a function of allocated HSDPA power per cell.

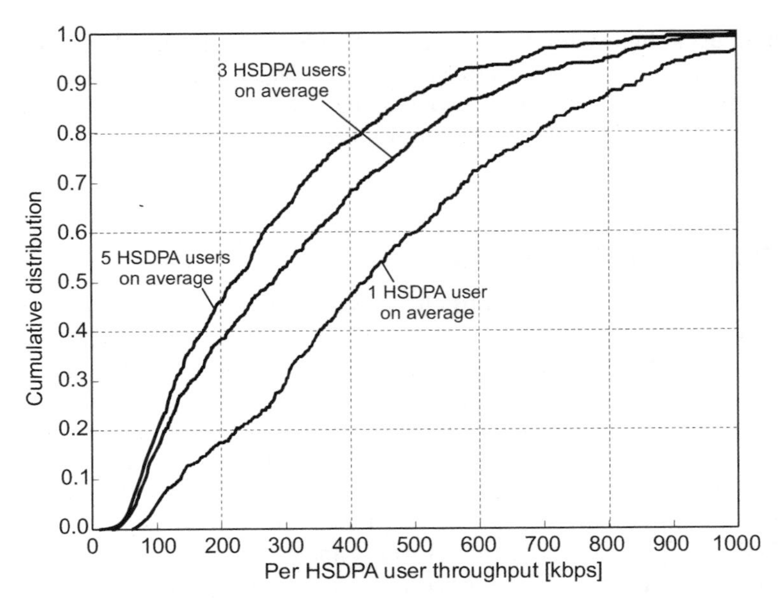


**Figure 7.16** Cumulative distribution function of HSDPA throughput experienced per user for different numbers of HSDPA users per cell. The HSDPA power allocated equals 7 W.

### *7.3.5 User data rates*

HSDPA throughput experienced per user depends on the number of allocated HSDPA users per cell that are sharing the common HS-DSCH. The cumulative distribution function of the HSDPA throughput experienced per user is reported in Figure 7.16, assuming 7-W HSDPA power allocation and five HS-PDSCH codes, while the remaining power is used for transmission of Release 99 channels. As expected, HSDPA throughput decreases as more users are allocated per cell. The decrease in average HSDPA throughput per user is not inversely proportional to the number of allocated HSDPA users. As an example, HSDPA user throughput at the 50% quantile equals 400 kbps and 270 kbps for one and three HSDPA users per cell, respectively. The reason for this is that multiuser diversity gain increases with more users. Users experiencing high data rates are typically close to the serving HS-DSCH cell, while users served with lower data rates are typically at the cell edge. Although not shown here, HSDPA throughput experienced per user is also observed to depend significantly on the amount of allocated HSDPA transmission power, as was also observed in Figure 7.15 for average HSDPA cell throughput.

### *7.3.6 Impact of deployment environment*

Recall that the network performance results presented have been obtained from a large macro-cellular scenario with an ITU Vehicular A power delay profile. In the following, we give a brief discussion on how performance results will change if some of the essential simulation parameters are altered.

#### 7.3.6.1 Multipath profile

It is known from the numerous radio channel measurement campaigns reported in [25] that the ITU Vehicular A power delay profile is representative for macro-cells where Node B antennas are mounted above the rooftop level. However, users located in close vicinity of the Node B might experience a power delay profile with less temporal dispersion – such as the ITU Pedestrian A channel profile (see, e.g., Greenstein's distance-dependent models for delay spread [24]). The ITU Pedestrian A power delay profile is typically used for micro-cells where the Node B antenna is mounted below rooftop level. ITU Pedestrian A has less temporal dispersion, and therefore also a better downlink orthogonality factor [2], which results in less own-cell interference at the UE after de-spreading. The ITU Pedestrian A channel profile also has a larger fading dynamic range which results in larger multiuser diversity scheduling gain (see [7]). Figure 7.17 compares both Vehicular A and Pedestrian A in a three-sector macro-cell environment. Hence, for the macro-cellular setup under consideration, average HSDPA cell throughput is increased by 30–50% in ITU Pedestrian A compared with Vehicular A. Note that the use of performance enhancement methods – like the advanced receivers in Section 7.8 – offer further improved cell throughputs over the numbers shown in Figure 7.17.

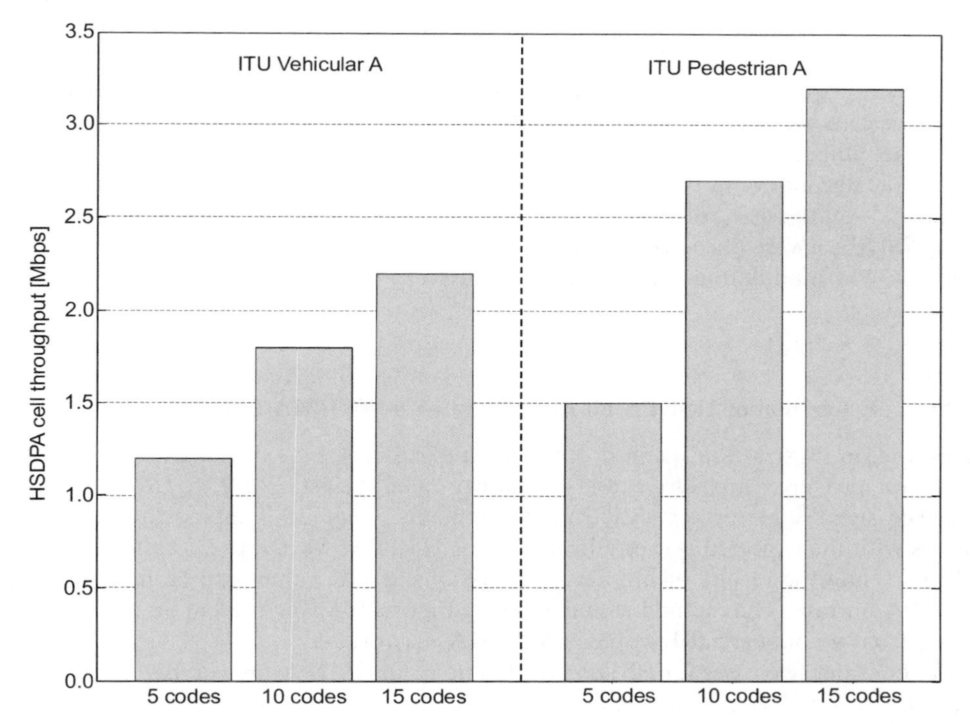


**Figure 7.17** Summary of average cell throughput for HSDPA in a macro-cellular environment depending on the assumed power delay profile and number of HS-PDSCH codes allocated.

#### 7.3.6.2 Cell topology

Cell topology does impact HSDPA performance. As an example, the results for micro-cells have been presented [1]. The micro-cellular setup is characterized by the higher isolation between neighbouring cells which results in less other-cell interference as well as less multipath propagation. The average HSDPA cell throughput is therefore found to increase from 2.2 Mbps for macro-cells to 4.8 Mbps for micro-cells. The larger HSDPA cell throughput in micro-cells is achievable because the HS-DSCH SINR at many of the users is sufficiently high to allow frequent transmission using 16QAM modulation.

#### 7.3.6.3 Indoor coverage

The other simulation results in this chapter have been obtained using an urban macro-cellular propagation model assuming users located outdoors at street level. Since quite a few users are located indoors in practice, the earlier simulations have been repeated with an additional indoor penetration loss of 20 dB. However, note that this additional path loss might be too pessimistic for users located on higher floors in high-rise buildings, since the path loss is typically observed to be less for elevated users [25], [26]. The added penetration loss of 20 dB tends to make thermal noise at users at the cell edge dominant over other-cell interference and, therefore, results in capacity loss when compared with the results for the outdoor-to-outdoor situation. For the scenario under consideration – with a 2.8-km site-to-site distance – the 20-dB penetration loss results in a 15% throughput decrease for HSDPA. However, the throughput decrease for Release 99 is observed to equal 40%. This clearly demonstrates that HSDPA provides an additional coverage gain over Release 99. Using HARQ with soft combining of retransmissions and fast scheduling mainly facilitates this coverage gain.

A dedicated indoor in-building system can also be used to significantly improve indoor coverage – that is, using distributed antenna systems or an indoor pico-cell. As studied in [16], HSDPA is very attractive for such scenarios since HS-DSCH SINR values are high and 16QAM modulation can frequently be used to offer high user data rates and cell capacities.

#### 7.3.6.4 Estimation of HSDPA bit rates based on a WCDMA drive test

This section illustrates a method of estimating HSDPA bit rates based on WCDMA drive test measurements. Drive tests typically provide at least CPICH $E_c/N_0$ and CPICH received signal code power (RSCP) measurements. Once we combine those measurements with the expected power allocations for HSDPA, we can predict the HS-DSCH SINR. Using the results of link level simulations, SINR values can be mapped to an HSDPA bit rate. The method is illustrated in Figure 7.18. We present an example case study from a commercial European WCDMA network.

We consider two cases for HSDPA deployment: an HSDPA-only carrier and a shared carrier with WCDMA. Assumed power allocations are shown in Figure 7.19. The shared carrier assumes HS-DSCH power of 7 W and the dedicated carrier 15 W. The network load was low during the measurements and average Node B transmission power was

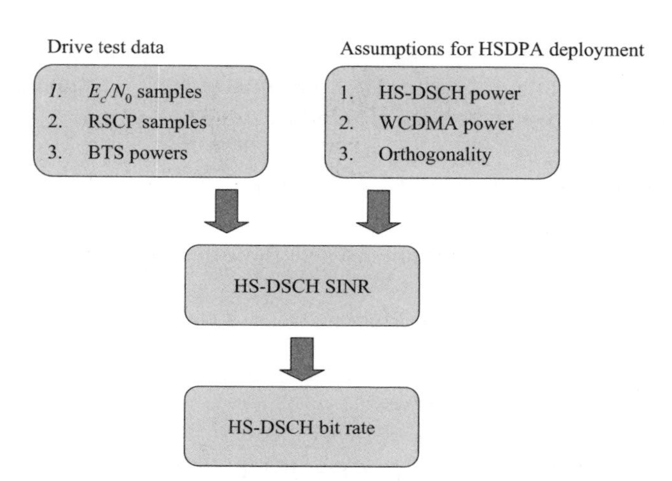


**Figure 7.18** Method of estimating HS-DSCH bit rate from WCDMA drive test measurements.

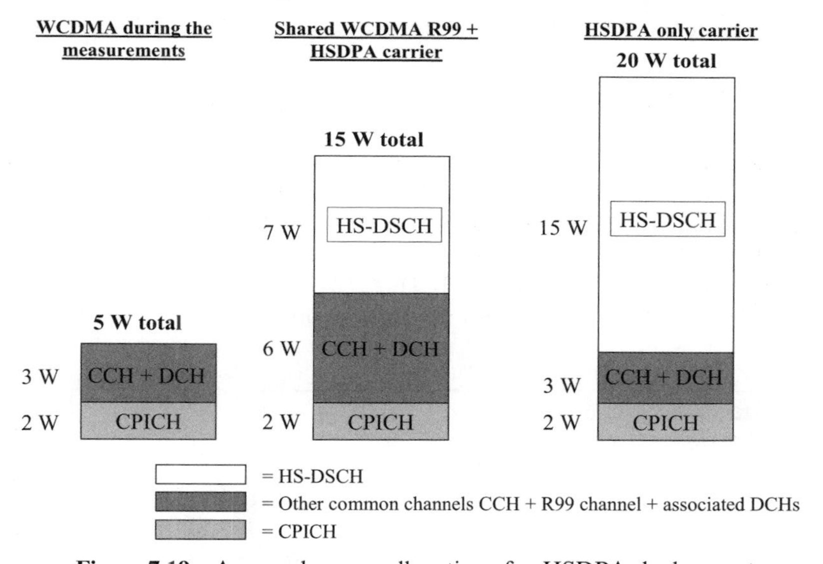


**Figure 7.19** Assumed power allocations for HSDPA deployment.

typically 5 W where most of the power was caused by the CPICH – that is, 2 W – and by the other common channels.

The interference and signal values from the drive tests are converted to corresponding values when HSDPA is deployed using the assumption from Figure 7.19. Total received interference with HSDPA is scaled according to the increase in total BTS power:

$$\text{Total interference } I_o = RSCP - \frac{E_c}{I_0} + 10\log_{10}\left(\frac{P_{tot,HSDPA}}{P_{tot,beforeHSDPA}}\right) \tag{7.6}$$

where *RSCP* is received CPICH signal code power, $E_c/I_0$ is CPICH $E_c/N_0$, $P_{tot,HSDPA}$ is the base station power using HSDPA, and $P_{tot,beforeHSDPA}$ that before HSDPA. This calculation assumes that the HS-DSCH is continuously transmitted from all cells, which is pessimistic for low-traffic scenarios. Own-cell interference is estimated from received CPICH power and BTS power using HSDPA:

$$\text{Own-cell interference } I_{or} = RSCP + 10\log_{10}\left(\frac{P_{tot,HSDPA}}{P_{CPICH}}\right) \tag{7.7}$$

Other-cell interference can be obtained when the other interference components are known:

$$\text{Other-cell interference } I_{oc} = I_0 - I_{or} - N_0 \tag{7.8}$$

where $N_0$ is the terminal receiver thermal noise level. The received energy per chip for the HS-DSCH can be obtained as follows:

$$HS\text{-}DSCH\, E_c = RSCP + 10\log_{10}\left(\frac{P_{HS\text{-}DSCH}}{P_{CPICH}}\right) \tag{7.9}$$

Finally, we calculate the HS-DSCH SINR using the signal and interference values above:

$$HS\text{-}DSCH\, SINR\,[\text{dB}] = 10\log_{10}\left(16\frac{HS\text{-}DSCH\, E_c}{(1-\alpha)I_{or} + I_{oc} + N_0}\right) \tag{7.10}$$

where $\alpha$ is the orthogonality of own-cell interference. Average orthogonality was assumed to be 0.6 in this case study, which corresponds to a typical urban macrocellular environment. If advanced terminal receivers with equalizers are deployed (see Section 7.8.1), orthogonality would be higher.

The distribution of the measured CPICH $E_c/N_0$ from the drive tests is shown in Figure 7.20. The median $E_c/N_0$ is −6.1 dB and the 90% value is −9.6 dB.

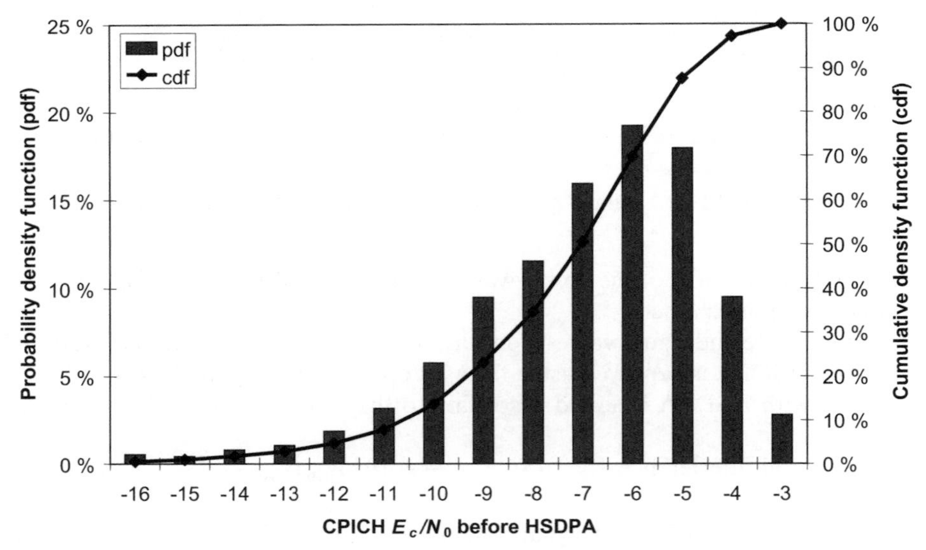


**Figure 7.20** Drive test data before HSDPA.

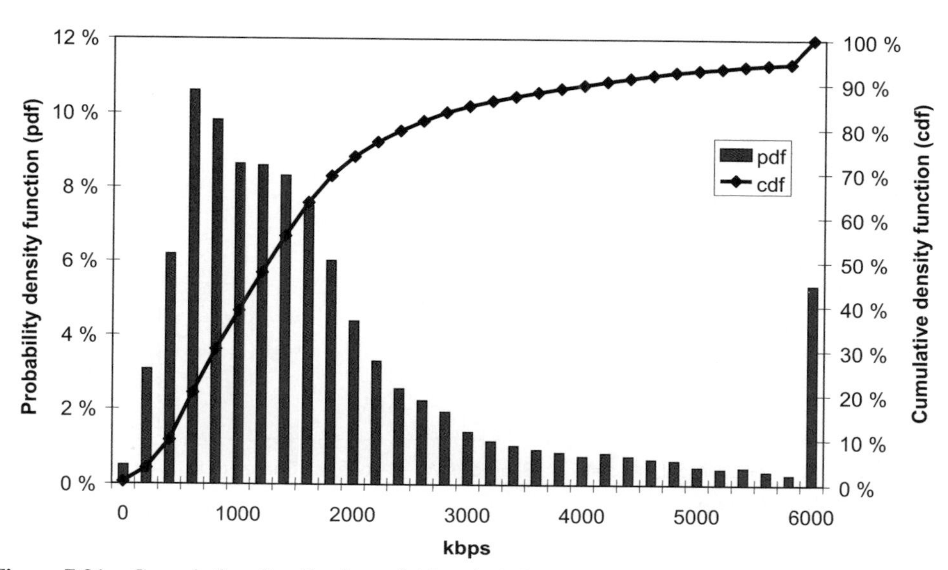


**Figure 7.21** Cumulative distribution of 15-code HSDPA bit rate (plot truncated at 6 Mbps).

We apply the calculation method to the measurement samples and obtain an HSDPA bit rate distribution for the 15-code HSDPA shown in Figure 7.21. The median bit rate expected is 1460 kbps and the 10% and 90% values are 600 kbps and 4200 kbps.

The bit rates for 5, 10, and 15 codes using a dedicated HSDPA carrier are shown in Figure 7.22. The estimated HSDPA data rates are similar to the simulation results given earlier in this chapter. The average data rates are 1200 kbps for 5-code HSDPA,

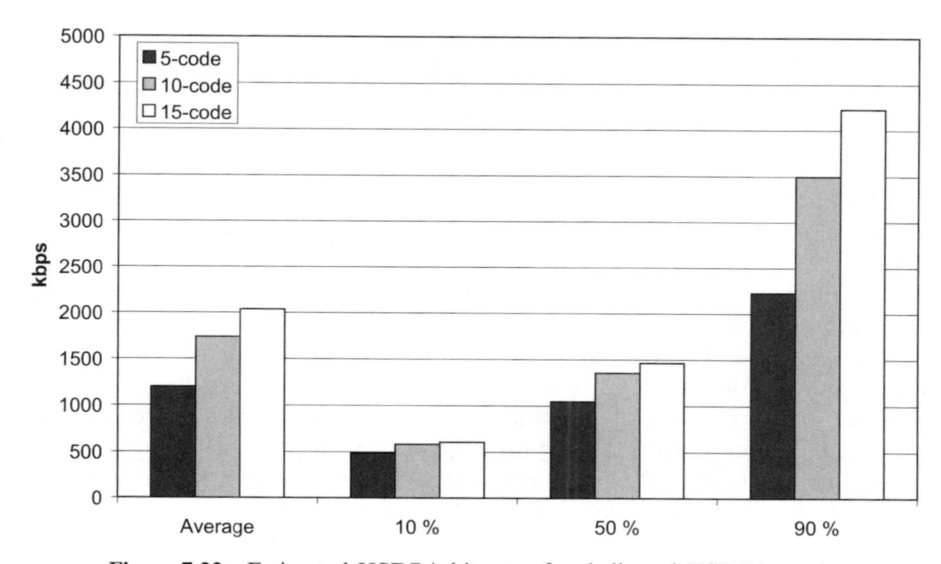


**Figure 7.22** Estimated HSDPA bit rates for dedicated HSDPA carrier.

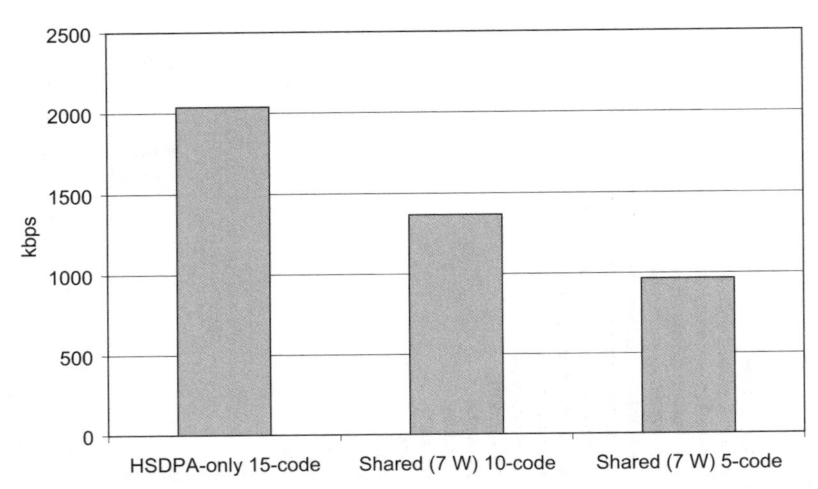


**Figure 7.23** Average HSDPA bit rates with dedicated HSDPA carrier and with shared carrier.

1740 kbps for 10-code, and 2040 kbps for 15-code HSDPA. A total of 90% of the drive test locations would provide at least 500–600 kbps.

Figure 7.23 illustrates the average bit rates achievable using dedicated and shared carriers. A dedicated 15-code HSDPA carrier can increase throughput by 50% compared with a 10-code HSDPA shared carrier and over 100% compared with a 5-code HSDPA shared carrier.

This section has shown that simple measurements from an operator's WCDMA network can be used to obtain quick estimates of the HSDPA data rates expected in that network.

### *7.3.7 HSDPA capacity for real time streaming*

In the previous sections, we presented performance results for best-effort traffic without any strict QoS constraints. In this section, we present results for constant bit rate streaming with a play-out buffer of 5 sec. These results are obtained using the simulation methodology and scenario described in Section 7.7.3.1, assuming that all traffic in the cell is transmitted on the HS-DSCH with up to ten HS-PDSCH codes. For additional details see [18].

Constant bit rate streaming services at 128 kbps are simulated using radio link control (RLC) acknowledged mode. New streaming calls are generated according to a homogeneous Poisson process with a call length of 40 sec. The play-out buffer functionality at the user is explicitly simulated. Once the amount of buffered data reaches 640 kilobits in the play-out buffer, play-out at 128 kbps is started. Hence, with a source bit rate of 128 kbps the minimum initial buffering time equals 5 sec. If it happens that the play-out buffer runs empty during a streaming call, then a re-buffering event occurs, where the play-out is stopped until the buffer reaches 640 kilobits. A satisfied user is defined as a user with an initial buffering time of less than 8 sec without experiencing any re-buffering events. An unhappy user is a user with either an initial buffering time longer than 8 sec, or

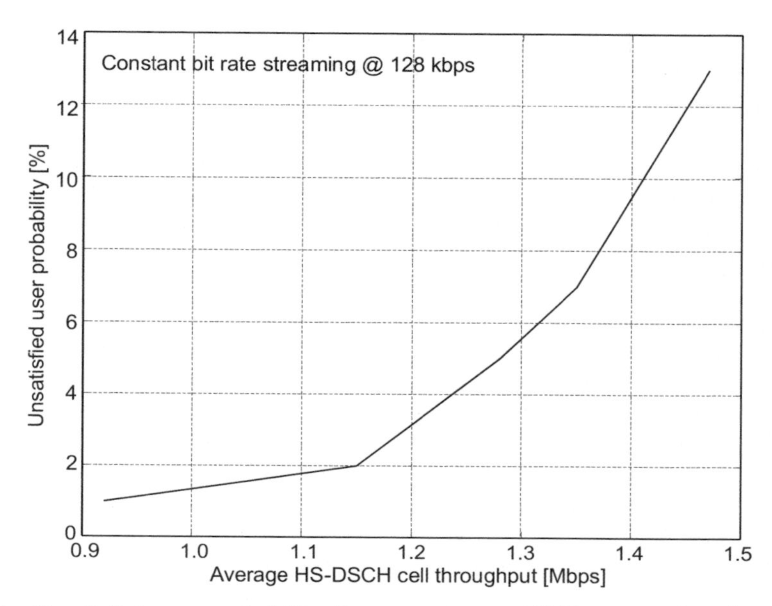


**Figure 7.24** Unsatisfied user probability for streaming at 128 kbps as a function of carried HS-DSCH throughput.

a user who experiences re-buffering events. Finally, users subject to initial buffering times or re-buffering times longer than 10 sec are dropped. Unsatisfied users are the sum of dropped and unhappy users. These QoS definitions for streaming services are similar to the ones used in [17]. Given these requirements for constant bit rate streaming, quality-based HSDPA admission control may be needed if there are too many streaming users entering the cell. Such algorithms are studied in [18] and [19]. The quality-based HSDPA access algorithm proposed in [18] is used for the results presented in Figure 7.24, where unsatisfied user probability is plotted vs carried HS-DSCH cell throughput. These results are obtained by running a series of simulations with different QoS targets for the quality-based HSDPA access algorithm. Assuming a maximum allowed unsatisfied user probability of 5%, HSDPA cell capacity at 128-kbps streaming equals approximately 1.3 Mbps, or equivalently ten users per cell.

Advanced streaming clients and servers are able to adapt their data rate to the available radio bit rate. Those adaptive streaming codecs are considered in Chapter 9.

## 7.4 Iub transmission efficiency

One of the significant operational expenses for many WCDMA operators is Iub transmission between the RNC and Node Bs, especially if leased E1/T1 lines are used. This section shows how HSDPA technology can improve Iub transmission efficiency and help reduce the cost per transmitted bit on HSDPA compared with that on WCDMA. Today, the majority of operators use asynchronous transfer mode (ATM) transmission on the

Iub, so this will be our working assumption for the subsequent analysis, even though Internet Protocol (IP) based transmission is also supported by 3GPP specifications. The improved Iub efficiency for HSDPA comes from the following factors [37]:

- Fast dynamic sharing of the HSDPA Iub bandwidth allocated between active HSDPA users. This is achieved by using fast MAC-hs flow control. With Release 99 the Iub bandwidth is typically allocated separately per user, which makes it difficult to dynamically share excess capacity during periods with low activity – for example, during short reading times for web browsing applications.
- Buffering of data in the Node B means that transmission with high peak data rates at the air interface can be supported with HSDPA without requiring a similar higher Iub bandwidth. It also implies that short temporary periods with Iub congestion do not necessarily result in unused HSDPA air interface capacity.
- HSDPA does not need soft handover, so the data transmitted to one HSDPA user is only sent once over a single Iub. WCDMA Release 99 relies on soft handover, where multiple Iub links are required for users in soft handover. As an example, results from [13] show that the soft handover overhead equals approximately 40% for WCDMA users, meaning that on average a Release 99 UE receives data from 1.4 cells.

Assuming a standard closed loop Transmission Control Protocol (TCP) web browsing traffic model and a three-sector macro-cell environment, it is reported in [37] that Iub efficiency is improved by a factor of 2 for HSDPA compared with that for WCDMA. This is a significant improvement, which basically means that twice as many user bits can be transmitted through the same Iub with HSDPA compared with using Release 99 WCDMA. In order to further address Iub dimensioning for HSDPA, the simulations from Section 7.7.3.4 are repeated using Iub bandwidth restrictions and high-speed fast medium access control (MAC-hs) flow control as illustrated in Figure 7.25. Recall from

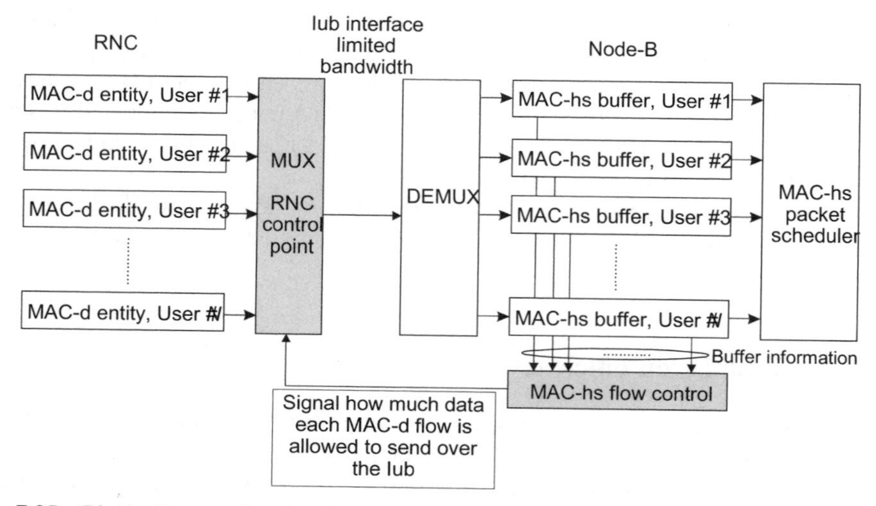


**Figure 7.25** Block diagram for simulation setup with limited Iub bandwidth and MAC-hs flow control.

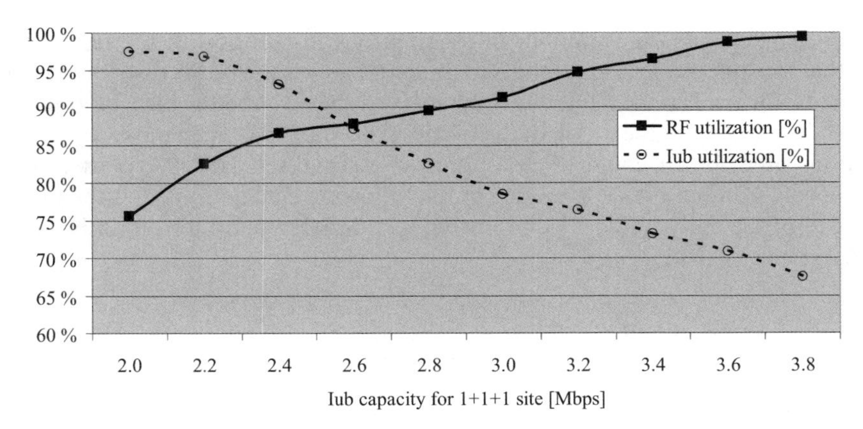


**Figure 7.26** Trade-off between HSDPA RF vs Iub efficiency.

Chapter 3 that MAC-hs flow control is operated by sending so-called credits to the RNC for each MAC-d flow, depending on the amount of buffered data in the Node B. The number of credits expresses the maximum number of payload data units (PDUs) that the RNC is allowed to send over the Iub during the next HS-DSCH interval of 10 ms. Note that the frame-length for the HS-DSCH frame protocol equals 10 ms. If Iub congestion occurs so the RNC cannot transmit all the PDUs according to the received credits, a simple congestion control scheme is applied, where the PDU flow is reduced equally for all active HSDPA MAC-d flows.

The trade-off between air interface RF vs Iub efficiency is reported in Figure 7.26. Note from Section 7.7.3.4 that without any Iub bandwidth limitations the average HSDPA cell capacity equals approximately 900 kbps, which is 2.7 Mbps per three-sector Node B assuming five HS-PDSCH codes and 7 W for HSDPA transmission. The remaining power and some channelization codes are used for simultaneous Release 99 transmission. The results in Figure 7.26 show that in order to achieve the full 100% HSDPA radio frequency (RF) capacity the HSDPA Iub bandwidth required equals 3.8 Mbps. An HSDPA Iub bandwidth of 3.2 Mbps is sufficient to achieve 95% of the maximum available HSDPA RF capacity. Such Iub dimensioning corresponds to 20% over-dimensioning of the Iub. This 20% over-dimensioning is needed to account for temporal variations in the available HSDPA RF capacity caused by, for example, HSDPA users being close to the serving cell and variations in the traffic model. Note that the resulting 20% margin will be reduced when the number of active HSDPA users per site is increased in high-capacity sites. In that case the peak site capacity will be closer to the average site capacity due to the smaller dominance of a few users in extreme conditions.

## 7.5 Capacity and cost of data delivery

The previous sections presented total cell throughput in Mbps, but this section estimates the achievable data capacity in total downloaded gigabytes per subscriber per month.

Data capacity depends on a number of assumptions in addition to the traffic models: What is the 'burstiness' of the data transmission and how is the traffic distributed over the day. Voice traffic tends to be relatively constant over the day and the busy hour may carry even less than 10% of the total daily traffic. The data part depends clearly on the applications used, but it could be more bursty, especially if HSDPA is used for fixed wireless access from homes. We assume in the following calculations that the busy hour carries 20% of the daily traffic. The following additional assumptions have been used in HSDPA capacity calculation:

- A 2 + 2 + 2 HSDPA configuration.
- Spectral efficiency of 2 Mbps/cell using an HSDPA terminal with single antenna and Rake receiver (see Figure 7.14).
- Spectral efficiency of 4 Mbps/cell using an HSDPA terminal with antenna diversity and equalizer receiver (see Figure 7.38).
- Busy hour utilization 80%.
- Busy hour share of the daily traffic is 20%.

These assumptions are applied to calculate the total traffic in GB per month as follows:

$$\text{Capacity [GB/month]} = \frac{(2+2+2)\dfrac{2\,\text{Mbps}}{8} \cdot \dfrac{3600\,\text{s}}{\text{hour}} \cdot 80\% \cdot 30\text{ days}}{20\%\text{ busy hour share}} \tag{7.11}$$

Total carried traffic is 650 GB with a single-Rake HSDPA terminal and 1170 GB with a two-equalizer HSDPA terminal per month per site. Figure 7.27 illustrates the maximum HSDPA capacity per subscriber per month. These results show that a 2–4-GB/subscriber/month capacity is feasible with 300 broadband subscribers per site.

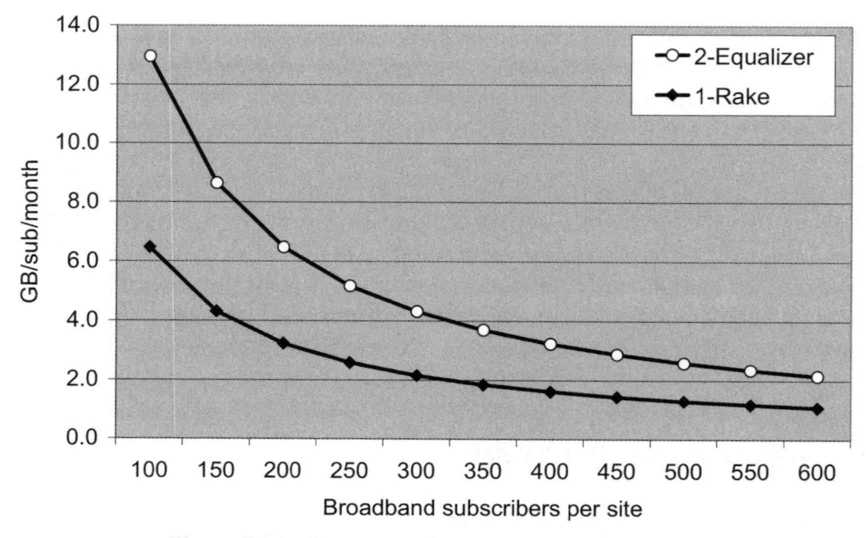


**Figure 7.27** Data capacity per subscriber per month.

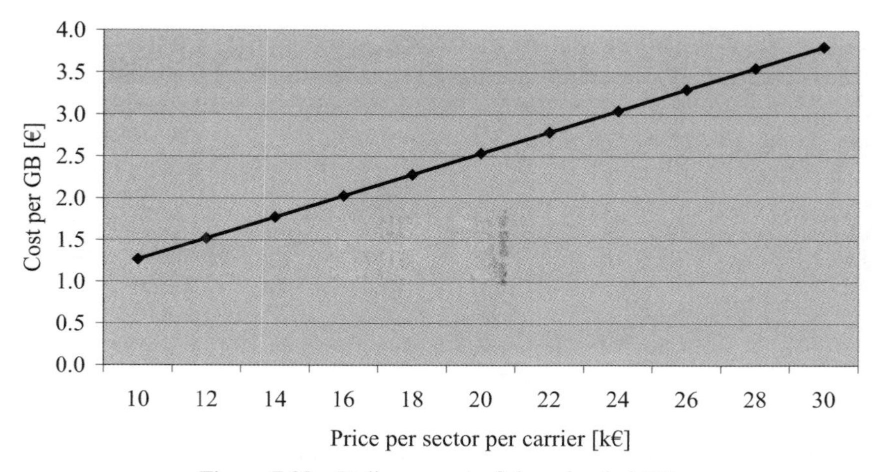


**Figure 7.28** Delivery cost of downloaded GB.

Another challenge – besides radio capacity – is the cost of data delivery for high-volume data. We give below a simplified view of the cost in which only depreciation of the network equipment capital expenditure is considered over 6 years. The cost of GB delivery is shown as a function of network equipment price. The price in Figure 7.28 is shown per sector per carrier including base station, RNC, and core network elements. The following formula is used to calculate the cost:

$$\text{Cost}\frac{€}{\text{GB}} = \frac{\text{TRX price} \times 10^9}{\frac{2\text{Mbps}}{8} \times 3600 \times \frac{80\%}{20\%} \times 365 \times 6 \text{ years}} \tag{7.12}$$

If network pricing is below €16 000 per sector per carrier, the data delivery cost with HSDPA is below €2/GB. That number can be considered as the lower bound of delivery cost when new capacity is added to the system. The busy hour utilization of 80% cannot be achieved in practice in all cells since the network must be designed to provide coverage also in those areas where the traffic density is not as high. If the average busy hour network utilization is 40% instead of 80%, the delivery cost is double that shown in Figure 7.28.

This calculation excludes such cost factors as Iub transmission cost or other network operation costs, marketing, and customer acquisition.

## 7.6 Round trip time

While the maximum data rate is typically used to benchmark radio systems, it is not a sufficient measure alone to describe radio performance from the application point of view. The pure data rate is relevant if we consider the download time of a large file, but there are a number of applications that are happy with a relatively low data rate –

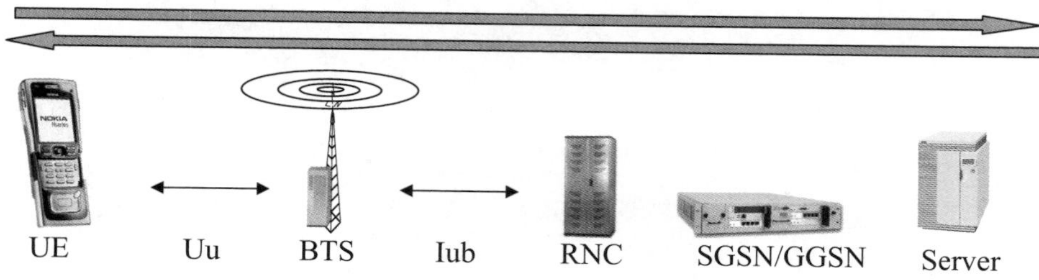


**Figure 7.29** Definition of round trip time (RTT).

say, 10–30 kbps – but require very low latency. Such applications include Voice-over-IP, push-to-talk and real time gaming. Also, interactive applications like web browsing benefit from low latency. This section presents typical WCDMA/HSDPA network latency. Application performance is considered in more detail in Chapter 9.

Latency can be measured as round trip time (RTT), which is defined as the time an IP packet takes to travel from the terminal through all network elements to the application server, and back. If the IP packet size is small, the data rate does not affect the delay, but the delay is then only defined by system frame structures and by element processing and interface delays. RTT is defined pictorially in Figure 7.29.

For RTT estimation, we use the following assumptions:

- A UE delay of 10–25 ms.
- A Node B delay of 10–15 ms.
- An air interface delay, including uplink buffering, of 43–53 ms for Release 99, 20 ms for HSDPA, and 10 ms for HSUPA. The WCDMA Release 99 data rate 64/64 kbps uplink/downlink assumes a 20-ms TTI while 128/384 kbps assumes a 10-ms TTI. High-speed uplink packet access (HSUPA) uplink assumes a 2-ms TTI.
- An Iub delay of 20–40 ms for Release 99, 10 ms for HSDPA, and 5 ms for HSUPA.
- An RNC delay of 20 ms for Release 99 and 10 ms for HSDPA/HSUPA.
- An Iu plus core network delay of 3 ms.

RTT evolution is illustrated in Figure 7.30. RTT for Release 99 is expected to be 110–150 ms, for HSDPA 70 ms, and for HSUPA <50 ms. The delay is mainly caused by the radio network elements and interfaces while the core network delay is very short. The MAC layer RTT with HSDPA + HSUPA is very short, approximately 10 ms, and total RTT mainly depends on processing delays in the network elements and in the UE.

Example ping measurements using both WCDMA Release 99 and HSDPA are shown in Figure 7.31. HSDPA RTT is between 70 and 85 ms in this particular radio network while WCDMA RTT is on average 130 ms, but there are a few large values caused by RLC retransmissions.

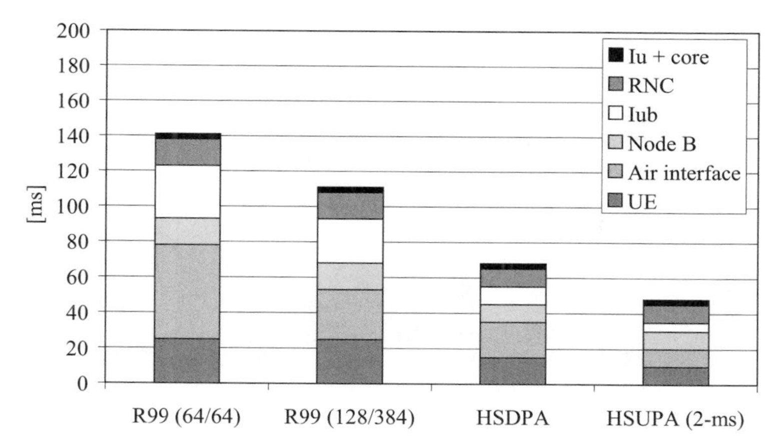


**Figure 7.30** Estimated round trip time (RTT) evolution.

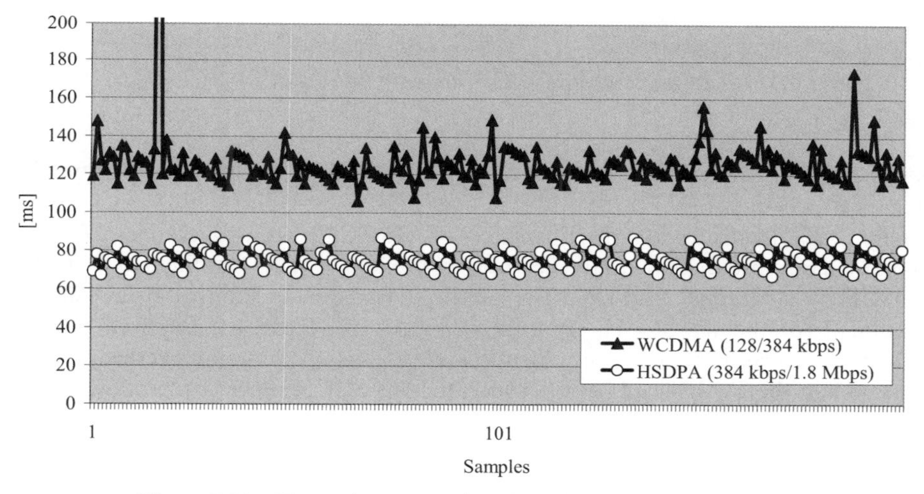


**Figure 7.31** Example measured WCDMA and HSDPA RTTs.

## 7.7 HSDPA measurements

This section first presents a few example measurement results from laboratory and further outdoor drive test results. The measurements were done using a Category 12 terminal with a maximum L1 bit rate of 1.8 Mbps and a maximum RLC L2 bit rate of 1.6 Mbps.

Figure 7.32 illustrates the functioning of HS-DSCH link adaptation. The distribution of the selected transport block size with different common pilot channel (CPICH) $E_c/N_0$ levels is shown. These $E_c/N_0$ levels are recorded when HSDPA power was off. With $E_c/N_0$ between $-3$ and $-6$ dB the largest possible transport block size of 3440 bits is used 94–99% of the time. When the $E_c/I_0$ value reduces to $-9$ dB the most typical transport block size is decreased from the maximum 3440 to 2404 to provide more channel coding.

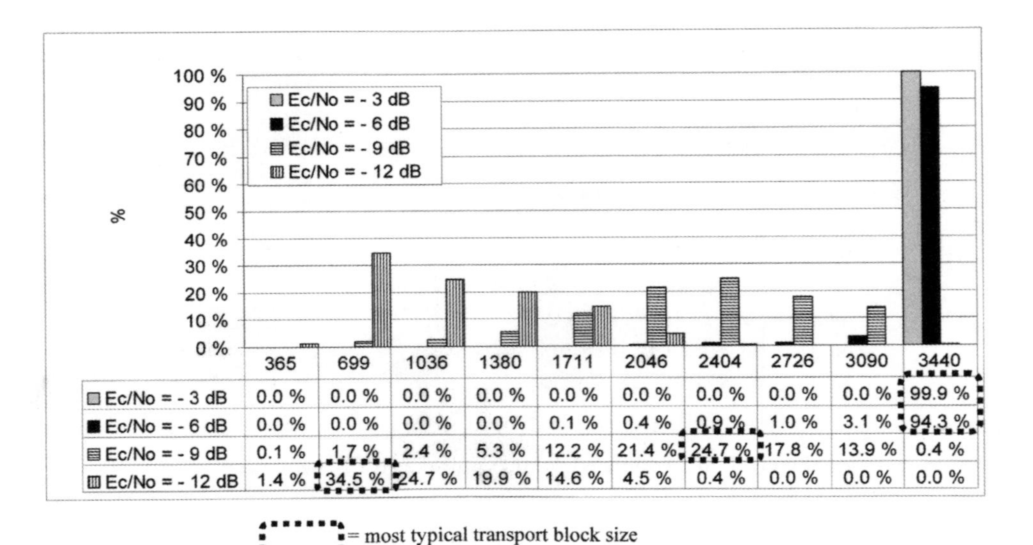


| | 365 | 699 | 1036 | 1380 | 1711 | 2046 | 2404 | 2726 | 3090 | 3440 |
|---|---|---|---|---|---|---|---|---|---|---|
| Ec/No = - 3 dB | 0.0 % | 0.0 % | 0.0 % | 0.0 % | 0.0 % | 0.0 % | 0.0 % | 0.0 % | 0.0 % | 99.9 % |
| Ec/No = - 6 dB | 0.0 % | 0.0 % | 0.0 % | 0.0 % | 0.1 % | 0.4 % | 0.9 % | 1.0 % | 3.1 % | 94.3 % |
| Ec/No = - 9 dB | 0.1 % | 1.7 % | 2.4 % | 5.3 % | 12.2 % | 21.4 % | 24.7 % | 17.8 % | 13.9 % | 0.4 % |
| Ec/No = - 12 dB | 1.4 % | 34.5 % | 24.7 % | 19.9 % | 14.6 % | 4.5 % | 0.4 % | 0.0 % | 0.0 % | 0.0 % |

**Figure 7.32** Selected transport block sizes with different CPICH $E_c/I_0$ levels.

When the $E_c/N_0$ value further reduces to $-12$ dB, the most typical transport block size decreases to 699 bits.

Figure 7.33 illustrates the performance of HS-SCCH power control. The radio network was configured so that the total combined power for the HS-SCCH and HS-DSCH was targeted at 10 W. The quality-based power control from Chapter 6 was applied for the HS-SCCH. At a good $E_c/N_0$ of $-3$ dB, the required power for the HS-SCCH was very low (only 80 mW), thus causing minimal overhead in interference. When the $E_c/N_0$ level becomes low, the HS-SCCH power control increases the

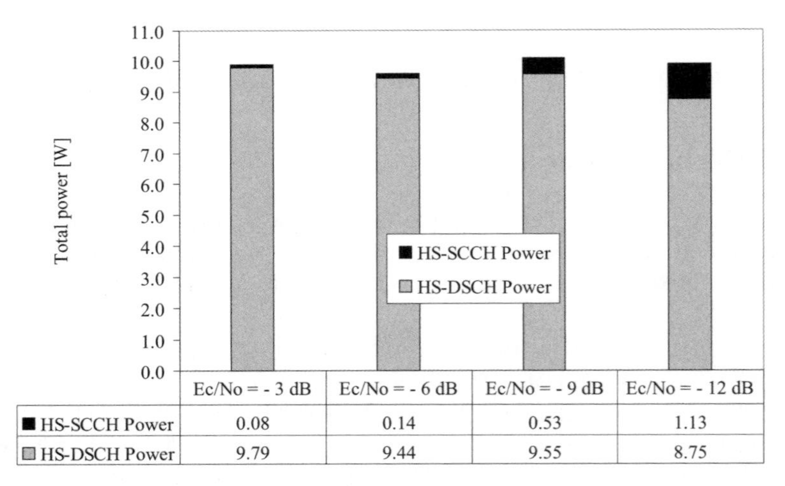


| | Ec/No = - 3 dB | Ec/No = - 6 dB | Ec/No = - 9 dB | Ec/No = - 12 dB |
|---|---|---|---|---|
| HS-SCCH Power | 0.08 | 0.14 | 0.53 | 1.13 |
| HS-DSCH Power | 9.79 | 9.44 | 9.55 | 8.75 |

**Figure 7.33** Power control for the HS-SCCH with different CPICH $E_c/I_0$ levels.

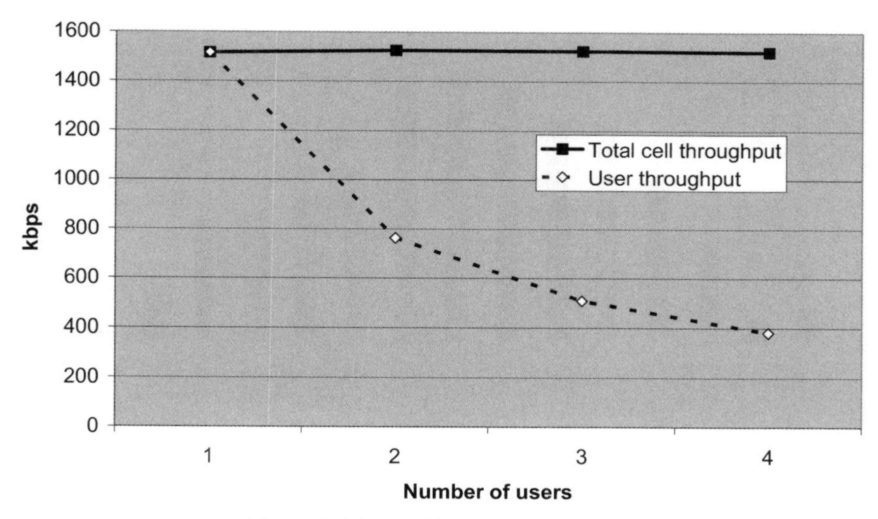


**Figure 7.34** Multiuser measurements.

power level to maintain reliable signalling quality. With an $E_c/N_0$ of −6, −9, and −12 dB the HS-SCCH power was 140 mW, 530 mW, and 1130 mW. A typical cell edge $E_c/N_0$ is approximately −9 dB corresponding to 0.5 W of required power for the HS-SCCH. This measurement result is comfortably in line with the HS-SCCH simulations shown earlier in the chapter.

The HS-SCCH power control measurements illustrate, first, that there is a need for HS-SCCH power control and, second, that the HS-SCCH power control needs to be optimized to minimize the interference from HS-SCCH transmission, especially for the case when multiple HS-SCCHs are used with code-multiplexing.

Figure 7.34 presents measurement results for one to four active HSDPA users. The results are obtained in good channel conditions using a round robin scheduler. Therefore, each user gets the same data rate. The total cell throughput for FTP file download was constant – 1.55 Mbps – regardless of the number of users. User throughput was total cell throughput divided by number of users – for example, each user gets 380 kbps with four parallel HSDPA users.

The following figures present the performance obtained from outdoor drive test measurements. The measurement environment was a large urban macro-cell with a maximum path loss of >160 dB and the lowest RSCP values below −115 dBm. The distribution of RSCP is shown in Figure 7.35. The median RSCP value was −87 dBm.

HSDPA power allocation was 6 W and the total base station power was 10 W during the measurements. The tested application was an FTP file download. Throughput was recorded for each download together with the RSCP and $E_c/N_0$ values. The distribution of the bit rate is shown in Figure 7.36. The maximum bit rates are 1.5–1.6 Mbps, which is limited by the maximum RLC bit rate of 1.6 Mbps with this terminal category.

The measured bit rates are compared with the simulation results in two different ways. The first way is to use the recorded RSCP and $E_c/N_0$ values to calculate the HS-DSCH SINR. The bit rate can be estimated from the SINR based on link level simulation

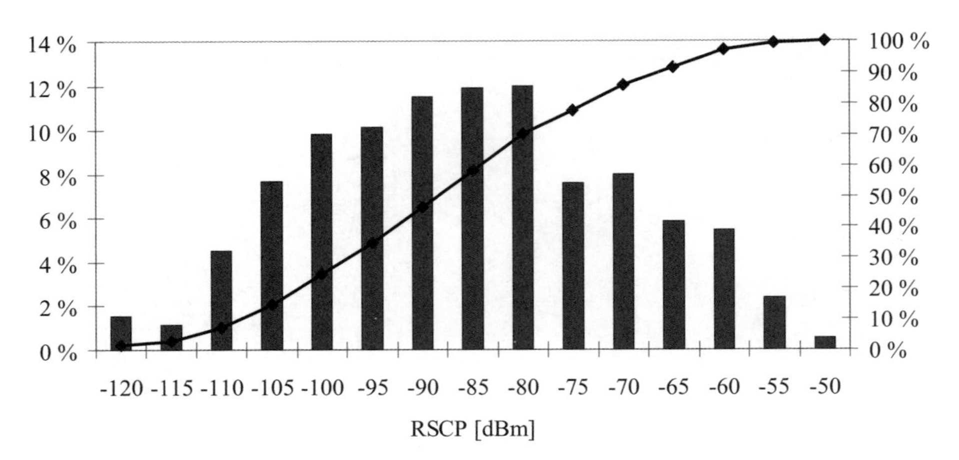


**Figure 7.35** Distribution of pilot power RSCP received in drive tests.

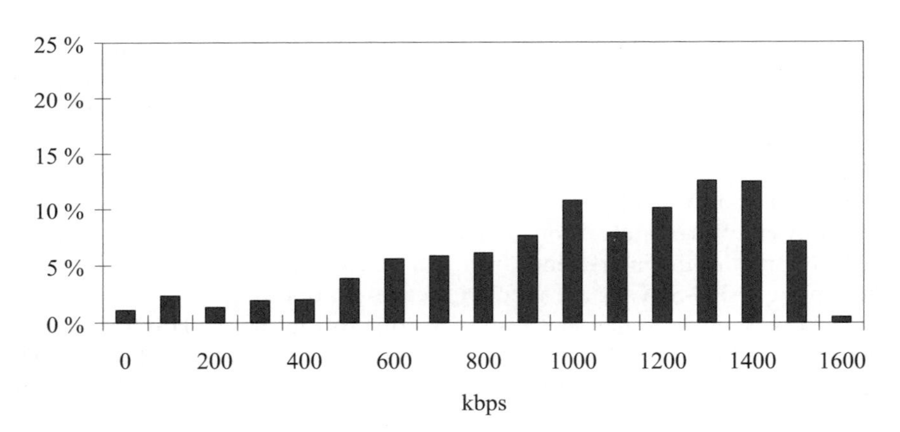


**Figure 7.36** Distribution of the application level bit rate in drive tests.

results. The other comparison is done using a standard macro-cell environment in the system simulator. The maximum path loss was assumed to 160 dB in the simulation. The bit rates from those two simulation methods are compared with the measurement results in Figure 7.37.

The median bit rates are fairly similar at 1.0–1.1 Mbps both in the field measurements and in the simulations. There are more samples in the low-throughput area in the measurements than in the simulations. That difference can be explained by the fact that there were a few samples with a very low signal level – that is, an RSCP of −120 dBm – in the measurements. Still, more than 90% of the measurement samples have at least 500 kbps, which is better than the highest WCDMA Release 99 bit rate of 384 kbps.

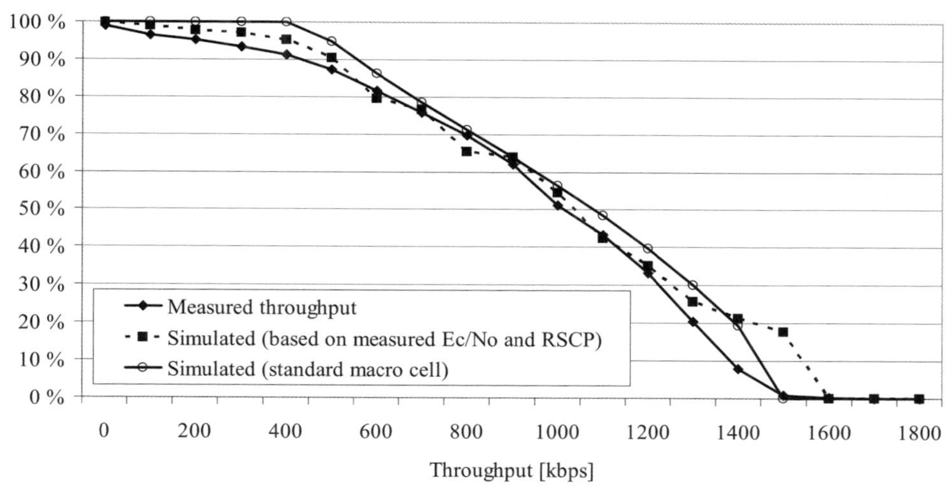


**Figure 7.37** Cumulative distribution of the application level bit rate in HSDPA drive tests – results comparison.

## 7.8 HSDPA performance evolution

The last few sections have considered the performance of HSDPA for a baseline setup. To accommodate an expected growth in traffic demand and to provide higher end user data rates, there are various enhancements readily available within the 3GPP framework to further increase capacity and service performance. The most significant enhancement techniques are discussed in the following.

### *7.8.1 Advanced UE receivers*

Advanced UE receivers are able to obtain a higher SINR, which results in higher user bit rates and cell capacities. No new algorithms are required in the network to take benefit of advanced UEs since these UEs will automatically report higher CQI values, so the Node B scheduler will automatically allocate higher bit rates. The following three types of advanced UEs are considered:

- A Rake receiver with receiver diversity (Enhanced Type I in 3GPP Release 6).
- A single-equalizer receiver (Enhanced Type II in 3GPP Release 6).
- An equalizer receiver with receiver diversity (Enhanced Type III in 3GPP Release 7).

The type prefix used here follows 3GPP terminology. 3GPP Release 6 includes performance requirements for a number of different enhanced UEs. The gain from using linear minimum mean-squared error (LMMSE) receivers predominantly comes from equalizing time-dispersive radio towards the serving HS-DSCH cell. Equalizing the radio channel results in a better equivalent downlink orthogonality factor and, therefore, less own-cell interference. The gain from using LMMSE is, therefore, primarily available to users that are dominated by own-cell interference while experiencing relatively high

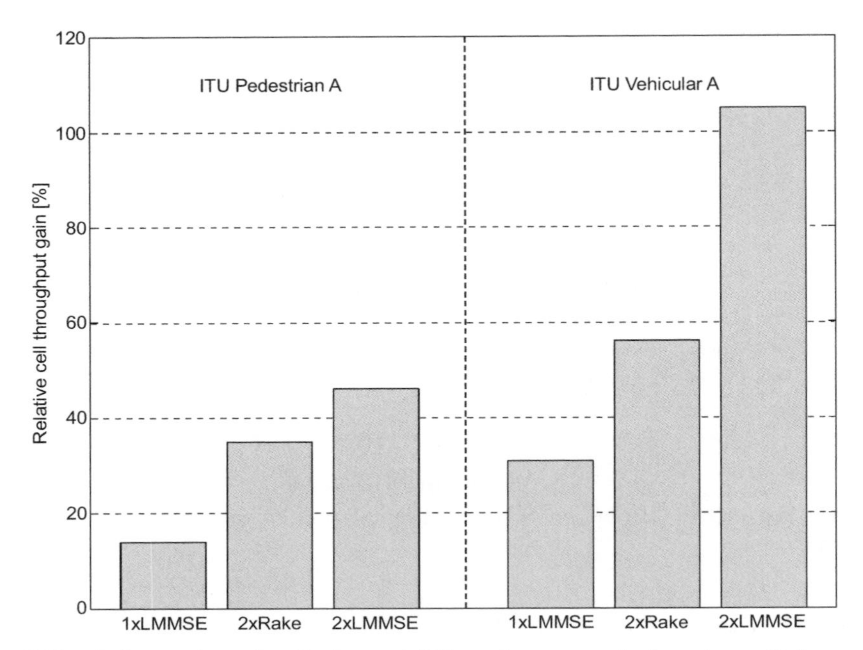


**Figure 7.38** Cell capacity gain of advanced UE receivers over single-antenna Rake receiver in macro-cellular environments with PF scheduling and best effort traffic.

temporal dispersion – that is, users in clear dominance areas that have a Vehicular A multipath profile. The gain from using two-antenna Rake receivers consists of two factors: (i) a coherent combining gain of ideally 3.0 dB, and (ii) an antenna diversity gain. Mapping from the gain in the experienced HS-DSCH SINR to a throughput gain depends strongly on the link adaptation operating point. As an example, if most users are already using all the HS-PDSCH codes and 16QAM modulation, then there is just a minor gain to be had from advanced UEs. On the other hand, if most users are operating at the lower part of the link adaptation range with only a few HS-PDSCH codes and QPSK modulation, then a 3-dB increase in the SINR maps to nearly a 100% gain in throughput.

The average cell capacity of using advanced terminal receivers is reported in Figure 7.38 for a macro-cellular environment with similar simulation settings to those in Section 7.7.3.1. These results are reproduced from [33]–[34]. As expected, the gain from using LMMSE is larger for the Vehicular A than for the Pedestrian A channel. An average capacity gain of approximately 100% is achieved in Vehicular A if all terminals are using two antennas and an LMMSE receiver compared with having single-antenna Rake receivers. This is equivalent to an average cell throughput of 4.4 Mbps. It is furthermore observed from Figure 7.38 that the gain from using two antennas and a Rake receiver – $2 \times$ Rake – is larger for the Vehicular A than for the Pedestrian A channel. The latter is observed because most users are already operating in the higher end of the link adaptation range in Pedestrian A. In Vehicular A, users are operating lower in the link adaptation dynamic range due to inter-path interference, which results in larger throughput gain from increasing the HS-DSCH SINR experienced.

The results presented in Figure 7.38 are obtained from dynamic network simulations using proportional fair scheduling. If a blind scheduler – such as round robin – is used, then the gain from using advanced terminals increases significantly. As an example, for round robin scheduling and Vehicular A, the average gain in cell capacity is reported to equal 178% for a 2 × LMMSE [33]. This is observed because, first, users typically operate lower in the link adaptation dynamic range when round robin scheduling is used, and, second, because the dual-antenna diversity gain is larger when no multiuser scheduling diversity gain is present. Additional simulation results and considerations for advanced HSDPA user terminals can be found in [29], [33], [34].

### 7.8.2 Node B antenna transmit diversity

As described in Chapter 4, both open loop and closed loop dual-branch antenna transmit diversity schemes are supported for HSDPA. However, the open loop scheme – space time transmit diversity (STTD) – is typically found to provide no additional gain for HSDPA at low to moderate UE speeds if combined with proportional fair scheduling [29], [31]. However, STTD does provide a capacity gain of 10% to 20% when a blind round robin scheduler is applied, and when users are moving too fast for the proportional fair scheduler to track radio channel fast fading [29].

Closed loop transmit diversity provides higher gain than open loop diversity. The closed loop, however, cannot be applied with fractional DPCH, but requires an associated DCH to be used. Therefore, closed loop diversity is not seen as attractive for long-term HSDPA performance improvement.

If we have an associated DCH available, the transmit diversity scheme that provides the largest gain is closed loop mode-2 (CLTD2), where both amplitude and phase weighting is used. CLTD2 provides both a coherent combining gain and a diversity gain. However, in many cases the diversity gain is found to be marginal if a multiuser diversity scheduling gain is already available in the system. For typical scenarios where all HSDPA users apply ideal CLTD2s, the cell capacity gain is reported to be 20% to 30% [29], [33] for UE speeds lower than 30 km/h. However, note that CLTD2 only works optimally for HSDPA users with an active set size equal to 1, since the user would otherwise derive the feedback weight to the Node B from the combined link quality in all the soft handover legs, while HSDPA-only is being transmitted to the user from one cell.

### 7.8.3 Node B beamforming

As discussed in [27], for Release 99 WCDMA, conventional Node B beamforming techniques can bring higher capacity in macro-cellular environments, assuming low azimuthal dispersion at the Node B antenna array. A capacity gain around 150% is achievable by forming a grid of fixed beams with an antenna array consisting of four elements [27]. One of the advantages of using beamforming techniques is that multiple scrambling codes can be introduced at the Node B by assigning different codes to different beams. The latter is attractive since the resulting code space can be increased without jeopardizing downlink orthogonality as narrow directional beams separate different scrambling codes. This kind of technique is particularly attractive for HSDPA (as described in [28]), since it allows parallel transmission on more than 15

HS-PDSCH codes per cell to multiple users. The beamforming capacity gain over single-antenna transmission for HSDPA is reported to equal a factor of 2.5 to 4.7 for an antenna array of four and eight elements, respectively [28]. As a final remark, note that beamforming also helps to improve uplink performance.

### 7.8.4 *Multiple input multiple output*

The use of multiple transmit and receive antennas form a so-called 'multiple input multiple output' (MIMO) scenario, which can be used to create parallel virtual orthogonal information channels from the Node B to the UE so that the data rate can be further increased. This concept is typically referred to as 'information MIMO'. The parallel virtual information channels are obtained by using signal processing algorithms to adaptively adjust the used antenna weights at the transmitter and receiver as a function of radio channel impulse response between the different antenna ports. The maximum number of parallel virtual channels equals $L = \text{Min}\{R, N, M\}$, where $N$ is the number of transmit antennas, $M$ is the number of receive antennas, and $R$ is the rank of the $N \times M$ radio channel matrix [35], [36]. In principle, the peak data rates from using information MIMO can be increased by a factor of $L$ compared with the case with one transmit and one receive antenna. However, the creation of $L$ parallel virtual channels requires first a relatively high SINR since the $N \times M$ radio channel matrix should be accurately estimated to determine the antenna weights that form the $L$ parallel virtual channels. Second, the SINR in the $L$ parallel virtual channels should be higher than or equal to the SINR in the reference $1 \times 1$ radio channel to experience an $L$-fold increase in the data rate. Many candidate information MIMO schemes have been investigated in 3GPP for HSDPA, with the emphasis on configurations with two antennas at the Node B and at the UE [38]. Simulation results show rather small gains for highly loaded macro-cellular environments. This is mainly because realistic $2 \times 2$ MIMO macro-cell scenarios typically only offer one dominant virtual channel with a high SINR, while the second virtual channel offers a poor SINR with little potential for further increase of the HSDPA data rate. However, HSDPA MIMO schemes for smaller micro- and pico-cells with higher experienced SINR are expected to show more promising gains. HSDPA MIMO is, therefore, a work item for 3GPP Release 7 as discussed in Chapter 2.

## 7.9 Conclusions

WCDMA Release 99 provides, in practice, 384 kbps as the maximum data rate in the downlink while HSDPA pushes the peak data rate initially to 1.8 and 3.6 Mbps, and later up to 10 Mbps. Simulations and measurements show that the HSDPA link performance is robust due to fast L1 retransmissions, incremental redundancy, and link adaptation. Field measurements show median data rates of 1 Mbps with a terminal maximum capability of 1.8 Mbps. Practical data rates will improve when higher bit rate UE categories are available.

HSDPA reduces network latency compared with WCDMA Release 99 DCH. The calculations and measurements show that average RTTs can be clearly below 100 ms with HSDPA downlink and Release 99 DCH in the uplink. With HSUPA uplink the RTT is expected to be below 50 ms.

HSDPA improves cell throughput by up to 1 Mbps/MHz/sector in three-sector macrocells due to advanced link level techniques, enhanced terminal receivers, and multiuser diversity with fast scheduling. Fast proportional fair scheduling is shown to improve cell throughput by over 30% with low mobile speeds. HSDPA was shown to improve the capacity of best effort packet and guaranteed bit rate streaming.

Simulations show that HSDPA can coexist with WCDMA on the same carrier. The shared carrier provides an attractive performance, especially when the amount of WCDMA traffic is low and the HSDPA terminal has a limited bit rate of 1.8 Mbps. For a high-capacity and high bit rate solution a dedicated HSDPA carrier can be used later.

HSDPA itself may require additional investments in Iub transport to facilitate the high peak bit rates, but HSDPA also improves the efficiency of Iub compared with WCDMA Release 99: clearly, more user bits can be carried with the same Iub capacity with HSDPA than with Release 99. The reason is, first, that HSDPA is transmitted only from one cell without soft handover and, second, that HSDPA users have a shared Iub resource instead of dedicated resources like in Release 99.

HSDPA bit rates and capacity can be further boosted by more than 100% by advanced terminal receivers and by more than 150% by base station beamforming with four-element antennas.

## 7.10 Bibliography


[1] T. E. Kolding, K. I. Pedersen, J. Wigard, F. Frederiksen, and P. E. Mogensen (2003), High speed downlink packet access: WCDMA evolution, *IEEE Vehicular Technology Society (VTS) News*, **50**(1), 4–10, February.

[2] K. I. Pedersen and P. E. Mogensen (2002), The downlink orthogonality factors' influence on WCDMA system performance, *IEEE Proc. Vehicular Technology Conference, September*, pp. 2061–2065.

[3] N. B. Mehta, L. J. Greenstein, T. M. Willis, and Z. Kostic (2002), Analysis and results for the orthogonality factor in WCDMA downlinks, *IEEE Proc. of Vehicular Technology Conference, May*, pp. 100–104.

[4] 3GPP, Technical Specification Group RAN, Medium Access Control (MAC) Protocol Specification, 3GPP TS 25.321 version 6.6.0.

[5] C. E. Shannon (1949), Communication in the presence of noise, *Proc. Institute of Radio Engineers*, **37**(1), 10–21, January.

[6] F. Frederiksen and T. E. Kolding (2002), Performance and modeling of WCDMA/HSDPA transmission/H-ARQ schemes, *IEEE Proc. Vehicular Technology Conference, September*, pp. 472–476.

[7] T. E. Kolding (2003), Link and system performance aspects of proportional fair packet scheduling in WCDMA/HSDPA, *IEEE Proc. Vehicular Technology Conference, September*, pp. 1717–1722.

[8] K. I. Pedersen, T. F. Lootsma, M. Støttrup, F. Frederiksen, T. E. Kolding, and P. E. Mogensen (2004), Network performance of mixed traffic on high speed downlink packet

access and dedicated channels in WCDMA, *IEEE Proc. Vehicular Technology Conference, September.*

[9] H. Holma and A. Toskala (eds) (2002), *WCDMA for UMTS: Radio Access for Third Generation Mobile Communications* (2nd edn), John Wiley & Sons, Chichester, UK.

[10] A. Das, F. Khan, A. Sampath, and H. Su (2002), Design and performance of downlink shared control channel for HSDPA, *Proc. PIMRC, September*, pp. 1088–1091.

[11] 3GPP, Technical Specification Group RAN, User Equipment (UE) Transmission and Reception (FDD), 3GPP TS 25.101 version 6.7.0.

[12] 3GPP, Technical Specification Group RAN, Multiplexing and Channel Coding (FDD), 3GPP TS 25.212 version 5.3.0 (R1-01-0430).

[13] K. I. Pedersen, A. Toskala, and P. E. Mogensen (2004), Mobility management and capacity analysis for high speed downlink packet access in WCDMA, *IEEE Proc. Vehicular Technology Conference, September.*

[14] 3GPP, Technical Specification Group RAN, Spreading and Modulation (FDD), 3GPP TS 25.213 version 6.3.0, June, 2005.

[15] T. E. Kolding, F. Frederiksen, and P. E. Mogensen (2002), Performance aspects of WCDMA systems with high speed downlink packet access (HSDPA), *IEEE Proc. Vehicular Technology Conference, Fall.*

[16] K. Hiltunen, B. Olin, and M. Lundevall (2005), Using dedicated in-building systems to improve the HSDPA indoor coverage and capacity, *IEEE Proc. Vehicular Technology Conference, Spring, June, 2005.*

[17] M. Lundevall, B. Olin, J. Olsson, J. Eriksson, and F. Eng (2004), Streaming applications over HSDPA in mixed service scenarios, *IEEE Proc. Vehicular Technology Conference, Fall, Los Angeles, September, 2005.*

[18] K. I. Pedersen (2005), Quality based HSDPA access algorithms, *IEEE Proc. Vehicular Technology Conference, Fall, September, 2005.*

[19] P. Hosein (2003), A class-based admission control algorithm for shared wireless channels supporting QoS services, *Proceedings of the Fifth IFIP TC6 International Conference on Mobile and Wireless Communications Networks, Singapore, October.*

[20] W. Bang, K. I. Pedersen, T. E. Kolding, and P. E. Mogensen (2005), Performance of VoIP on HSDPA, *IEEE Proc. Vehicular Technology Conference, Stockholm, June.*

[21] F. Poppe, D. de Vleeschauwer and G. H. Petit (2000),Guaranteeing quality of service to packetized voice over the UMTS air interface, *Eighth International Workshop on Quality of Service, June*, pp. 85–91.

[22] F. Poppe, D. de Vleeschauwer, and G. H. Petit (2001), Choosing the UMTS air interface parameters, the voice packet size and the dejittering delay for a voice-over-ip call between a umts and a pstn party, *IEEE INFOCOM*, **2**, 805–814, April.

[23] R. Cuny and A. Lakaniemi (2003), VoIP in 3G Networks: An end-to-end quality of service analysis, *IEEE Proc. Vehicular Technology Conference, Spring, April*, Vol. 2, pp. 930–934.

[24] L. J. Greenstein, V. Erceg, Y. S. Yeh, and M. V. Clark (1997), A new path-gain/delay-spread propagation model for digital cellular channels, *IEEE Trans. on Veh. Technol.*, **46**(2), 477–485, May.

[25] L. Correia (ed.) (2001), *Wireless Flexible Personalised Communications: COST 259 Final Report*, John Wiley & Sons, Chichester, UK, May.

[26] F. Frederiksen, P. E. Mogensen, and J.-E. Berg (2000), Prediction of path loss in environments with high-raised buildings, *IEEE Proc. Vehicular Technology Conference, September*, pp. 898–903.

[27] K. I. Pedersen, P. E. Mogensen, and J. Ramiro-Moreno (2003), Application and performance of downlink beamforming techniques in UMTS, *IEEE Communications Magazine*, 134–143, October.

[28] K. I. Pedersen and P. E. Mogensen (2003), Performance of WCDMA HSDPA in a beamforming environment under code constraints, *IEEE Proc. Vehicular Technology Conference, Fall, October*.

[29] J. Ramiro-Moreno, K. I. Pedersen, and P. E. Mogensen (2003), Network performance of transmit and receive antenna diversity in HSDPA under different packet scheduling strategies, *IEEE Proc. Vehicular Technology Conference, April*.

[30] A. G. Kogiantis, N. Joshi, and O. Sunay (2001), On transmit diversity and scheduling in wireless packet data, *IEEE Proc. International Conference on Communications*, Vol. 8, pp. 2433–2437.

[31] L. T. Berger, T. E. Kolding, J. Ramiro Moreno, P. Ameigeiras, L. Schumacher, and P. E. Mogensen (2003), Interaction of transmit diversity and proportional fair scheduling, *IEEE Vehicular Technology Conference, Jeju, Korea, April*, Vol. 4, pp. 2423–2427.

[32] 3GPP, Technical Specification Group RAN, Base Station Conformance Testing (FDD), 3GPP TS 25.141 version 6.9.0, March 2005.

[33] T. Nihtilä, J. Kurjenniemi, M. Lampinen, and T. Ristaniemi (2005), WCDMA HSDPA network performance with receiver diversity and LMMSE chip equalization, *IEEE Proc. Symposium on Personal, Indoor and Mobile Radio Communications (PIMRC), Berlin, September 11–14*.

[34] J. Kurjenniemi, T. Nihtilä, M. Lampinen, and T. Ristaniemi (2005), Performance of WCDMA HSDPA network with different advanced receiver penetrations, *Proc. Wireless Personal Multimedia Communications (WPMC), Aalborg, Denmark, September 17–22*.

[35] J. Bach Andersen (2000), Array gain and capacity of known random channels with multiple element arrays at both ends, *IEEE Journal on Selected Areas in Communications*, **18**(11), 2171–2178, November.

[36] D. Gesbert, M. Shafi, D. Shiu, P. J. Smith, and A. Naguib (2003), From theory to practice: An overview of MIMO space-time coded wireless systems, *IEEE Journal on Selected Areas in Communications*, **21**(3), 281–302, April.

[37] A. Toskala, H. Holma, E. Metsala, K. I. Pedersen, and D. Steele (2005), Iub efficiency analysis for high speed downlink packet access in WCDMA, *Proc. WPMC, Aalborg, Denmark, September*.

[38] 3GPP, Technical Specification Group RAN, Technical Report on Multiple Input Multiple Output in UTRA, 3GPP TS 25.876 version 1.7.1.

# 8

# HSUPA bit rates, capacity and coverage

Jussi Jaatinen, Harri Holma, Claudio Rosa, and Jeroen Wigard

This chapter presents the performance of high-speed uplink packet access (HSUPA), including bit rates, capacity, and coverage. The outline is as follows: Section 8.1 contains a short summary of the main performance factors. This is followed by a study of link layer performance aspects in Section 8.2. Section 8.3 looks at the cell capacity gains from Node B based scheduling and L1 hybrid automatic repeat request (HARQ). Section 8.4 introduces further uplink enhancements with antenna and baseband solutions.

## 8.1 General performance factors

The performance of HSUPA depends highly on the selected scenario and on deployment and service parameters, in the same way as high-speed downlink packet access (HSDPA). Many of the essential parameters are similar to those explained for HSDPA. Here, the most significant are briefly mentioned:

- *Network algorithms* – HSUPA-specific algorithms like the Node B based packet scheduler, HSUPA resource allocation, and the serving cell change algorithm.
- *Deployment scenario* – the available interference levels in the cell defined by factors such as propagation loss, other-cell interference, and multipath propagation.
- *User equipment (UE) transmitter capability* – maximum supported bit rates and maximum transmit power.
- *Node B receiver performance and capability* – number of receive antennas, receiver type, and implementation margins.
- *Traffic* – mixture of dedicated channel (DCH) and HSUPA suitable traffic, traffic type like streaming vs messaging, and number of active users.

*HSDPA/HSUPA for UMTS* Edited by Harri Holma and Antti Toskala

In this chapter, we will attempt to adhere closely to the basic assumptions agreed in the Third Generation Partnership Project (3GPP). During the discussion of results, it will be indicated which factors in the above play the most important roles.

The following performance metrics are used in this chapter:

- *Average cell throughput* – the average amount of data from all users in a cell that can be received during a certain amount of time.
- *10% packet call throughput* – representing the end user experience. It indicates the 10% outage of user throughput (i.e., 90% of end users will experience a better packet call throughput than this value).
- $E_b/N_0$ – the required power per information bit over noise in order to receive a certain bit rate with a certain block error probability. This has a direct impact on the load of the system [16].
- $E_c/N_0$ – the required power per chip bit over noise in order to receive a certain bit rate with a certain block error probability.

Other less general measures will be introduced in addition to the presented results.

## 8.2 Single-user performance

The link performance of an enhanced uplink dedicated channel (E-DCH) link is studied in this section. We use so-called 'fixed reference channels' (FRCs) in link level studies. FRCs are a set of E-DCH channel configurations defined in 3GPP for performance testing purposes. They are defined in [21] and shown in Table 8.1.

FRC1 to FRC3 use a 2-ms transmission time interval (TTI), whereas the other FRCs are specified for a TTI of 10 ms. The maximum bit rate varies with the number of codes with FRC3 providing a peak bit rate of more than 4 Mbps. The 3GPP performance requirements using these FRCs are specified in [19]. The requirements specify a minimum performance under certain conditions.

For each FRC, there exists one pair of suggested power ratios between the enhanced dedicated physical data channel (E-DPDCH), the enhanced dedicated physical control

**Table 8.1** Fixed reference channels (FRCs) defined for E-DCH.

| *FRC* | *TTI length* (ms) | *Transport block size* (bits) | *Codes* | *Coding rate* | *Maximum bit rate* (kbps) | *UE category* |
|---|---|---|---|---|---|---|
| 1 | 2 | 2 688 | 2*SF4 | 0.71 | 1353 | 2 |
| 2 | 2 | 5 376 | 2*SF2 | 0.71 | 2706 | 4 |
| 3 | 2 | 8 064 | 2*SF4 + 2*SF2 | 0.71 | 4059 | 6 |
| 4 | 10 | 4 800 | 1*SF4 | 0.53 | 508 | 1 |
| 5 | 10 | 9 600 | 2*SF4 | 0.51 | 980 | 2 and 3 |
| 6 | 10 | 19 200 | 2*SF2 | 0.51 | 1960 | 4 and 5 |
| 7 | 10 | 640 | 1*SF16 | 0.29 | 69 | 1 |

**Table 8.2** Beta factors defined for fixed reference channels.

| | *E-DPDCH/DPCCH* (dB) | *E-DPCCH/DPCCH* (dB) |
|---|---|---|
| FRC1 | 9 | 2.05 |
| FRC2 | 10 | 4.08 |
| FRC3 | 6 | 0 |
| FRC4 | 9 | −1.94 |
| FRC5 | 9 | −1.94 |
| FRC6 | 10 | −5.46 |
| FRC7 | 6 | 0 |

channel (E-DPCCH), and the DPCCH. The power ratios are called 'beta factors'. The suggested beta factor is defined as an average over multipath channels. Beta factors are summarized in Table 8.2. Here the values have been rounded to the signalling values available, and for FRC3 the value given corresponds to SF4 codes. SF2 code powers are derived from this by multiplying by $\sqrt{2}$. These beta factors are designed for antenna diversity reception while 3 dB should be added for single-antenna reception.

In the following the single-link performance of FRC2, 5, and 6 are presented under the assumptions listed in Table 8.3. FRC5 represents a first-phase HSUPA UE, whereas FRC2 and FRC6 represent future release UEs. Note that power control is off in these simulations. The performance requirements are defined for all FRCs with diversity reception, and all FRCs except 2 and 3 for single-antenna reception.

Figure 8.1 shows the data rate without power control as a function of received $E_c/N_0$ per base station antenna. The performance of FRC2 with a 2-ms TTI and FRC6 with a 10-ms TTI are fairly similar. FRC2 can provide higher peak data rates when $E_c/N_0$ is

**Table 8.3** Link simulation assumptions.

| | |
|---|---|
| HARQ combining | Incremental redundancy |
| Maximum number of transmissions | Four |
| Number of HARQ processes | Four for a 10-ms TTI, eight for a 2-ms TTI |
| Base station receiver | Two-antenna Rake |
| Turbo-codec | $R = 1/3$, $K = 4, 8$ iterations, Max-Log MAP |
| For E-DPDCH simulation:<br>Demodulation of E-DPCCH<br>Simulated parameters | <br>Off<br>Throughput as a function of total $E_c/N_0$ |
| For E-DPCCH simulation:<br>Demodulation of E-DPDCH<br>Simulated parameters | <br>Off<br>Missed detection and false alarm as a function of total $E_c/N_0$. False alarm margin to be used is 2.0 dB. |

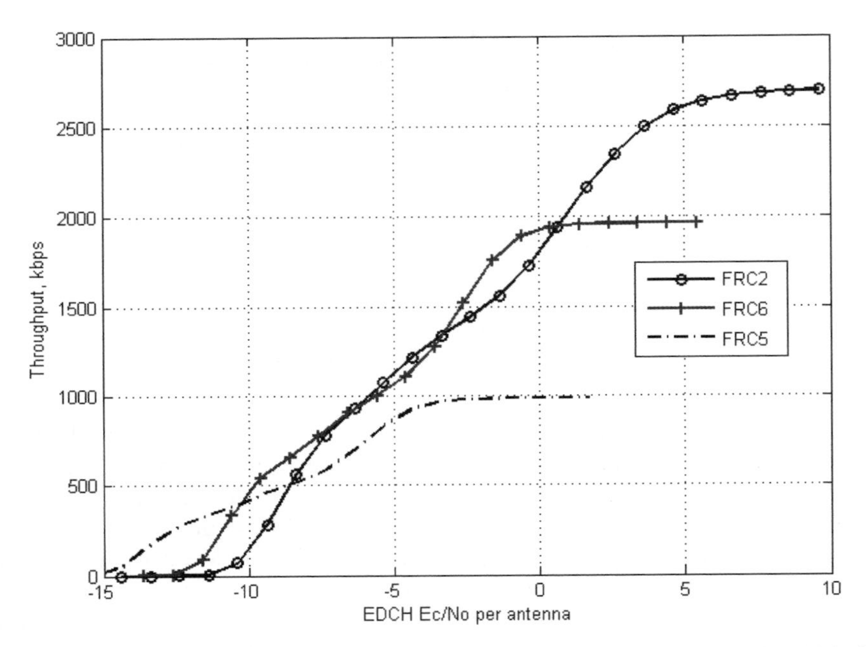


**Figure 8.1** HSUPA throughput in Vehicular A at 30 km/h, no power control used (similar to 3GPP test case).

high enough. In order to achieve data rates beyond 2 Mbps, the received $E_c/N_0$ needs to be higher than approximately 0 dB in this case.

The reliability of the control channel E-DPCCH must be reasonably high for the proper decoding performance of E-DPDCH. If the E-DPCCH is missed or the codeword received is incorrect, the HARQ buffers are easily corrupted by inappropriate soft-combining. The probability of missed detection should be low, preferably <0.2%. That refers to the case where the UE has transmitted data but the Node B fails to receive the E-DPCCH correctly. Figure 8.2 illustrates the missed detection probability in different multipath profiles. Since the E-DPCCH carries no CRC encoding, a threshold-based approach to channel power detection is necessary. The threshold is defined as the 'false alarm rate', which refers to the case where no data are transmitted by the UE but an E-DPCCH codeword is erroneously detected at the Node B receiver.

Link simulations indicate that the received $E_c/N_0$ needs to be relatively high to achieve high data rates. Figure 8.3 shows the uplink noise rise that is caused by a single user as a function of received $E_c/N_0$. If the received $E_c/N_0$ is 0 dB, the signal and noise powers are equal causing a 3-dB noise rise in total. If the required $E_c/N_0$ is 5 dB, the single user alone would cause a 6-dB noise rise. When the noise rise increases, the cell coverage area shrinks for other simultaneous users on the same carrier. In order to limit noise rise and to guarantee the coverage area, the highest $E_c/N_0$ values may not be practical in large macro-cells.

The rest of this section presents link level performance results without packet scheduling. The assumptions are listed in Table 8.4.

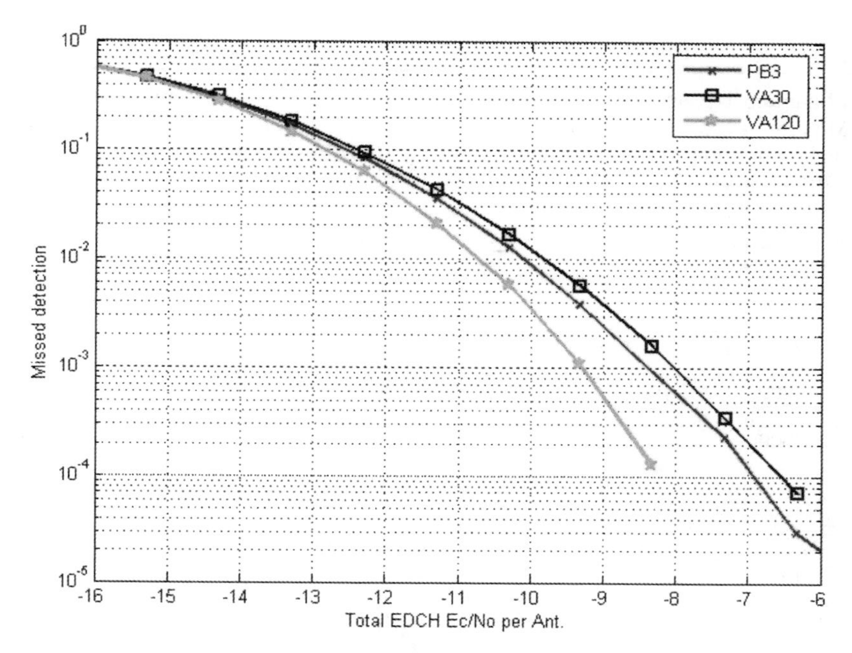


**Figure 8.2** E-DPCCH DTX detection performance for a 2-ms TTI.

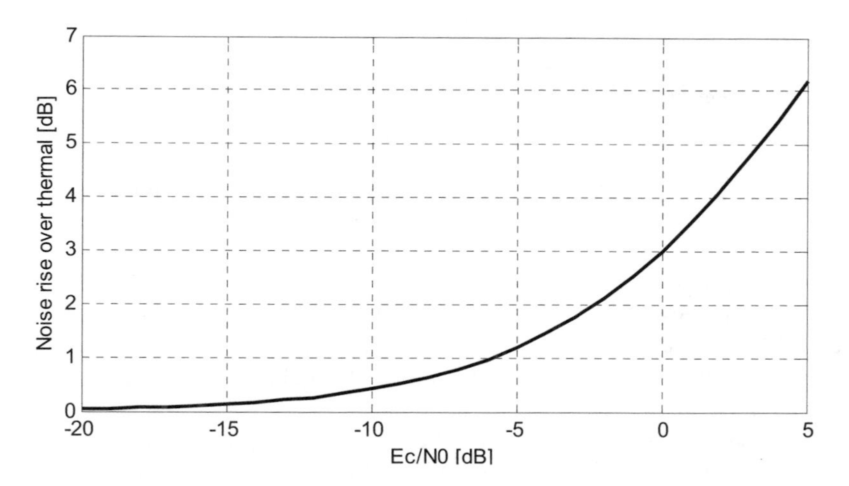


**Figure 8.3** Uplink noise rise caused by single user as a function of required $E_c/N_0$.

Figure 8.4 shows single-user uplink throughput on the radio link control (RLC) level in kbps vs the received signal code power (RSCP) of the common pilot channel (CPICH) in the UE. A lower RSCP means a larger path loss and, thus, a lower uplink throughput. The throughput is assumed to be limited by the maximum UE power level – capacity and packet scheduling aspects are not considered. The peak bit rate of 1.4 Mbps that is

**Table 8.4** Assumptions for user bit rate analysis.

| | |
|---|---|
| Maximum UE bit rate | Category 3 with maximum 1.45 Mbps |
| Maximum UE transmission power | Class 3 with maximum 24 dBm |
| Base station noise level | −105 dBm |
| CPICH power level | 33 dBm |
| RLC overhead | 5% |
| Slow fading standard deviation | 8 dB |

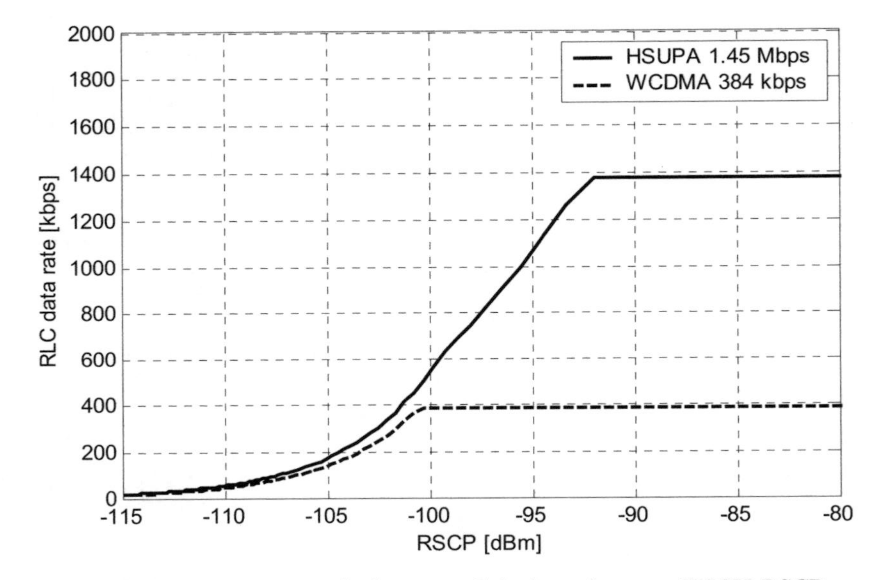


**Figure 8.4** Instantaneous single-user uplink throughput vs CPICH RSCP.

characteristic of a HSUPA Category 3 UE can be achieved with an RSCP of −92 dBm. HSUPA provides a clearly higher data rate than a wideband code division multiple access (WCDMA) 384-kbps UE when the RSCP level is higher than approximately −100 dBm. HSUPA also provides some gain in throughput with lower RSCP levels due to faster L1 HARQs.

Figure 8.5 shows the cumulative distribution of the uplink single-user data rate over the area of a large macro-cell including shadow fading and soft handover. The cell size is defined so that 64 kbps can be provided with a 95% coverage probability. An uplink rate of 384 kbps can be offered with 70% probability and 1.4-Mbps HSDPA with approximately a 40% probability from the UE power point of view.

The average bit rate and the 95% bit rate from Figure 8.5 are illustrated in Table 8.5. HSUPA can take benefit of the higher peak bit rate than that of WCDMA and can provide a 2.7× higher average bit rate than WCDMA. There is some gain from HSUPA at the 95% probability point, but that gain is clearly less than the gain in terms of the average data rate.

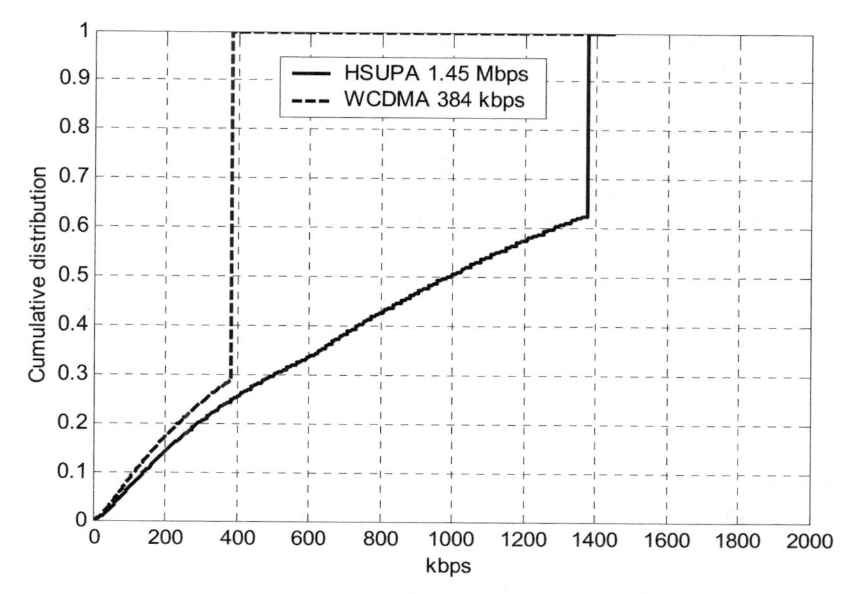


**Figure 8.5** User bit rate distribution.

**Table 8.5** Average and 95% outage user bit rates in the cell.

| | *WCDMA 384 kbps* (kbps) | *HSUPA 1.4 Mbps* (kbps) | *Gain from HSUPA* |
|---|---|---|---|
| Average user bit rate in the cell | 324 | 881 | 2.7× |
| 95% outage user bit rate in the cell | 64 | 82 | 1.3× |

## 8.3 Cell capacity

While the previous section considered single-link performance, this section considers interference-limited cell capacity. The capacity gain from HSUPA is mainly obtained using L1 HARQ and Node B based scheduling. Those two aspects are considered in the following two subsections.

### *8.3.1 HARQ*

The advanced physical layer retransmission schemes employing soft combining gain, incremental redundancy, and low retransmission delay are some of the central features of HSUPA. The introduction of L1 HARQ schemes has two main advantages:

1. Retransmissions are faster at L1 in HSUPA than L2 RLC-based retransmissions in Release 99.
2. Soft combining of retransmissions is used in HSUPA.

In other words, the retransmissions cause less delay and delay variations with L1-controlled retransmission schemes (located in the Node B) than with L2-based retransmission schemes (located in the RNC). Faster retransmissions allow us to have a higher retransmission probability while maintaining the same user delay performance. This leads to a decrease in required $E_b/N_0$ and, hence, an increase in spectral efficiency. Combining techniques – such as chase combining (CC) and incremental redundancy (IR) – can further improve the performance of Node B controlled L1 retransmission schemes.

In order to investigate the impact of L1 retransmission schemes and combining techniques on user and cell throughput performance, the following definitions are introduced:

1. The block error probability (BLEP) is the probability of failed detection of a data frame at first transmission. It is a function of the target $E_b/N_0$ used in the power control algorithm, as well as of the propagation and mobility environment.
2. The frame erasure rate (FER) is calculated over a sufficiently long time period and is defined as follows:

$$\mathrm{FER} = 1 - \frac{\text{Correctly received data frames}}{\text{Total transmitted data frames}} \tag{8.1}$$

3. Finally, the effective $E_b/N_0$ is defined as the ratio of required $E_b/N_0$ to corresponding frame erasure rate.

$$\text{Effective } E_b/N_0 = \frac{E_b/N_0}{1 - \mathrm{FER}} \tag{8.2}$$

Figure 8.6 illustrates the impact of HARQ techniques on user-experienced data rate.

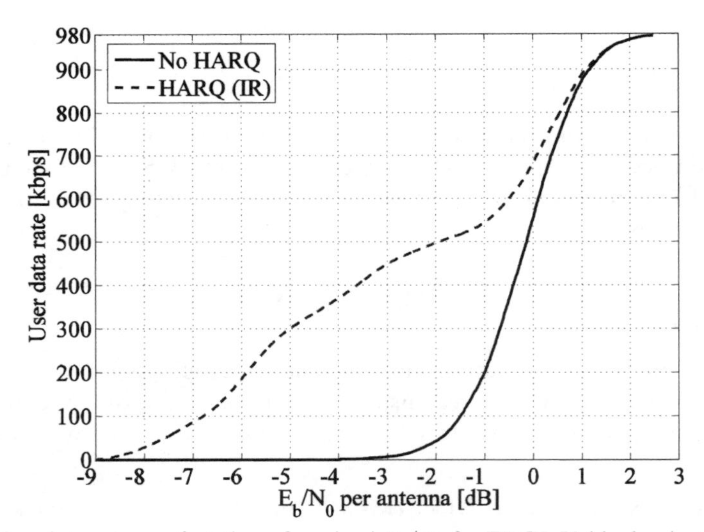


**Figure 8.6** User data rate as a function of received $E_b/N_0$ for FRC5, Vehicular A at 30 km/h, with and without HARQ.

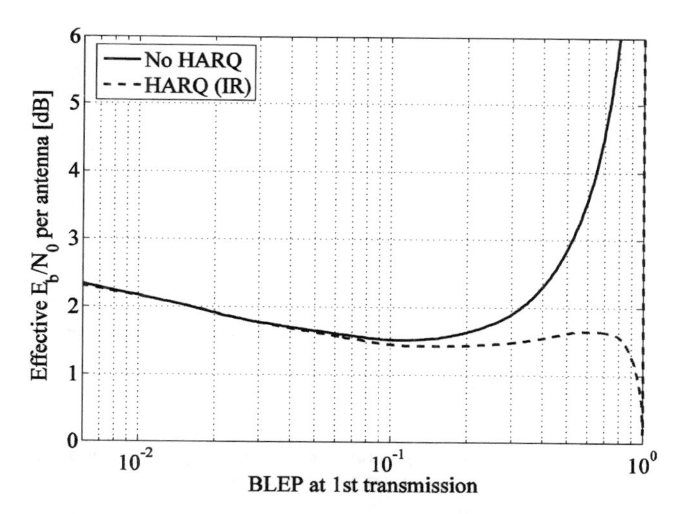


**Figure 8.7** Effective $E_b/N_0$ vs BLEP at first transmission for FRC5, Vehicular A at 30 km/h, with and without HARQ.

For a given received energy per bit-to-noise ratio, incremental redundancy provides a lower frame erasure rate than is the case with no HARQ, thus leading to higher user throughput. The gain from incremental redundancy (HARQ) clearly depends on the system operation point: the higher the required $E_b/N_0$, the lower the target BLEP at first transmission and the lower the gain from HARQ.

Following the definition of cell throughput in [16], uplink spectral efficiency is inversely proportional to the effective $E_b/N_0$. The optimal operation point typically depends on the propagation and mobility scenario, as well as on whether or not combining techniques are used. The results reported in Figure 8.7 show how uplink spectral efficiency can be improved by increasing the BLEP at first transmission, which leads to a lower $E_b/N_0$ requirement. This is possible since the low round trip delay with L1 retransmission schemes allows fast recovery from transmission errors. For instance, spectral efficiency gain from increasing the BLEP at first transmission from 1% to 10% is between 0.6 and 0.7 dB, corresponding to a cell throughput increase of approximately 15–20%.

Incremental redundancy can be successfully deployed to further increase uplink spectral efficiency, although the gain from HARQ is really appreciable only for FER values above 50%. In Figure 8.8, the effective $E_b/N_0$ with incremental redundancy is minimized at 300 kbps, which corresponds to an FER of approximately 70%, and to a BLEP at first transmission higher than 90% in Figure 8.7. Assuming that with RLC retransmissions the physical channel is operated at an error probability of 1%, the maximum spectral efficiency gain from L1 HARQ is approximately 2 dB, corresponding to a cell throughput increase of 60%. However, the price to pay is significantly increased transmission delay, reduced user data rate, as well as more processing power at the base station due to multiple retransmissions. Assuming a more realistic operation region for the BLEP at first transmission between 20% and 30%, the cell throughput gain from L1 HARQ is approximately 15% to 20%.

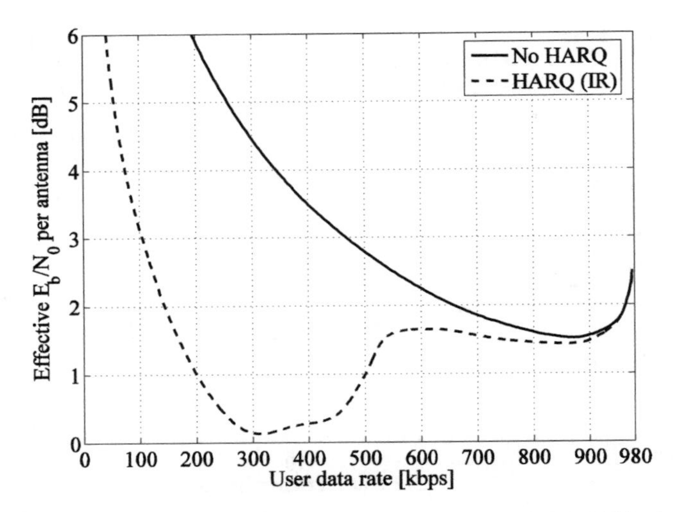


**Figure 8.8** Effective $E_b/N_0$ as a function of the user data rate for FRC5, Vehicular A at 30 km/h, with and without HARQ.

### *8.3.2 Node B scheduling*

Node B based scheduling with L1 control signalling can operate much faster than RNC-based scheduling with L3 control signalling. It can provide the system with two main advantages:

1. *Tighter control of total received uplink power* is possible with fast Node B packet scheduling, which allows faster adaptation to interference variations. The required power headroom can be reduced and the system can be operated with higher average interference levels leading to higher cell throughput.
2. Node B scheduling allows *faster reallocation of radio resources* between users. The long scheduling delays associated with the transmission of capacity requests from the UE to the radio network controller (RNC) and the transmission of capacity grants from the RNC to the UE represent the main disadvantage of RNC packet scheduler implementation. To reduce such delays, the Node B scheduling algorithm considered in this section is based on utilization of allocated radio resources, and does not require explicit signalling of a capacity request from the UE to the Node B. Conceptually, the scope of the proposed scheduling algorithm is to dynamically take resources from users with low utilization of allocated radio resources and to redistribute them between users with high utilization. More details on the packet scheduling algorithm can be found in [1] and [2]. The packet scheduling parameters set up for both RNC and Node B scheduling are reported in Table 8.6. The scheduling and modification period define the frequency of packet scheduling on both a cell and a user basis, respectively.

With both RNC and Node B scheduling, the Node B is assumed to deploy a two-branch antenna diversity with Rake processing of the received signals. Both inner loop and outer loop power control algorithms are implemented in the system level simulator. The model

**Table 8.6** Packet scheduling parameter setup for simulation of RNC and Node B scheduling.

| | *Node B scheduling* (ms) | *RNC scheduling* |
|---|---|---|
| Packet scheduling period | 100 | 200 ms |
| User data rate modification period | 100 | 600 ms |
| Allocation delay | 40 | 100 ms |
| Inactivity timer<br>After having been inactive for this time period a UE is downgraded to the minimum allowed data rate | 100 | 2 s |

also includes soft and softer handover, with selection diversity performed at the RNC for soft handover, and maximal ratio combining at the Node B for softer handover.

The model used to simulate packet data traffic is the modified gaming model described in [5]. As illustrated in Figure 8.9, the state of a simulated user with respect to the generation and transmission of data packets can be divided into three main phases:

1. In the first phase of the packet call, source data are generated in the user RLC buffer. During this first phase, data are transmitted based on both buffer occupancy and the data rate allocated by the packet scheduler.
2. During the second phase, the generation of source data in the user RLC buffer is terminated. At the same time, the terminal continues transmitting the data it has accumulated during the packet call period. This second phase is ended whenever all the buffered data have been successfully transmitted.
3. Finally, during the third phase no data are generated in the user RLC buffer. This phase is considered 'reading time'. Since the buffer is empty when the third phase starts, during reading time no data are transmitted.

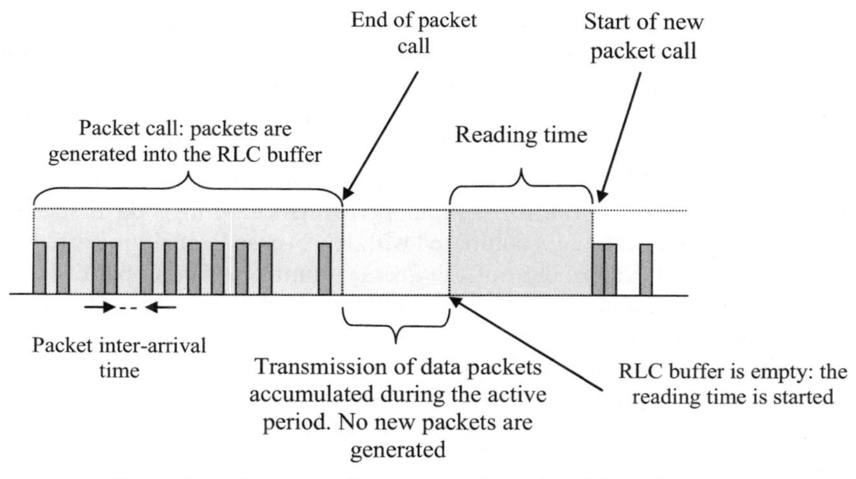


**Figure 8.9** Source traffic model and its closed loop feature.

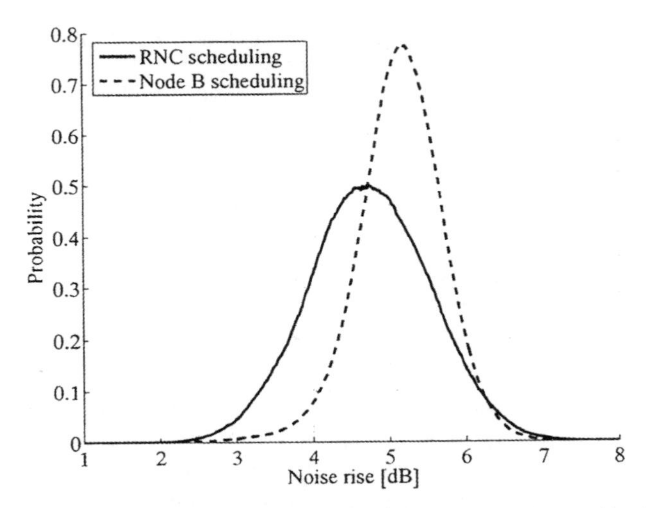


**Figure 8.10** Noise rise distribution for RNC and Node B scheduling, Vehicular A at 3 km/h, an average of 20 users per cell, and a 5% noise rise outage of 6 dB.

The average source data rate during packet calls is 250 kbps. The detailed information on the system level simulator can be found in [2].

The performance of the proposed Node B scheduling algorithm is compared with that of a Release 99 compliant RNC packet scheduler implementation. In order to focus on packet scheduling performance, L1 retransmission without soft combining is assumed. Therefore, the system is assumed to operate with the BLEP at first transmission equal to 10%. The idea is to compare the two scheduling scenarios when operating the physical channel at the same error probability, thus focusing on the efficiency of the resource allocation algorithm.

The impact of fast Node B scheduling on system performance can be seen in the noise rise distributions for RNC and Node B scheduling in Figure 8.10. Both distributions have a 5% noise rise outage equal to 6 dB – that is, the probability of exceeding a noise rise of 6 dB is 5%. Fast Node B packet scheduling allows the system to more efficiently control the total received uplink power than is the case with RNC scheduling, and the required power headroom to prevent the system from entering an unwanted load region is consequently reduced. Hence, the average uplink load to meet the specified noise rise outage constraint can be increased, and cell throughput consequently improved. The improvement in the average cell throughput performance with Node B scheduling is illustrated in Figure 8.11. The gain compared with RNC-based scheduling is between 6% and 9%, and is almost independent of the average number of users per cell.

In order to also give an indication on the improved level of service in the system with Node B scheduling, Figure 8.11 reports average cell throughput as a function of 10% packet call throughput outage. Packet call throughput is a measure of the average air interface throughput experienced by a user during a packet data session. Only 10% of packet data sessions experience a packet call throughput lower than the 10% packet call throughput outage. Figure 8.12 shows how Node B scheduling can be deployed to either increase cell throughout performance or to improve the overall quality of service in the

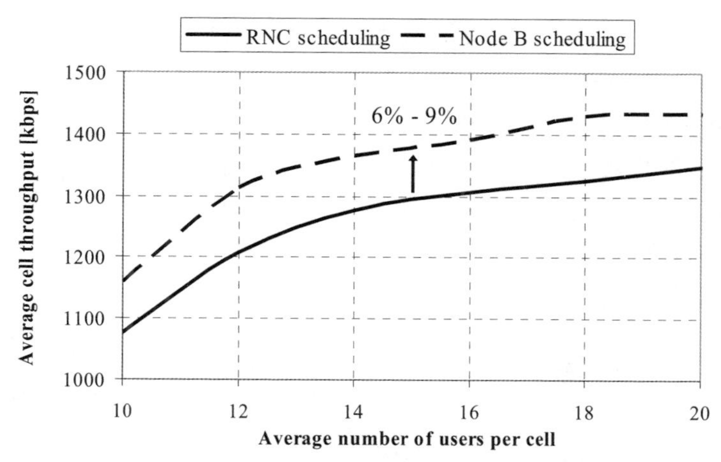


**Figure 8.11** Average cell throughput as a function of average number of users per cell for both RNC and Node B based scheduling and a 5% noise rise outage of 6 dB.

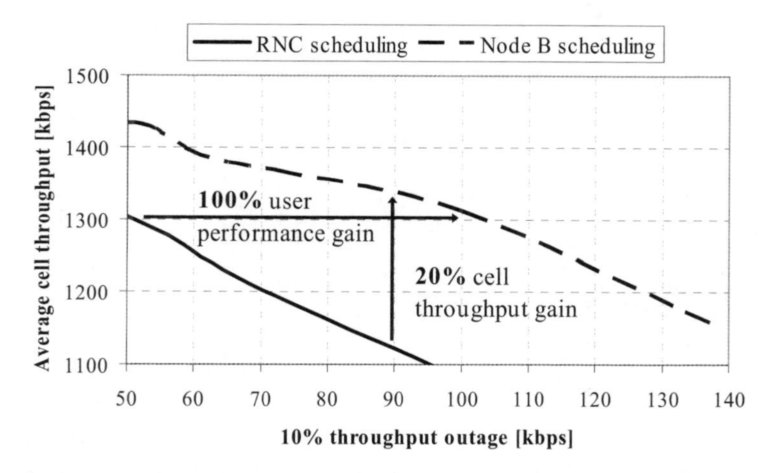


**Figure 8.12** Average cell throughput as a function of 10% throughput outage for both RNC and Node B scheduling.

system. As an example, for a 10% packet call throughput outage of 90 kbps the cell throughput gain from Node B scheduling over RNC scheduling is about 20%. For a cell load of 1.3 Mbps, the 10% outage point increases from approximately 100% from 50 kbps with RNC scheduling to more than 100 kbps with Node B scheduling.

Finally, the combined gain of jointly deploying fast L1 retransmission schemes and fast Node B scheduling is estimated. Different scenarios are simulated: first, WCDMA Release 99 with a BLER target of 1% and RNC-based slow scheduling and retransmissions and, second, HSUPA with a BLER target of 10% and 20%, Node B based fast scheduling, and fast retransmissions, both scenarios with soft combining (L1 HARQ) and without soft combining (L1 ARQ). The capacity results are shown in Figures 8.13 and 8.14 for a Vehicular A channel profile at 3 and 50 km/h, respectively.

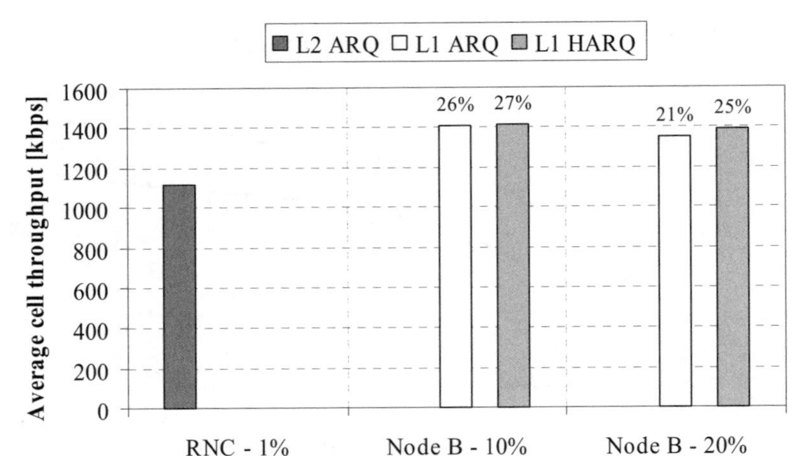


**Figure 8.13** Average cell throughput with Vehicular A at 3 km/h, 5% noise rise outage of 6 dB, and 10% throughput outage of 64 kbps.

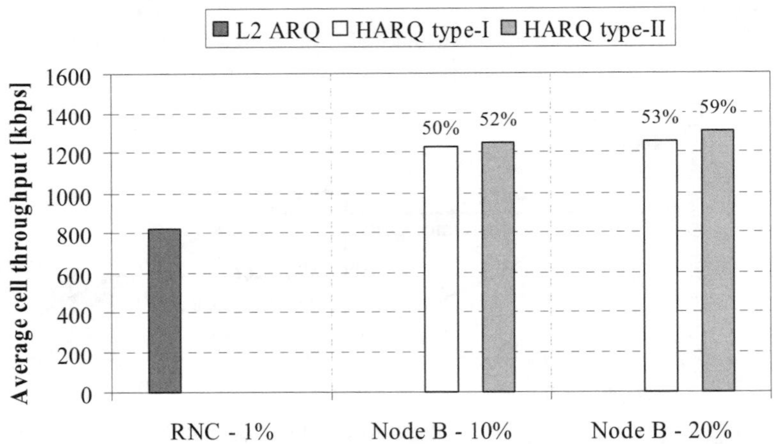


**Figure 8.14** Average cell throughput with Vehicular A at 50 km/h, 5% noise rise outage of 6 dB, and 10% throughput outage of 64 kbps.

Under the assumptions considered, the cell throughput gain of HSUPA compared with WCDMA Release 99 is 25% to 60%, depending on the mobility scenario. The reason for a higher cell throughput gain with a UE speed of 50 km/h can be explained by the higher spectral efficiency gain when increasing the BLEP target at first transmission from 1% to 10%. The reason is that a low BLEP is more difficult to achieve at high mobile speed since power control accuracy is not as good as at low mobile speed.

Most of the gain from L1 retransmission schemes is obtained from operating the physical channel at a higher error probability, rather than from deploying soft combining techniques. Node B based scheduling provides a cell throughput increase of approximately 15% to 20% on top of the gain from L1 HARQ.

## 8.4 HSUPA performance enhancements

Uplink spectral efficiency and data rates can be improved with antenna and baseband solutions. This section briefly introduces antenna diversity, beamforming, interference rejection combining, and multiuser detection (MUD).

*Multibranch antenna diversity and beamforming* – antenna arrays can be operated in two distinct modes: (i) diversity mode or (ii) beamforming mode. Diversity techniques rely on the statistical independence between antenna elements to reduce the likelihood of deep fades, and are additionally able to provide an average signal-to-interference-plus-noise ratio (SINR) gain by coherently combining the signals received at each antenna element [9]. Beamforming techniques direct narrow beams towards each user in such a way that multiple access interference is reduced by means of spatial filtering and potential suppression of interfering signals [11]. The basic choice is whether to maximize diversity or antenna gain. In [13] performance evaluation of different antenna array approaches for the WCDMA uplink is presented. It is shown how diversity and beamforming concepts perform similarly with a small number of antennas per sector, while the conventional beamforming approach starts to overcome diversity techniques with a large number of antennas and low azimuth spread. However, diversity techniques are shown to perform better in low-mobility environments since longer averaging periods can be applied for better channel estimates. Also, in [15] it is shown that for channel environments with a large angular spread, space diversity schemes with interference cancellation represent an attractive solution compared with beamforming in terms of complexity and robustness, against calibration errors of the antenna array.

The choice of antenna solution must consider both uplink and downlink requirements. See the downlink HSDPA performance discussion in Chapter 7. Beamforming mode looks attractive for enhancing downlink performance.

*Interference rejection combining* – traditional antenna diversity combining uses maximal ratio combining (MRC) where interference is assumed to be spatially white. Interference rejection combining (IRC) exploits the spatial properties of interference [28]. MRC aims to maximize the received signal power while IRC aims to maximize the received signal-to-interference ratio. IRC can improve performance substantially in the case of dominant interferers. Since HSUPA enables high peak data rates, thus possibly leading to interference scenarios characterized by dominant interferers, IRC has the potential to provide significant capacity and coverage gains. In case of low data rate users in WCDMA, there are no dominant interferers, and the IRC receiver cannot provide any gains. Implementation of an IRC receiver is relatively simple compared with employing the MUD algorithm. IRC algorithms are used in Global System for Mobile Communications (GSM) base station receivers to improve performance since dominant interferers are typical in the GSM system.

*Multiuser detection* – although multiple access interference (MAI) is often approximated as additive white Gaussian noise (AWGN) [11], in reality it consists of the distinct received signals of code division multiple access (CDMA) users. As a consequence,

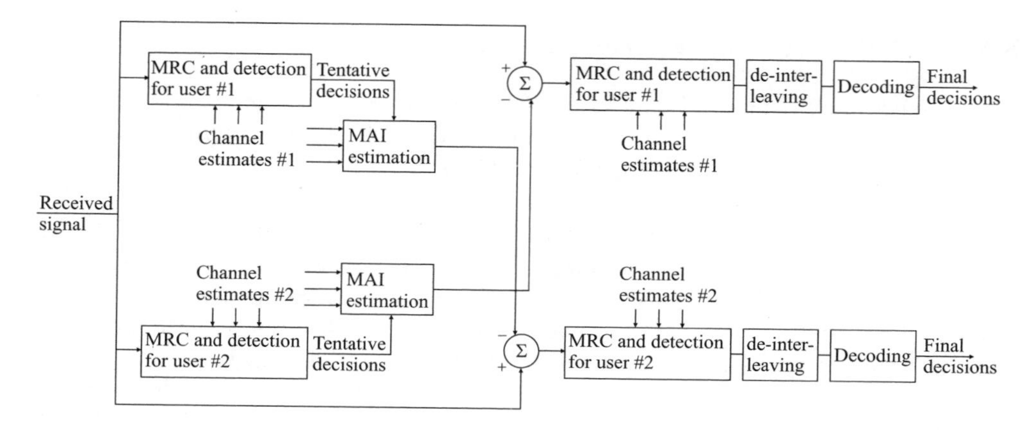


**Figure 8.15** Parallel interference cancellation receiver for two users [16].

multiple access interference possesses some structure and can be taken into consideration in the receiver. Verdú was able to demonstrate that CDMA systems are not inherently interference-limited, but rather limited by conventional matched filter receivers [13]. However, the optimal multiuser detector in [12] presents a major drawback: its implementation complexity is an exponential function of the number of users, and as a consequence it is not feasible for most practical CDMA receivers. A survey of different MUD techniques for DS-CDMA systems is presented in [14] and [15], where the trade-off between performance and complexity of multiuser detectors is also discussed. One way to perform MUD is through interference cancellation: the basic principle is re-creation of CDMA interference in order to subtract some or all of the multiple access interference seen by each user. Figure 8.15 depicts the simplified structure of a single-stage parallel interference cancellation receiver for two users.

Advanced base station receivers have the potential for significantly boosting end user bit rates. In this perspective, four-branch antenna diversity and interference cancellation can provide HSUPA with a higher gain than is the case with Release 99, since HSUPA enables much higher data rates than Release 99. This is clearly illustrated in Table 8.7, which reports user data rates in a scenario with five users. It is assumed that interference cancellation is able to remove 70% of intra-cell interference. Due to the high peak data rates available with HSUPA, interference cancellation can improve HSUPA user data

**Table 8.7** User data rates for a scenario with five users/cell.

| | | *User data rates* (kbps) | *Gain from IC* (%) |
|---|---|---|---|
| Release 99 | No interference cancellation | 375 | — |
| | 70% interference cancellation efficiency | 384 | 2 |
| HSUPA | No interference cancellation | 414 | — |
| | 70% interference cancellation efficiency | 583 | 41 |

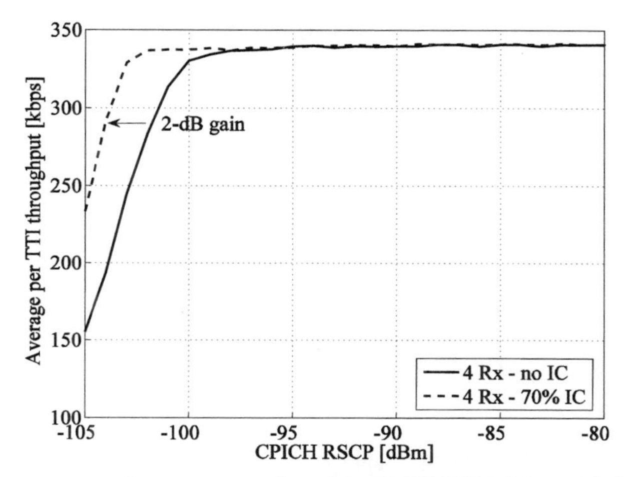


**Figure 8.16** Average user data rate as a function of CPICH RSCP with four-branch antenna diversity, with and without interference cancellation (the CPICH RSCP is calculated assuming a downlink cable loss of 4 dB and a CPICH transmission power of 33 dBm).

rates by 40%, while the gain is almost negligible with Release 99 data rates. Release 99 user data rates are limited by the peak bit rate of 384 kbps.

Interference cancellation techniques can also be successfully deployed to improve uplink coverage. Cell coverage is illustrated in Figure 8.16; the link budget is improved by 2 dB which means the same data rate can be provided at a 2-dB lower signal level with an interference cancellation base station receiver than with a Rake receiver.

Cell throughput gain from interference cancellation is presented in [3]. It is shown that starting from the well-known uplink load equation in [16], cell throughput gain from interference cancellation can be approximated as follows:

$$G \cong \frac{1+i}{1+i-\beta \cdot \eta_{UL}} \tag{8.3}$$

In (8.3) $i$ is the other-to-own-cell-interference ratio, $\eta_{UL}$ is the uplink fractional load, and $\beta$ is the interference cancellation efficiency [17], [18]. Cell throughput gain from interference cancellation increases with MUD efficiency $\beta$ and decreases with $i$, since interference cancellation is only effective for intra-cell interference. Also, the impact of MUD efficiency on cell throughput gain is scaled by the uplink fractional load $\eta_{UL}$. The reason for this is that the gain from multiuser detection clearly increases with the amount of interference to be cancelled.

Figures 8.17 and 8.18 plot the theoretical cell throughput gain from interference cancellation for $\beta$ equal to 0.3 and 0.7, respectively, as a function of the uplink fractional load $\eta_{UL}$, and for different values of the other-to-own-cell-interference ratio $i$. In a fully loaded isolated cell ($\eta_{UL} = 1, i = 0$) the theoretical capacity gain is 40% for a cancellation efficiency of 30%, and approximately 230% for an efficiency of 70%. However, assuming a cell load around 75% and an other-to-own-cell-interference ratio in the range from 0.4 to 0.6, cell throughput gain from interference cancellation significantly reduces to approximately 20% and 50% for $\beta$ equal to 0.3 and 0.7, respectively.

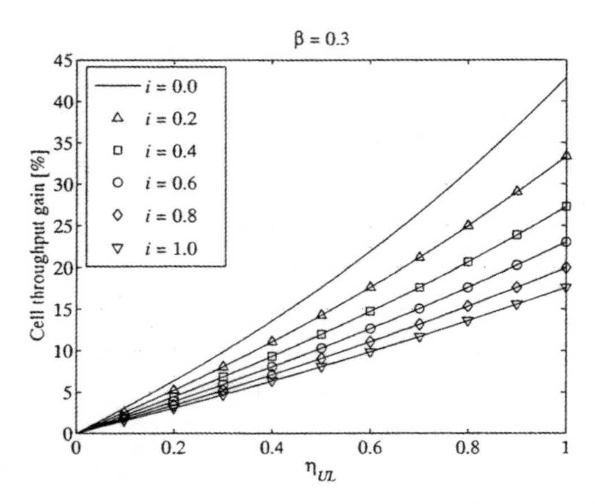


**Figure 8.17** Theoretical cell throughput gain from IC as a function of $\eta_{UL}$ and for $\beta = 0.3$.

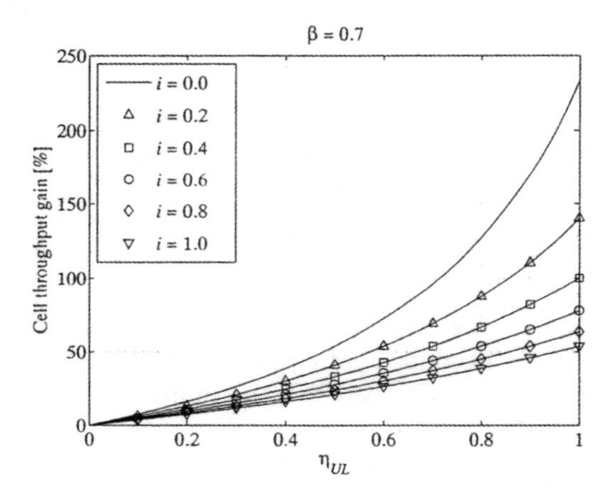


**Figure 8.18** Theoretical cell throughput gain from IC as a function of $\eta_{UL}$ and for $\beta = 0.7$.

The combined performance gain from HSUPA with four-branch antenna diversity and interference cancellation compared with the basic WCDMA Release 99 are illustrated in Figure 8.19. Gain amounts are given relative to a reference Release 99 scenario with two-branch antenna diversity, matched filter receiver, 1% BLER target, and L2 ARQ retransmissions. Overall cell throughput enhancement is between 200% and 300%, depending on the efficiency of the interference cancellation receiver deployed.

## 8.5 Conclusions

WCDMA Release 99 provides, in practice, 384 kbps as the maximum data rate in the uplink while HSUPA pushes the peak data rate initially to 1–2 Mbps and later up to 3–5 Mbps. The higher data rates in HSUPA are achieved without higher order modula-

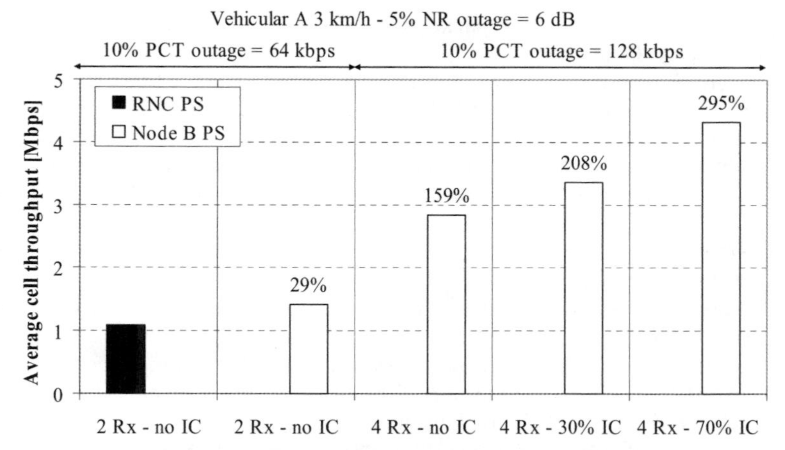


**Figure 8.19** Average cell throughput for different antenna diversity and IC scenarios with Node B PS and L1 HARQ, and compared with the performance of a reference Release 99 scenario (RNC PS and L2 ARQ).

tion which enables efficient mobile power amplifier operation and facilitates good coverage for high data rates. From the coverage point of view, high HSUPA bit rates above 1 Mbps are even available in a large macro-cell at a 50% probability.

HSUPA uses advanced techniques to also improve cell capacity and spectral efficiency in loaded cells. The techniques include fast L1 retransmissions with HARQ and fast Node B based scheduling. Fast retransmissions allow a higher error rate in HSUPA than in the Release 99 DCH while still maintaining the same end-to-end delay. If the block error rate is increased from 1% to 10%, cell capacity gain is 15–20%, and with higher block error rate levels even beyond 50%. Fast Node B scheduling allows using higher average interference levels in HSUPA than in the Release 99 DCH since fast scheduling makes signal-to-interference distribution narrow and reduces the interference outage probability. The capacity gain from fast scheduling is 10–20% depending on the required outage probability. In total, the expected capacity gain from HSUPA is 25–60%.

HSUPA cell capacity can be further improved by using base station antenna beamforming, four-branch reception, interference rejection combining, and interference cancellation. All solutions are supported by the current 3GPP HSUPA standard. These techniques can reduce the amount of interference and, thus, provide higher data rates and higher cell capacity.

## 8.6 Bibliography

[1] C. Rosa, J. Outes, K. Dimou, T. B. Sørensen, J. Wigard, F. Frederiksen, and P. E. Mogensen (2004), Performance of fast Node B scheduling and L1 HARQ schemes in WCDMA uplink packet access, *IEEE Proceedings of the 59th Vehicular Technology Conference, May.*

[2] C. Rosa (2004), Enhanced uplink packet access in WCDMA, Ph.D. thesis, Aalborg University (Denmark), December.

[3] C. Rosa, T. B. Sørensen, J. Wigard, and P. E. Mogensen (2005), Interference cancellation and 4-branch antenna diversity for WCDMA uplink packet access, *IEEE Proceedings of the 61st Vehicular Technology Conference, May/June.*

[4] 3GPP, Technical Specification Group RAN, Feasibility Study for Enhanced Uplink for UTRA FDD, 3GPP TR 25.896 version 6.0.0, Release 6, March 2004, available at *www.3gpp.org*

[5] T. Hytönen (2001), *Optimal Wrap-Around Network Simulation*, Helsinki University of Technology Report, A432.

[6] R. H. Clarke (1968), A statistical theory of mobile-radio reception, *Bell Systems Technical Journal*, **47**, 957–1000.

[7] W. C. Jakes (1974), *Microwave Mobile Communications*, IEEE Press, Los Alametos, CA.

[8] J. Ramiro-Moreno (2003), System level performance analysis of advanced antenna concepts in WCDMA, Ph.D. thesis, Aalborg University (Denmark), July.

[9] J. Ylitalo and E. Tiirola (2000), Performance evaluation of different antenna array approaches for 3G CDMA uplink, *IEEE Proceedings of the 51st Vehicular Technology Conference, May*, Vol. 2, pp. 883–887.

[10] D. Dahlhaus and Zhenlan Cheng (2000), Smart antenna concepts with interference cancellation for joint demodulation in the WCDMA UTRA uplink, *IEEE Proceedings of the 6th International Symposium on Spread Spectrum Techniques and Applications, September*, Vol. 1, pp. 244–248.

[11] J. G. Prakis (1995), *Digital Communications*, McGraw-Hill, London.

[12] S. Verdú (1986), Minimum probability of error for asynchronous Gaussian multiple-access channels, *IEEE Transactions on Information Theory*, **32**(1), 85–96, January.

[13] S. Verdú (1998), *Multi-user Detection*, Cambridge University Press, New York.

[14] A. Duel-Hallen, J. Holtzman, and Z. Zvonar (1995), Multiuser detection for CDMA systems, *IEEE Personal Communications*, **2**(2), 46–58, April.

[15] S. Moshavi (1996), Multi-user detection for DS-CDMA communications, *IEEE Communications Magazine*, **34**(10), 124–136, October.

[16] H. Holma and A. Toskala (2004), *WCDMA for UMTS: Radio Access for Third Generation Mobile Communications* (3rd edn), John Wiley & Sons, Chichester, UK.

[17] B. Hagerman, F. Gunnarsson, H. Murai, M. Tadenuma, and J. Karlsson (2004), WCDMA uplink interference cancellation performance: Field measurements and system simulations, *Nordic Radio Symposium, Session 10 – WCDMA Enhancements, August.*

[18] S. Hämäläinen, H. Holma, and A. Toskala (1996), Capacity evaluation of a cellular CDMA uplink with multiuser detection, *IEEE Proceedings of the 4th International Symposium on Spread Spectrum Techniques and Applications, September*, Vol. 1, pp. 339–343.

[19] 3GPP, Technical Specification Group RAN, Base Station Radio Transmission and Reception, 3GPP TR 25.104 version 6.9.0, Release 6, June 2005, available at *www.3gpp.org*

[20] 3GPP, Technical Specification Group RAN, Physical Layer – Measurements (FDD), 3GPP TS 25.215 version 6.3.0, Release 6, June 2005, available at *www.3gpp.org*

[21] 3GPP, R4-050315, EUL UL Simulation Assumptions, May 2005, available at *www.3gpp.org*

[22] D. Astely and A. Artamo (2001), Uplink spatio-temporal interference rejection combining for WCDMA, *IEEE Third Workshop on Signal Processing Advances in Wireless Communications (SPAWC '01), 20–23 March*, pp. 326–329.

# 9

# Application and end-to-end performance

Chris Johnson, Sandro Grech, Harri Holma, and Martin Kristensson

This chapter analyses high-speed packet access (HSPA) radio access from an end user service perspective. The introduction in Section 9.1 contains a representative list of packet services that illustrates the spectrum of characteristics that a system based on HSPA radio access needs to be capable to accommodate. Section 9.2 covers always-on connectivity using HSPA radio access and associated signalling and typical end-to-end delays. The expected performance of the applications over an HSPA network is estimated in Section 9.3. Section 9.4 presents how network loading affects application performance and provides guidelines for network dimensioning.

## 9.1 Packet application introduction

This section briefly introduces a few packet-based applications. The list concentrates on mobile applications that have already been launched either in the fixed or in the mobile domain. The list is not intended to be exhaustive, but it illustrates some key applications that can be run over HSPA networks.

- *Browsing* – users on the move want to have access to information and entertainment including news headlines, weather reports and sports results. The service may also be location-specific. Browsing includes both mobile-specific Wireless Application Protocol (WAP) browsing as well as public Internet browsing either via a mobile phone (or, more generally, a handheld device) or using a laptop computer. The browsing service is interactive and service response times should be fast to make the user experience attractive.
- *Music and game download* – in addition to providing basic mobile telephony, today's mobile terminals are capable of also serving as music players and gaming devices as well as mobile phones. The user can download songs or games over cellular networks.

*HSDPA/HSUPA for UMTS* Edited by Harri Holma and Antti Toskala

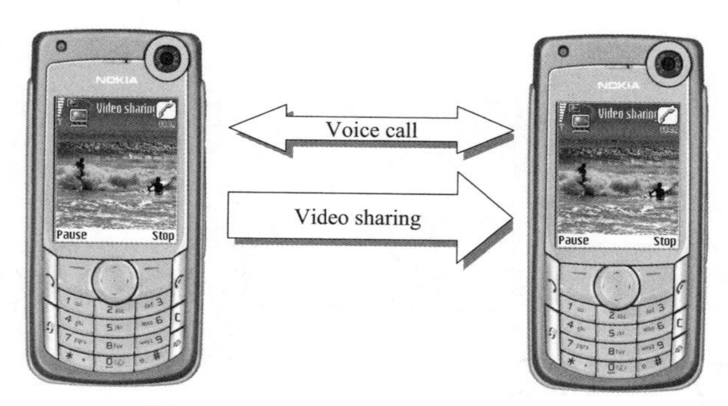


**Figure 9.1** Real time video sharing.

Download time should preferably be below 1minute in order to keep the user satisfied. When the file sizes increase, the required data rates must increase correspondingly. Data rates higher than 500 kbps are required in order to download a digitally encoded record track, with typical coding rates, in less than 1 minute.

- *Mobile-TV streaming* – mobile streaming can include TV news, movie trailers, music videos, etc. A few tens of kbps can offer basic news while good-quality streaming requires clearly higher data rates, typically slightly over 100 kbps on an average-sized mobile terminal screen. The data rate needs to be maintained with reasonable stability during the streaming session in order to avoid re-buffering events. In today's mobile networks mobile-TV is delivered via point-to-point streaming connections, while in the future it will also be possible to provide them via broadcast solutions like the Third Generation Partnership Project (3GPP) Release 6 Multimedia Broadcast and Multicast Service (MBMS).
- *Real time video sharing* – streaming video can also be shared from mobile to mobile in addition to the content-to-person distribution format outlined above. A one-way real time video sharing service can be initiated during a normal voice call for instantaneous person-to-person video sharing (as illustrated in Figure 9.1). The service can be further enhanced with a two-way packet-based video connection (i.e., a packet-switched video call).
- *Push-to-talk* – this application allows users to call a pre-defined group of people by pressing just one button, similar to a walkie-talkie service. Push-to-talk calls can be made to individuals or to groups and one-way call connection establishment should preferably be very fast. The push-to-talk service can be extended by means of one-way video sharing to create a push-to-show service.
- *Push e-mail* – traditional e-mail services operate in pull mode where incoming mails are fetched interactively by the end user, or in an automated fashion by the mail client at regular pre-defined polling intervals. In the case of push e-mail the server sends the e-mail subject field and part of the e-mail to the terminal as soon as the e-mail arrives at the server. Push-to-talk and push e-mail are the first always-on services where the terminal must have an IP connection open all the time. An example mail application is shown in Figure 9.2.
- *On-line gaming* – mobile gaming has been around for years, but on-line network gaming can enhance the service: real time games can be played between users over

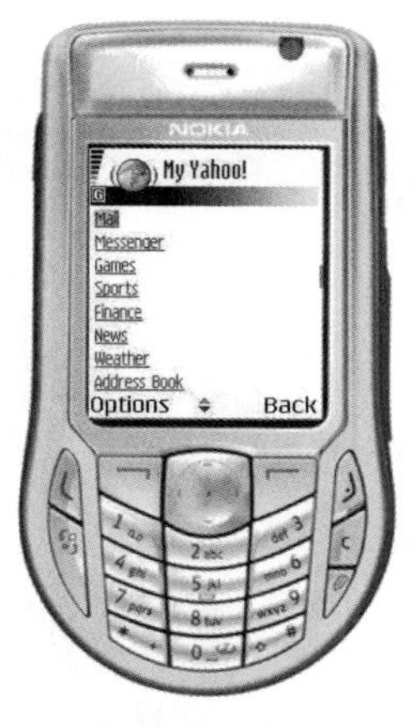


**Figure 9.2** E-mail application on a Symbian Series 60 terminal.

**Figure 9.3** Example real time games.

mobile networks. The gaming experience can be boosted with a simultaneous voice call or a push-to-talk connection. Example real time games are shown in Figure 9.3.

- *Mobile weblog* – this is a mixture between a multimedia diary and a weblog. The user can upload the pictures and videos taken by the camera phone to his weblog to be shared with his friends. Uplink data rates need to be sufficient to allow fast upload of multimedia content from the mobile.
- *Wireless broadband access* – Internet access from a laptop is a service where the operator provides a 'bit pipe' to the end user, preferably with high bit rate and low latency. Broadband access could be a fully mobile solution (e.g., when the user is sitting in a train or a bus travelling fast through the cells). Another use case is portable access where the user is typically stationary when using a laptop, but he/she still wants to take the laptop to a hotel, coffee shop, home, office, and access the Internet. This latter use case could be compared with wireless local area network (WLAN) hot spot access. Fixed broadband access is yet another use case where the laptop is located in the same place all the time, but instead of using a fixed line digital subscriber line (DSL) connection the user may opt to use wireless data. HSPA radio

**Figure 9.4** Use cases for broadband data access: (a) fixed broadband access (wireless DSL); (b) portable broadband access (similar to hot spots); and (c) mobile broadband access (cellular data access).

networks can provide data rates that are similar to fixed line broadband access. The use cases are illustrated in Figure 9.4.

## 9.2 Always-on connectivity

### *9.2.1 Packet core and radio connectivity*

Always-on connectivity is required to provide push services, like push e-mail and push-to-talk, from the server to the terminal. The always-on capability also improves the end user performance of interactive services by avoiding setup delays. It is not practical to maintain the physical radio connection all of the time, but the network must be able to maintain a logical connection without physical resource allocation.

An always-on application can preserve its transport layer connections if it can keep the same Internet Protocol (IP) address despite its mobility. The IP address in a Universal Mobile Telecommunications System (UMTS) network is allocated as part of the Packet Data Protocol (PDP) context during GPRS session management [3]. Mobility management must be maintained in the Serving GPRS Support Node (SGSN) which knows the location of the mobile with routing area accuracy. Mobility information can also be maintained in the radio network controller (RNC) which may know the location of the mobile with cell or with UTRAN registration area (URA) accuracy. The mobility management connection between the user equipment (UE) and the SGSN is created by means of General Packet Radio Service (GPRS) attach signalling [3]. The radio resource management connection to the RNC is obtained by radio resource control (RRC) connection setup signalling. The different levels of connectivity are illustrated in Figure 9.5.

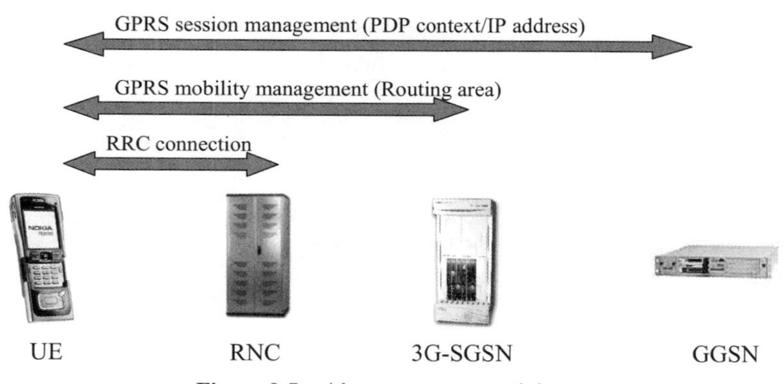


**Figure 9.5** Always-on connectivity.

When a UE is RRC-connected to an RNC, it can utilize four different RRC states:

- Cell_DCH state – where the UE can make use of high-speed downlink/uplink packet access (HSDPA/HSUPA) data rates. When there is no activity or very little activity, the UE is moved to Cell_FACH state.
- Cell_FACH state – where the UE can access the random/forward access (RACH/FACH) common channels and can transmit small amounts of data (up to a few hundred bytes). If there are more data to be transmitted, the UE is moved to Cell_DCH state. If there is no activity, the UE is moved to Cell_PCH state or to URA_PCH state.
- Cell_PCH state – where the UE only receives paging messages, but cannot transmit any data before moving to Cell_FACH state.
- URA_PCH state is similar to Cell_PCH state but the location of the UE is known only at the URA level while in Cell_PCH state the location is known at the cell level. State transition from Cell_PCH to URA_PCH state can be triggered based on UE mobility.

RRC state transitions are illustrated in Figure 9.6.

When a UE is RRC idle, the Iu connection between the RNC and SGSN is released and the RNC is unaware of the UE. If data arrive for the UE, the SGSN pages the UE and an RRC connection is created.

RRC states CELL_FACH, CELL_PCH, and URA_PCH are only applicable to packet-switched services. Circuit-switched services are limited to using RRC idle and the CELL_DCH state.

PCH states are useful in terms of saving network resources and UE power consumption compared with Cell_DCH state. An indicative power consumption in different RRC states is illustrated in Figure 9.7.

Cell_DCH state requires the largest amount of power since both the UE receiver and UE transmitter are running continuously. With a typical power consumption in the order of 200–300 mA and battery capacity of 1000 mAh, talk-times can be 3–5 hours.

In Cell_FACH state the UE transmitter runs only when it has data to be sent. The UE can consume less power at the transmitter than in Cell_DCH state but the receiver chain

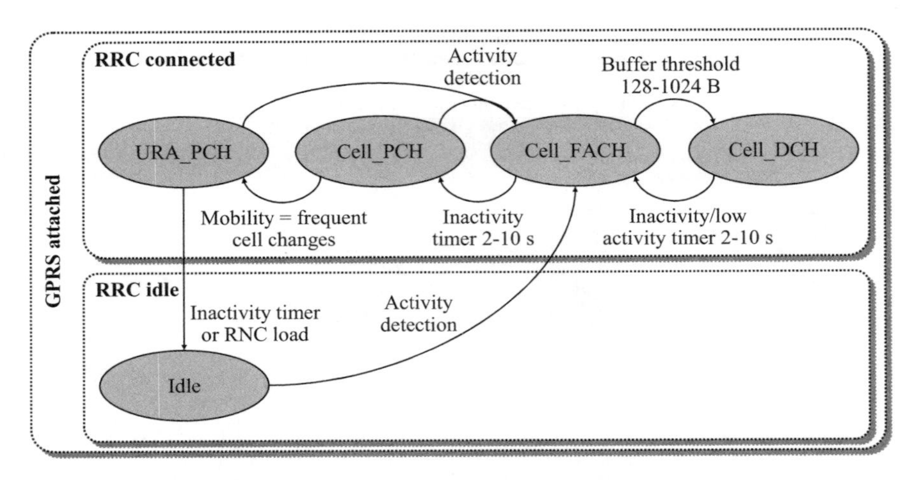


**Figure 9.6** RRC states for packet-switched services.

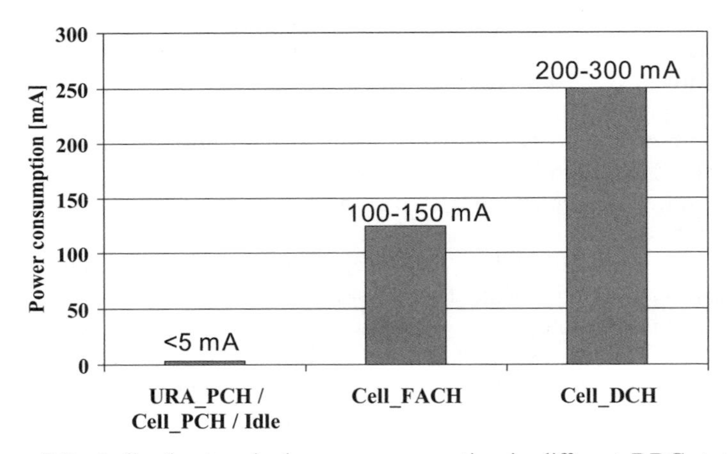


**Figure 9.7** Indicative terminal power consumption in different RRC states.

must be on in order to decode any FACH messages. Therefore, power consumption in Cell_FACH state is still relatively higher than in paging channel (PCH) states where the UE only receives paging messages and can utilize sleep mode. The UE should be moved relatively quickly to PCH state when the data transmission is over in order to minimize UE power consumption.

RRC state changes are controlled by the operator using RNC parameters. A UE cannot decide to make state changes itself.

When a UE goes beyond the wideband code division multiple access (WCDMA)/ HSPA coverage area during packet data transmission, the RNC will command the UE to make an inter-system cell change to a GPRS cell. The IP address is preserved despite the cell change, and any active packet applications remain connected since the same Gateway GPRS Support Node (GGSN) is used both for HSPA and for GPRS. The inter-system

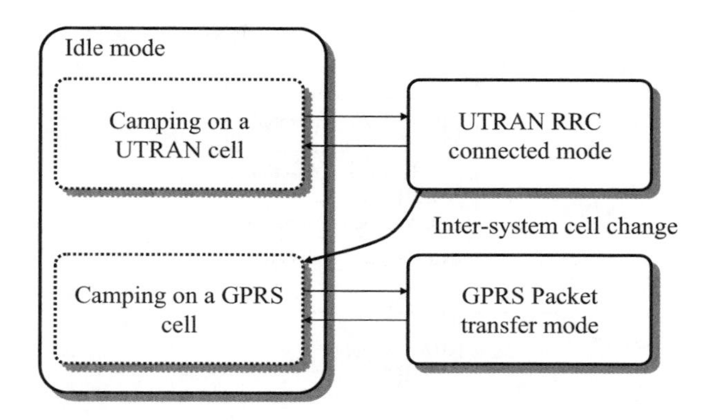


**Figure 9.8** Inter-system cell change from HSPA to GPRS.

cell change from UTRAN RRC connected mode to a GPRS cell is illustrated in Figure 9.8.

A detailed description of signalling related to the packet session setup, RRC state change, inter-system cell change, and analysis of the associated delays are given in the following sections.

### *9.2.2 Packet session setup*

This section, first, presents the signalling required for packet session setup and, second, shows measurement results from a commercial WCDMA network illustrating typical setup delays.

The packet session establishment procedure can be divided into the following phases:

1. Radio resource connection (RRC) setup.
2. GPRS mobility management = GPRS attach.
3. GPRS session management = PDP context activation.
4. Radio resource allocation.

The following figures describe the signalling required for each step. The RNC and UE signal one another using the RRC protocol specified by 3GPP in [4]. The RNC and Node B signal one another using the Node B Application Part (NBAP) protocol specified by 3GPP in [5]. The RNC and core network signal one another using the Radio Access Network Application Part (RANAP), which is specified by 3GPP in [6].

The first phase of a mobile-originated packet-switched data session is to establish an RRC connection with the RNC. From the UE perspective, RRC connection establishment involves three RRC messages and air–interface synchronization. From the network perspective, RRC establishment involves RRC, NBAP, and Access Link Control Application Part (ALCAP) signalling as well as air–interface synchronization. The signalling used to establish an RRC connection is presented in Figure 9.9.

The UE sends an RRC Connection Request message to the RNC using the RACH transport channel which is encapsulated by the physical random access channel

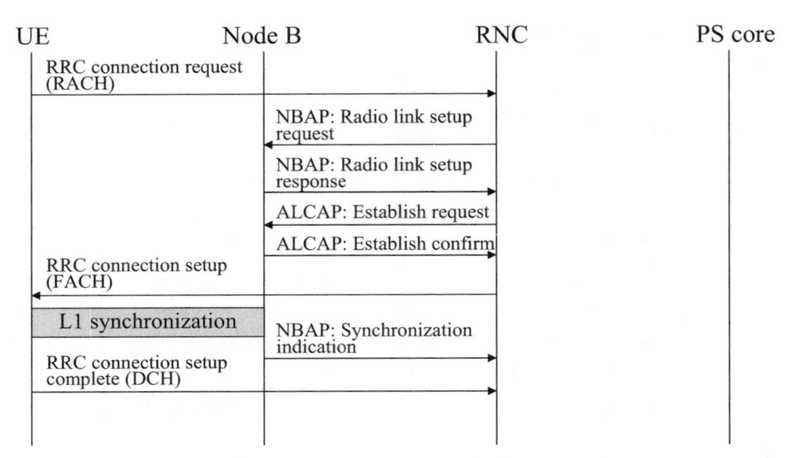


**Figure 9.9** Radio resource control (RRC) connection setup.

(PRACH). Transmission on the PRACH physical channel must be preceded by the transmission of uplink PRACH power control pre-ambles and their subsequent acknowledgement by the Node B. Once the PRACH power control pre-ambles have been acknowledged, the RRC Connection Request message can be sent. The RRC Connection Request message is relatively small and requires only a single transport block which is typically transmitted using a 20-ms transmission time interval (TTI) and transparent mode radio link control (RLC). The RRC Connection Request message can be retransmitted if necessary based upon 3GPP parameters N300 and T300 [5]. These parameters are broadcast to the UE in system information block 1.

Once the RNC has received the RRC Connection Request message it uses NBAP signalling to request a radio link at the relevant Node B. The Node B starts to transmit a dedicated physical control channel (DPCCH) for the new radio link. The transmit power is based upon a downlink open loop power control calculation. The RNC then reserves an associated set of Iub transmission resources using the ALCAP Establish Request and Establish Confirm messages. The RNC responds to the UE's RRC Connection Request message using an RRC Connection Setup message. This message is transmitted using the FACH transport channel. The RRC Connection Setup message is relatively large and typically requires seven transport blocks. These transport blocks have a size of 168 bits and are transmitted using a 10-ms TTI and unacknowledged mode RLC. The transport block set size is typically defined such that a maximum of two transport blocks can be sent per 10-ms TTI. If the UE is in an area of poor downlink coverage, then it is possible that it does not receive one or more of the RRC Connection Setup message transport blocks. In this case the RNC is required to retransmit the complete set of transport blocks from layer 3. The parameters defining retransmission of the RRC Connection Setup message are not specified by 3GPP. The RNC databuild may include implementation-dependent parameters defining the period between retransmissions and the maximum allowed number of retransmissions.

Once the UE has received the RRC Connection Setup message, it attempts to achieve air–interface synchronization using the DPCCH transmitted by the Node B. The rate at which a UE achieves air–interface synchronization depends on the initial DPCCH power level and on the 3GPP specified parameters T312 and N312 [4]. Layer 1 of the UE must generate N312 in-sync primitives within a time period defined by T312. A typical value for N312 is 4 and a typical value for T312 is 6 sec. Once the UE achieves air–interface synchronization, it starts to transmit the uplink DPCCH. This allows the Node B to achieve air–interface synchronization.

Once the Node B has achieved air–interface synchronization, it informs the RNC using an NBAP Synchronization Indication message. Meanwhile, the UE responds to the RRC Connection Setup message using an RRC Connection Setup Complete message. This message typically requires two 148-bit transport blocks. There are two possible signalling radio bearer bit rates: 3.4 kbps with a 40-ms TTI and 13.6 kbps with a 10-ms TTI. The RNC databuild typically includes a parameter to define the choice of bit rate. The 13.6-kbps bit rate allows faster call setup but also has a greater Iub transmission resource requirement. If the 13.6-kbps signalling bit rate is used, then the bit rate is reduced to 3.4 kbps once call setup has been completed in order to save network resources.

The RRC Connection Setup Complete message and all subsequent RRC signalling are transmitted using acknowledged mode RLC and any necessary retransmissions can be completed relatively rapidly at layer 2. At this stage the UE has established an RRC connection.

In the above example, the RRC connection setup used a dedicated channel (DCH). Another option is to run the RRC connection setup (including the RRC Connection Setup Complete message) on the RACH/FACH. In that case the subsequent GPRS mobility and session management signalling would take place on the RACH/FACH as well. DCH allocation including base station and Iub resource reservation would then take place during the radio bearer reconfiguration phase.

The second phase of a mobile-originated packet-switched data session involves GPRS mobility management signalling with the core network. This signalling is illustrated in Figure 9.10.

The UE sends a GPRS Attach Request message to the packet-switched core network via the RNC. The RNC relays the contents of the message to the core network using a RANAP Initial UE Message. The RANAP message is combined with a Signalling Connection Control Part (SCCP) Connection Request message which is used to request Iu signalling resources. The core network replies with a Connection Confirm message to

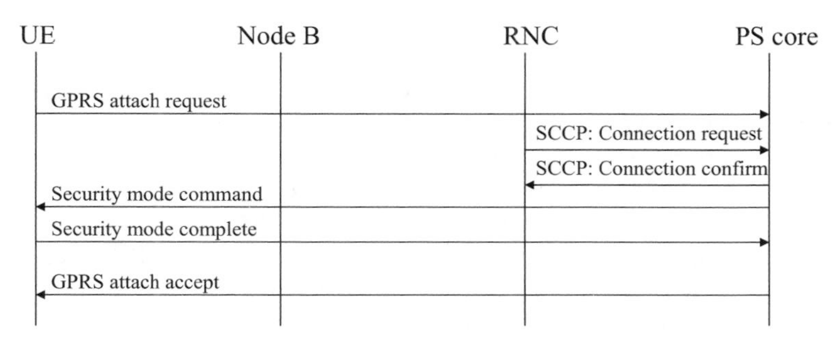


**Figure 9.10** GPRS mobility management.

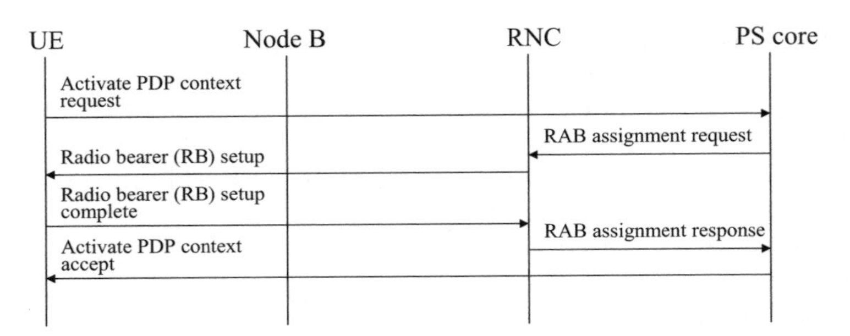


**Figure 9.11** GPRS session management.

acknowledge that a signalling connection has been established across the Iu. The core network then completes the security mode procedure. There may also be a requirement to complete the authentication and ciphering procedure. It is usually possible to configure whether the authentication and ciphering procedure is required for all connections or only a specific percentage of connections. Once the security mode procedure is complete, the core network is able to accept the GPRS Attach Request and the UE becomes registered for packet-switched services.

The third phase of a mobile-originated packet-switched data session involves GPRS session management signalling and radio access bearer (RAB) assignment. The signalling is illustrated in Figure 9.11.

The UE starts by sending the core network a PDP Context Request message. This message may include a specification of the quality of service requirements. Alternatively, the message may specify that the subscribed quality of service requirements defined for that UE within the HLR should be applied. The message also requests an IP address and specifies the access point name (APN) to which the UE wishes to connect. The core network then sends a RAB Assignment Request to the RNC. The RNC runs its admission control algorithm and sends a Radio Bearer Setup message to the UE which defines a set of physical, transport, and logical channel configurations. Some RNC implementations may assign a finite user plane bit rate at this stage (i.e., before the RNC has any knowledge of the quantity of traffic to be transmitted). Alternatively, the RNC assigns a zero bit rate to the user plane at this stage. The RNC then waits for a Capacity Request prior to assigning an appropriate bit rate. This approach helps to avoid assigning high bit rates when transferring relatively small quantities of data. The UE responds with the Radio Bearer Setup Complete message and the RNC informs the core network using the RAB Assignment Response message. At this stage the UE has established a radio link to the Node B, a radio bearer to the RNC and a RAB to the core network. The core network is then able to complete this phase by sending the Activate PDP Context Accept message. This message includes the quality of service attributes as well as the IP address that has been assigned to the UE.

The fourth phase of a mobile-originated packet-switched data session involves radio bearer and radio link reconfiguration. The signalling is illustrated in Figure 9.12.

The RNC starts the procedure by sending an RRC Measurement Control message to the UE. This message defines a traffic volume threshold which can be used to trigger an uplink Capacity Request according to 3GPP measurement reporting event 4a. The RNC

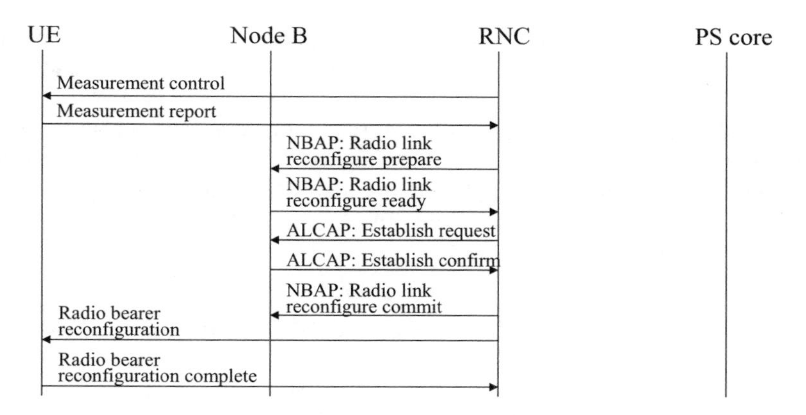


**Figure 9.12** Radio bearer reconfiguration for resource allocation.

has a similar threshold for the downlink allowing it to generate a downlink capacity request. Figure 9.12 illustrates the case of an uplink data transfer where the capacity request is triggered by the UE sending a measurement report. The measurement report includes information regarding the quantity of data waiting to be transmitted. This information helps the RNC to assign an appropriate bit rate. The RNC generates a capacity request from the measurement report and, assuming the capacity request is accepted, the RNC sends an NBAP Radio Link Reconfiguration Prepare message to the Node B. This message tells the Node B how its existing radio link will need to be reconfigured to be capable of supporting the packet-switched data session. If the 13.6-kbps signalling bit rate has been used, the Radio Link Reconfiguration Prepare message also lets the Node B know how this bit rate will have to be reduced when the packet-switched bearer becomes active. The Node B responds with the NBAP Radio Link Reconfiguration Ready message, but does so before applying the reconfiguration. A second set of Iub resources are reserved by the RNC using the same approach as during the first phase of signalling (i.e., using ALCAP Establish Request and Establish Confirm messages). Once the Iub resources have been successfully reserved, the RNC sends the NBAP Radio Link Reconfiguration Commit message to the Node B. This message informs the Node B of the connection frame number (CFN) to use so that it can start to apply the radio link configuration which includes the packet-switched radio bearer. The CFN must be defined such that the UE is able to apply the new configuration at the same time as the Node B. The UE is informed of the CFN within the subsequent Radio Bearer Reconfiguration message. This means that the CFN must be defined such that it occurs after the UE has received the Radio Bearer Reconfiguration message. The value of the CFN should thus be a function of the signalling bearer bit rate: a shorter activation time can be used with the 13.6-kbps than with the 3.4-kbps signalling bit rate. Some margin should also be allowed for layer 2 retransmissions and processing time. It is common for the RNC databuild to include a parameter that defines a configurable time offset for the CFN.

Once the CFN occurs and the new configuration has become active, the UE responds with the Radio Bearer Reconfiguration Complete message. This message and all subsequent signalling messages are transmitted across the air–interface using the 3.4-kbps bit rate even if the 13.6-kbps bit rate had been applied previously. The UE is now configured with a finite user plane bit rate towards the packet-switched core network and is ready to transfer data.

A set of packet session setup times as measured from a commercial WCDMA network are illustrated in Figure 9.13 [7]. Three case are shown: Case 1 when the mobile is not GPRS-attached, Case 2 when the mobile is already GPRS-attached, and Case 3 when the mobile has a PDP context already available, but no RRC connection. The required signalling procedures are as follows:

- Case 1 = RRC connection + GPRS attach + PDP context activation + Radio bearer setup.
- Case 2 = RRC connection + PDP context activation + Radio bearer setup.
- Case 3 = RRC connection + Radio bearer setup.

The total setup time including radio bearer reconfiguration is 3.5–4.0 sec for Case 1. When the UE is already GPRS-attached in Case 2, the setup delay is less than 2 sec with a 13.6-kbps signalling rate, and less than 2.5 sec if radio resource allocation is included. A detailed signalling delay break-down for Case 2 is shown in Table 9.1. If the PDP context is already active, the setup time is approximately 1 sec with 13.6 kbps plus the radio bearer reconfiguration delay.

If the UE already has an RRC connection, only the RRC state change to CELL_DCH may be required. The UE may already have an RRC connection due to an existing circuit-switched call or due to another packet-switched connection. If there is an ongoing

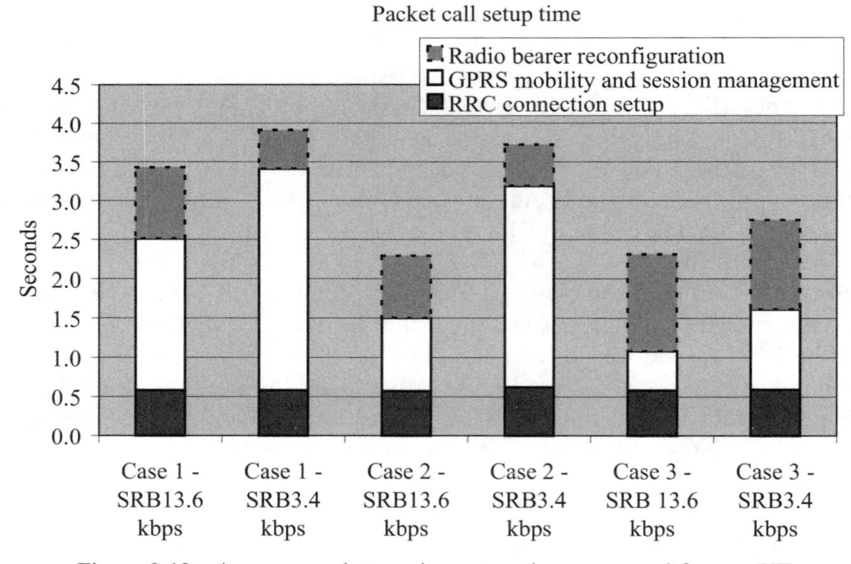


**Figure 9.13** Average packet session setup time measured from a UE.

**Table 9.1** Example packet session setup time recorded by a UE that is already GPRS-attached.

| | *Signalling radio bearer* | |
|---|---|---|
| | *3.4 kbps* (ms) | *13.6 kbps* (ms) |
| RRC Connection Request (uplink) | 0 | 0 |
| RRC Connection Setup (downlink) | 410 | 420 |
| RRC Connection Setup Complete (uplink) | 610 | 600 |
| Service Request (uplink) | 650 | 640 |
| Security Mode Command (downlink) | 1130 | 860 |
| Security Mode Complete (uplink) | 1140 | 860 |
| Activate PDP Context Request (uplink) | 1150 | 870 |
| Radio Bearer Setup (downlink) | 2060 | 1320 |
| Radio Bearer Setup Complete (uplink) | 2100 | 1360 |
| Activate PDP Context Accept (downlink) | 2690 | 1580 |
| Radio Bearer Reconfiguration (downlink) | 2940 | 1860 |
| Radio Bearer Reconfiguration Complete (uplink) | 3130 | 2280 |

circuit-switched call then the UE will be in CELL_DCH and will not require an RRC state change. RRC state changes are analysed in Section 9.2.3.

The RAN under investigation was configured for these measurements so that radio resources were allocated based only upon measurement reports indicating whether there were data in either the uplink or downlink buffers. Radio bearer reconfiguration takes place based upon the measurement report. Alternatively, the RAN can be configured so that radio resources are allocated directly after PDP context activation.

Packet session setup time can be optimised by refining the values assigned to specific parameters and by refining the way in which resources are allocated. The RACH procedure requires the Node B to acknowledge PRACH pre-ambles. An acknowledgement is sent once the Node B has successfully received a pre-amble. PRACH power control parameters are controlled by the RNC. If pre-ambles are transmitted with greater power then the Node B is more likely to receive them. However, if pre-ambles are transmitted with too much power then the Node B receiver will experience increased uplink interference. The time spent ramping up the transmit power of the pre-ambles can be reduced if the pre-amble step size is increased and if the initial pre-amble transmit power is configured to be relatively large.

The rate at which uplink and downlink synchronization is achieved across the air–interface depends upon the 3GPP parameters N312 and T312. If N312 is configured with a low value then it becomes easier for the UE to move into synchronized state (i.e., fewer in-sync primitives are required). This means that synchronization can be achieved more rapidly. The drawback is that the synchronization decision may be less reliable. The same argument can be applied to the equivalent uplink parameters at the Node B.

The rate at which the NBAP and AAL2 signalling procedures are completed is affected by the Iub bandwidth allocated to NBAP and AAL2. Separate virtual channel connections (VCCs) are typically assigned for common NBAP signalling, dedicated NBAP signalling, and AAL2 signalling. If the bandwidth assigned to each of these VCCs is increased then the rate at which each set of procedures can be completed increases. The drawback of assigning larger bandwidths is an increased Iub capacity requirement.

The activation time offset which is used to define the CFN at which the synchronized reconfigurations become active – for example, the radio link reconfiguration procedure during radio bearer reconfiguration – also has some room for optimization. This offset should be made as small as possible to reduce call setup delay. Making it too small means that the Node B may start to apply the new configuration before the UE has received the corresponding instruction. The offset should account for the time consumed by sending the RRC message to the UE and should also account for the possible necessity of completing RLC retransmissions. The time consumed by sending the RRC message is a function of the message size and also the signalling radio bearer bit rate. The activation time offset should thus be a function of these two variables. The time required for RLC retransmissions depends upon the bit rate and also the RLC polling time – that is, how long it takes the UE to inform the RNC that a retransmission is required. If the RLC polling time can be reduced, then the activation time offset can also be reduced. It is typical to allow for two RLC retransmissions when evaluating an appropriate offset.

3GPP Release 6 allows use of HSDPA and HSUPA for carrying the signalling radio bearer, which allows a faster call setup time. The faster setup time is achieved, first, because the signalling speed over HSDPA/HSUPA is fast and, second, because no activation time is required. 3GPP Release 6 can use an asynchronous procedure for setup and avoid the activation time since the physical channel configuration is not changed in the radio bearer setup. This is the case since both signalling and the user plane are carried over the same HSDPA/HSUPA channel. 3GPP Release 6 allows the packet session setup time to be reduced to below 1 sec. 3GPP Release 7 includes a work item to further reduce setup times.

### *9.2.3 RRC state change*

After a period of inactivity a UE is typically moved from Cell_DCH to Cell_PCH or URA_PCH state to save UE battery life and to save radio network resources. When data are present in the UE buffer ready to be sent, the UE is moved to Cell_FACH or to Cell_DCH state depending on the amount of data in the buffer. This section presents the signalling flow when a UE is first moved from Cell_PCH to Cell_FACH, and then from Cell_FACH to Cell_DCH state.

When a UE initiates, for example, a file download, it first has to establish a TCP connection with the server. TCP connection establishment involves small packets that can be carried by the RACH/FACH. The state transition from Cell_PCH to Cell_FACH is illustrated in Figure 9.14. Before the UE is allowed to send the RACH message, it has to obtain the necessary information from the system information block that is transmitted on the downlink common channel (i.e., the BCCH). Next, the UE starts with RACH pre-ambles followed by a Cell Update in the RACH message part. The RNC responds with a Cell Update Confirm to move the UE to CELL_FACH state. Cell Update Confirm also includes UTRAN mobility information, which allocates a new radio network temporary identifier (RNTI). Cell-RNTI is used as a UE identifier in Cell_FACH state. The UE acknowledges reception of that message with a UTRAN

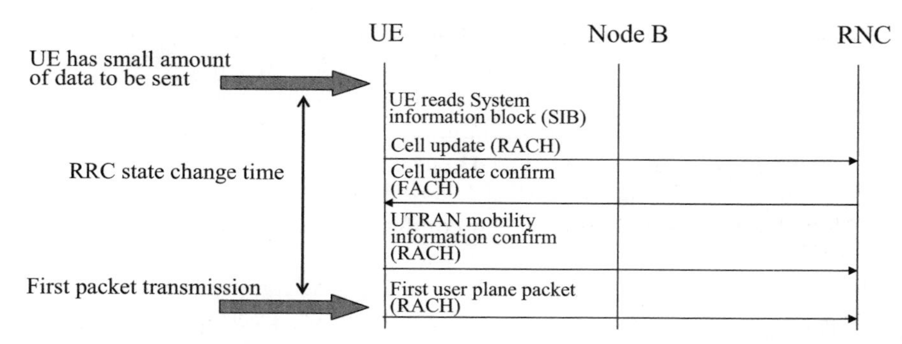


**Figure 9.14** RRC state change from PCH to FACH.

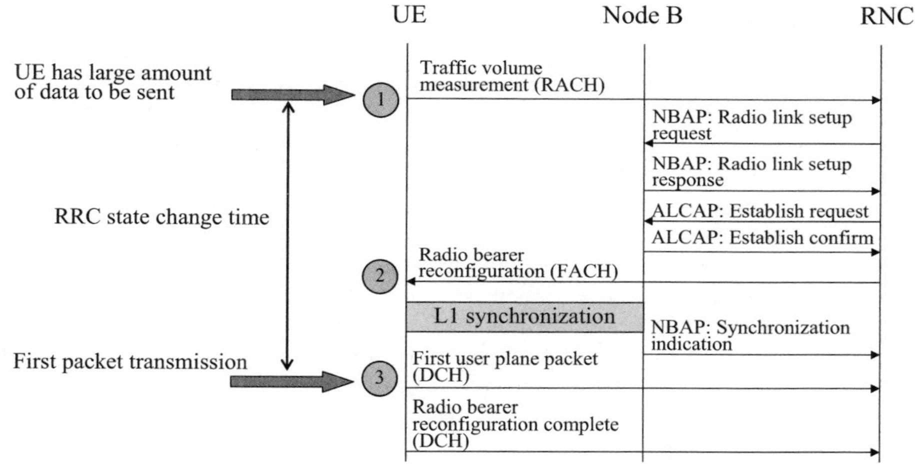


**Figure 9.15** RRC state change from FACH to DCH: ① = traffic volume measurement indicates the amount of data the UE has in its buffer; ② = radio bearer reconfiguration commands the UE to Cell_DCH state; and ③ = first user plane packet transmission.

Mobility Information Confirm. The UE is now ready to send user data on the RACH channel.

Once the UE is in Cell_FACH state and there is a large amount of data to be transmitted – such as large TCP packets – the UE is moved to Cell_DCH state. The state transition from Cell_FACH to Cell_DCH is illustrated in Figure 9.15. The UE starts with RACH pre-ambles followed by a RACH message part carrying a traffic volume measurement. The measurement report informs the RNC about the amount of data the UE has in its transmission buffer. If the state transition is caused by downlink data, the traffic volume report is generated internally within the RNC. Once the RNC has decided to allocate a DCH or the HS-DSCH for the UE, it has to reserve Node B and Iub transport resources with the Radio Link Setup Request and Establish Request messages. When the reservations are acknowledged, the RNC sends a Radio Bearer Reconfiguration message on the FACH informing the UE about the DCH parameters. The Node B starts downlink DPCCH transmission when it has received the Radio Link Setup

**Table 9.2** Example state transition from FACH to DCH recorded by UE.

| | *Message* (ms) |
|---|---|
| Traffic volume measurement (RACH uplink) | 0 |
| Radio bearer reconfiguration (FACH downlink) | 340 |
| Radio bearer reconfiguration complete (DCH uplink) | 540 |

Request. The UE starts the L1 synchronization procedure when it has received the Radio Bearer Reconfiguration message. Once the UE is synchronized to the downlink signal, it starts uplink transmission of the DPCCH. The UE is allowed to start the transmission of uplink user data on the DCH after a pre-defined period during which the Node B has acquired uplink synchronization.

An example measured RRC state transition from Cell_FACH to Cell_DCH is shown in Table 9.2. When the amount of data in the UE buffer exceeds the DCH allocation threshold, the UE sends a Traffic Volume Measurement, which is defined as the starting point in this UE log. The UE receives the Radio Bearer Reconfiguration from the RNC 340 ms later. The delay is explained by message transmission times and by radio network delay for the reservation of base station, transmission, and RNC resources for the new DCH. When the UE has received the reconfiguration message and has obtained downlink synchronization, it can start uplink transmission as soon as the pre-defined uplink timers have elapsed. In this case the Radio Bearer Reconfiguration Complete message was sent 200 ms after reception of the downlink reconfiguration message. The total delay in the transition from Cell_FACH to Cell_DCH was approximately 0.5 sec in this example.

The transition from Cell_PCH to Cell_DCH takes more time than from Cell_FACH to Cell_DCH in 3GPP Release 5 since the cell update procedure is required in addition. 3GPP Release 6 allows combination of the traffic volume measurement and the cell update in a single message, thus reducing the delay.

### *9.2.4 Inter-system cell change from HSPA to GPRS/EGPRS*

This section analyses the signalling procedures and the delay in cell change from HSPA to GPRS during data transmission. The procedure is shown in Figure 9.16. Once the RNC observes that HSPA coverage becomes weak, it requests the UE to make measurements of potential second-generation (2G) target cells. Once the best 2G target cell is identified, the RNC instructs the UE to move onto a specific 2G cell using the Cell Change Order message. This message also specifies the CFN at which the UE should make the transition. Once the UE has moved onto the 2G cell, it must read the 2G system information messages including routing and location area codes and the network mode of operation. The network mode of operation defines whether or not combined location area and routing area updates are possible. The combined location and routing area update mode of operation requires a Gs interface between the Mobile Switching Centre (MSC) and the 2G SGSN. The signalling presented in Figure 9.8 assumes network mode of operation II – that is, separate location area and routing area updates. The UE sends a

**Table 9.3** Example inter-system cell change recorded by a UE for a UDP-based service.

| *Message* | *Delay (relative to cell change order)* (ms) |
|---|---|
| *Last UDP packet sent on 3G* | *−20* |
| 3G Cell Change Order from UTRAN | 0 |
| 2G Channel Request | 1 720 |
| 2G Immediate Assignment | 1 840 |
| 2G Location Updating Request | 1 940 |
| 2G Classmark Change | 2 180 |
| 2G Authentication Request | 2 440 |
| 2G UTRAN Classmark Change | 2 890 |
| 2G Authentication Response | 3 120 |
| 2G Ciphering Mode Command | 3 320 |
| 2G Ciphering Mode Complete | 3 590 |
| 2G TMSI Reallocation Command | 3 840 |
| 2G TMSI Reallocation Complete | 4 060 |
| 2G Identity Request | 4 260 |
| 2G Identity Response | 4 530 |
| 2G Location Updating Accept | 5 210 |
| 2G Channel Release | 5 440 |
| 2G Routing Area Update Request | 7 590 |
| 2G Channel Request | 8 050 |
| 2G Immediate Assignment | 8 150 |
| 2G Authorization and Ciphering Request | 9 770 |
| 2G Authorization and Ciphering Response | 10 010 |
| *First UDP Packet sent on 2G* | *10 720* |
| 2G Routing Area Update Accept | 12 330 |
| 2G Routing Area Update Complete | 12 340 |

Routing Area Update Request to the 2G SGSN. This triggers the Serving Radio Network Subsystem (SRNS) context transfer procedure. The SRNS Context Request message informs the RNC to start buffering and not to send any further downlink data to the Node B. The procedure is completed by the 2G SGSN sending the SGSN Context Acknowledge message. This message indicates that the 2G SGSN is now ready to start receiving data belonging to the UE's packet-switched connection. The 3G SGSN then sends the SRNS Data Forwarding Command to the RNC. This message instructs the RNC to start tunnelling data towards the 2G SGSN. The 2G SGSN forwards the data to the UE via a 2G radio network. The UE location is then updated so that data are sent directly from the GGSN to the 2G SGSN. The 2G SGSN sends a Routing Area Update Accept message and the UE confirms this with an associated Routing Area Update Complete message.

Example inter-system cell change signalling is shown in Table 9.3 [8]. Once the UE was moved to GPRS it took 1.7 sec to read the 2G system information and initiate the location area update procedure. The location area update procedure takes about 3.5 sec.

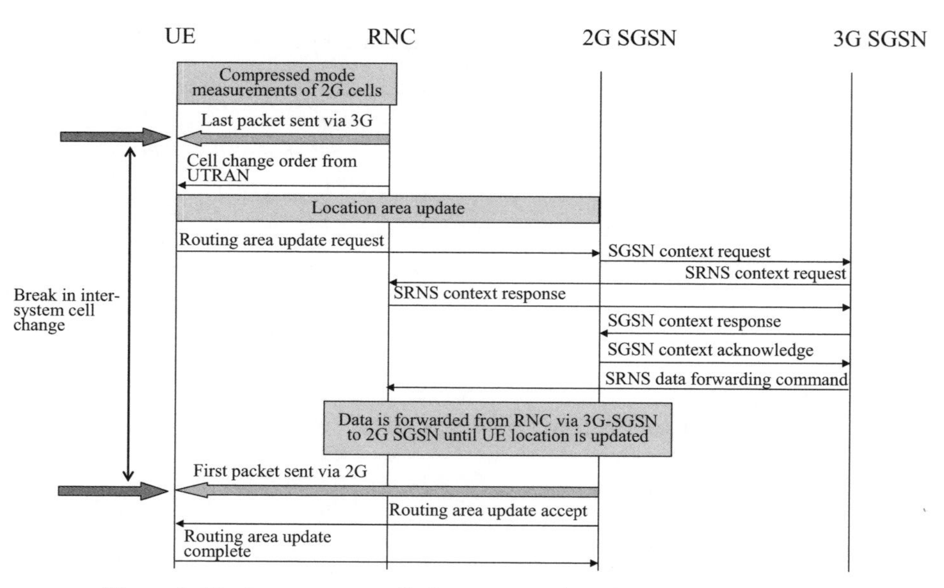


**Figure 9.16** Inter-system cell change procedure from HSPA to GPRS.

The UE then releases its channel and initiates a routing area update. The routing area update together with UE-related delays take 7 sec. The first User Datagram Protocol (UDP) packet is received on the 2G system before the routing area update procedure is complete. User plane data are sent as soon as the UE's location is updated within the packet-switched core network. The overall inter-system handover delay is 12.3 sec in this example when measured the Layer 3 signalling and 10.7 sec when measured from the end-to-end transport layer in the case of UDP.

Figure 9.17 illustrates these delays graphically.

The inter-system handover delay presented above is for a UDP-based service, such as streaming. The delay between the last item of user plane data sent on the 3G system and the first item of user plane data sent on the 2G system can be greater for a TCP-based service due to Transmission Control Protocol (TCP) retransmission timeout (RTO). The RTO defines the time interval for unacknowledged TCP segment retransmissions. It is

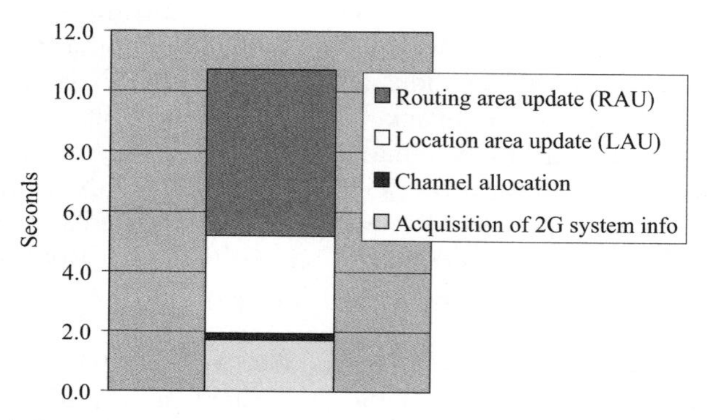


**Figure 9.17** Delay split for inter-system cell change from HSPA to GPRS.

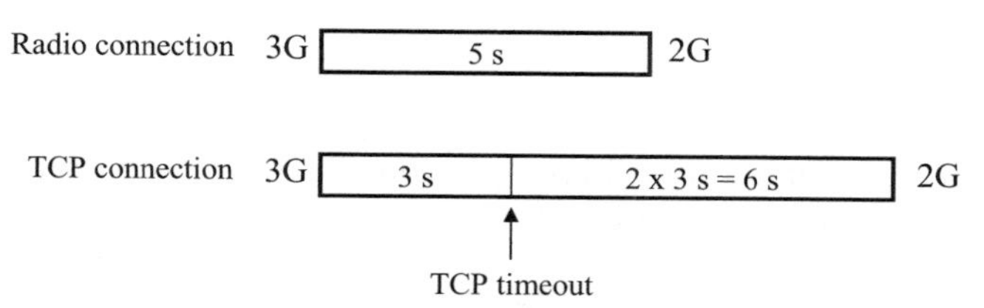


**Figure 9.18** TCP timeout during inter-system cell change.

updated dynamically based upon the estimated TCP round trip time (RTT). If the RTO expires, then the earliest unacknowledged TCP packet is retransmitted and the value of the RTO is doubled. The RTO is likely to expire during an inter-system cell change when the break is as long as in Table 9.3. If the RTO expires just before the data connection is re-established, the receiver will have to wait for the RTO to expire again before the TCP sender re-transmits any data. Figure 9.18 shows an example with a 5-sec break in the radio connection. The example assumes that the RTO value is 3 sec before the cell change. The RTO timer depends on the TCP parameters and on the buffering in the 3G network. TCP timeout takes place after 3 sec and the TCP data transfer continues after a further 9 sec even if the 2G radio connection is already available after 5 sec.

If the inter-system cell change delay is large, the RTO is likely to expire more often, causing more significant delays at the TCP layer. Reducing inter-system handover delays helps to minimize the impact of the TCP timeout. TCP performance can also be improved by forwarding any remaining packets from the RNC to the 2G SGSN to avoid any packet losses and to help the TCP layer to recover from the break in data transfer.

There are a number of ways to minimize an inter-system break. Introducing the Gs interface between the MSC and the 2G SGSN allows the use of combined location area and routing area updates. Measurements have shown that this reduces inter-system handover delay by approximately 5 sec. This represents a significant percentage of the overall inter-system handover delay.

The time consumed by the UE in reading 2G system information also represents a considerable part of inter-system handover delay. The delay can be minimized by broadcasting the relevant system information more often in a 2G system. 3GPP Releases 5 and 6 introduce inter-system network-assisted cell change (NACC). In this case the UE is able to acquire 2G system information while connected to the 3G system. Inter-system NACC together with combined RAU/LAU can push an inter-system cell change break to below 3 sec.

3GPP Release 6 also defines a packet-switched inter-system handover that allows the connection break to be reduced to less than 0.5 sec, thus making the transition from HSPA to enhanced data rate for global evolution (EDGE) practically seamless for packet applications.

## 9.3 Application performance over HSPA

This section presents the performance of a few applications over HSPA radio networks. The following are considered here: web browsing, TCP performance in general, push-to-talk and Voice-over-IP (VoIP), real time gaming, mobile TV streaming and push e-mail.

### 9.3.1 *Web browsing*

End users are increasingly accustomed to broadband web browsing performance through fixed Internet connections. In comparison, the browsing performance over early deployments of cellular systems supporting packet-switched services has suffered from limited throughput and rather significant latencies. Due to the interactive nature of web browsing, latency plays a particularly important role in determining perceived end user performance. This is particularly true for user-to-network interface link speeds above a few hundred kilobits per second, above which the effective end-to-end throughput seen by the end user will be limited by other network links along the path. In comparison, the effect of link RTT is cumulative over an end-to-end path. Thus, it is clearly the case that it is link RTT, rather than link speed, that becomes a performance limiting factor.

The role of RTT on perceived link performance can be verified using the network setup illustrated in Figure 9.19, where a network emulator is used to induce extrinsic delay on the end-to-end communication link. This setup can be used as a platform for carrying out subjective degradation category rating tests as defined in [2]. These tests are used for judgements of pairs of links, where one of the links does not bear any extrinsic delay and the other has been degraded by inducing some extrinsic delay. The links are compared and the degradation of the manipulated link is rated on a scale from 1 to 5, where 1 corresponds to very annoying and 5 corresponds to unnoticeable degradation.

Figure 9.20 shows the results obtained from a degradation category rating test carried out over a 1-Mbps/384-kbps link, emulating the typical throughput that can be expected

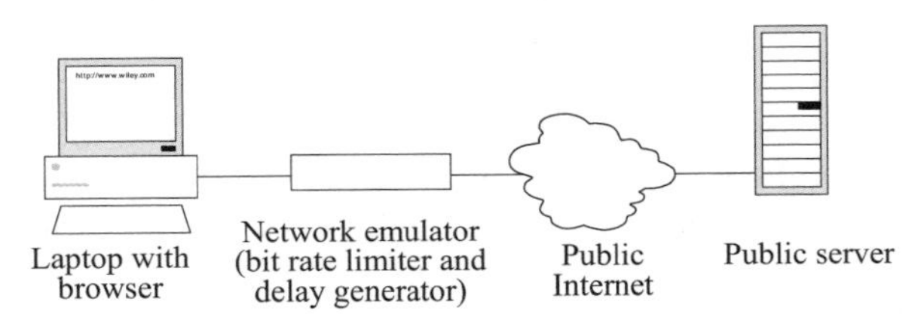


**Figure 9.19** Measurement setup for subjective web browsing measurements.

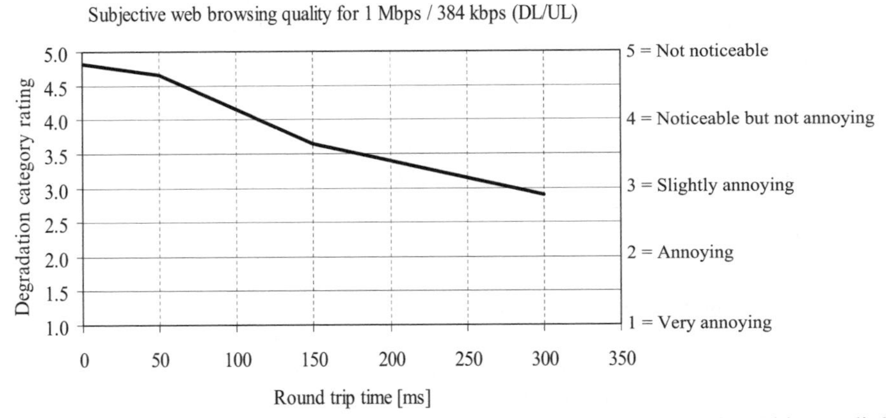


**Figure 9.20** Subjective web browsing quality with 1 Mbps downlink and 384 kbps uplink.

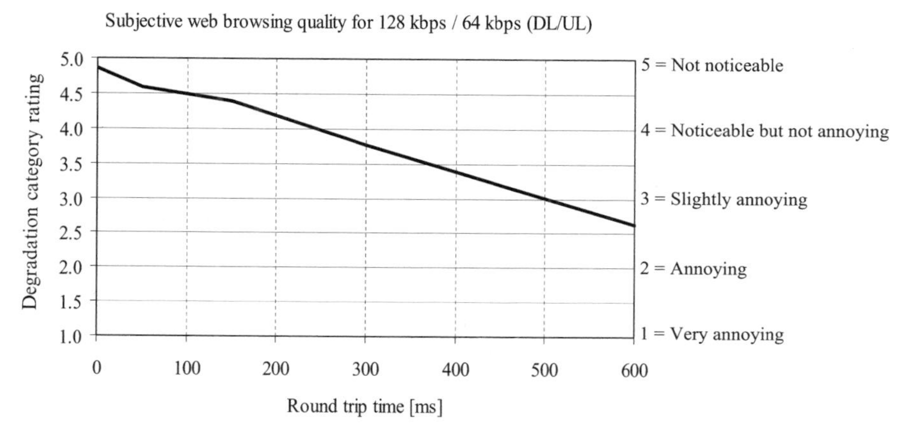


**Figure 9.21** Subjective web browsing quality with 128 kbps downlink and 64 kbps uplink.

from an HSPA link. The results show that – while there is an almost linear improvement in the performance rating as the link RTT is reduced and that at least some users still notice an improvement with RTTs lower than 100 ms – on average, users are not annoyed with RTTs up to 100 ms. In contrast, the degradation caused by RTTs approaching 300 ms tends to be graded as slightly annoying. A well-engineered HSPA access system should contribute fewer than 100 ms to overall end-to-end RTT. The effect of this contribution to end user browsing performance perception can be evaluated from the plot in Figure 9.20.

Figure 9.21 shows the results of the same test carried out on a 128-kbps/64-kbps link, emulating the typical throughput on an EGPRS link or WCDMA in a weak signal area. The results show once again an almost linear performance improvement with an upward offset of about 0.5–1 compared with the results in Figure 9.20. This suggests that low RTT is more important with high link speed than with low speed. Whereas with a 128-kbps/64-kbps link the noticeable, but not annoying, threshold was reached at an RTT of about 240 ms – the same threshold for a 1-Mbps/384-kbps link required approximately half of that value.

### *9.3.2 TCP performance*

TCP is the transport protocol that accounts for the vast majority of services over the Internet. It continuously controls the transmission rate when exchanging data between a sender and a receiver. The target is to fully utilize the available bandwidth but not to congest the network when multiple users share the resources: TCP should not send data faster than the bandwidth of the narrowest link in the transmission path. To cope with this, TCP typically goes through the adaptation states illustrated in Figure 9.22 and described below:

- *Synchronization* – in this phase the sender and receiver negotiate, for example, the sizes of the sender and receiver buffers.

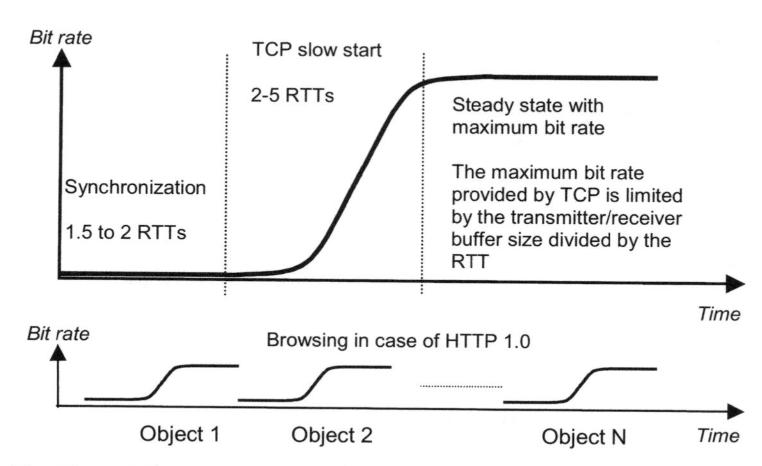


**Figure 9.22** Three different states of TCP/IP connection and the corresponding bit rate.

- *Slow start* – the sender gradually increases the transmission rate when it receives acknowledgements from the receiver.
- *Steady state* – the sender transmits new packets as soon as it has received acknowledgements from the receiver that the receiver is ready to accept more packets.

A short RTT speeds up both synchronization and the slow start because of faster information exchange. This improves end user experience because the shorter the synchronization and slow start phases, the faster the maximum bit rate of the connection can be used. During the steady state the RTT does not affect performance as long as the transmitter buffer size is large enough to keep all the unacknowledged packets 'travelling in the network' in its memory. It takes a couple of RTTs before the steady state bit rate (maximum bit rate) is reached. With RTTs typically less than 100 ms in fixed networks, the end user does not, in general, notice any effects of the TCP slow start. For example, if the RTT is around 100 ms – as is the case in HSDPA – then the slow start takes only around 200 to 500 ms, and thereafter the maximum bit rate is used. Decreasing the RTT further from the HSDPA level is obviously good, but the impact on end user experience is limited for the vast majority of TCP/IP applications. However, in 2G cellular radio networks, where RTTs are around 200–700 ms – and even longer – end user experience may be affected because then the slow start can take a number of seconds. The slow start process is particularly visible to the end user when browsing a web page consisting of multiple objects, each requiring a separate slow start in HTTP1.0 (illustrated in the lower part of Figure 9.22). To cope with this it is possible to use HTTP1.1 instead, where multiple objects may be transferred within the same TCP session.

One additional property of TCP is that it may time out in the event of a break in end-to-end connectivity. In the event of TCP timeout, the TCP sender stops transmitting and attempts to restart the connection from scratch going through the slow start phase one more time. If this happens frequently, it will have adverse effects on end user service perception. A common rule of thumb is that TCP timeout happens when RTT exceeds the average RTT of the connection + twice the variance of the RTT of the connection

[11]. Based on the formula it is unlikely that a TCP timeout would happen during ordinary operation of HSPA because the RTT variations are typically not large, even if the network load varies. TCP timeout, however, may happen during the inter-system cell change from HSDPA to GPRS due to the long break (see the discussion in Section 9.2.4).

### *9.3.3 Full duplex VoIP and Push-to-Talk*

In comparison with many other applications running over IP, the throughput required for full duplex VoIP is low, up to a few tens of kbps, but the latency requirements are, on the other hand, more demanding. Thus, once again, the RTT – and not the link throughput – is the factor that limits end user service performance and network capacity. The ITU recommendation [1] for one-way transmission time for full duplex voice indicates that users are satisfied with a mouth-to-ear delay of up to around 280 ms. For delays longer than 280 ms the interactivity of the voice connection decreases rapidly, and when the mouth-to-ear delay approaches 400 ms many voice users are dissatisfied with the interactivity. Note that the delays mentioned here are mouth-to-ear delays and, hence, include not only the transmission delays but also processing (coding/decoding) delays in the transmitter and in the receiver. The International Telecommunication Union (ITU) recommendation [1] also includes guidelines for coding/decoding delays. For most codecs relevant to the mobile domain these delays are from just below 50 ms up to 100 ms. Leaving room for these processing delays – both at the transmitter and the receiver end – results in end-to-end transmission delay for radio links and a backbone transmission network of well below 200 ms in total. When we compare this delay requirement with RTTs of below 200 ms in WCDMA Release 99 and below 100 ms in HSPA, it is clear that VoIP works well in both technologies. With the shorter RTT in HSPA radio links, the delay allowed in backbone transmission (and Internet) is larger in HSPA than in WCDMA Release 99. However, note that when the radio load increases in HSPA the RTT increases. There is in HSPA, hence, an inherent quality/capacity trade-off when carrying VoIP (this has been studied by simulation investigations in Chapter 10). With the DCHs in WCDMA Release 99 the RTT is, on the other hand, very stable and close to being independent of radio interface load.

For push-to-talk applications the mouth-to-ear delay requirements are less demanding than for full duplex VoIP. On the other hand, these applications set tight requirements on the setup times for radio connection. This is because each time the user requests to talk, the system must set up a radio connection and the time for doing this directly impacts the end user experience for the push-to-talk connection. The Open Mobile Alliance (OMA) push-to-talk requirement [9] specification has defined three quality of end user criteria for delays:

- The 'right-to-speak' response time during push-to-talk session establishment – this is the time it takes from the instant that the user initiates a push-to-talk session until he/she receives feedback (in the form of a beep or similar sound) from the terminal that he/she can start talking. The right-to-speak time should be less than 1.6 to 2 sec depending on the push-to-talk use case.

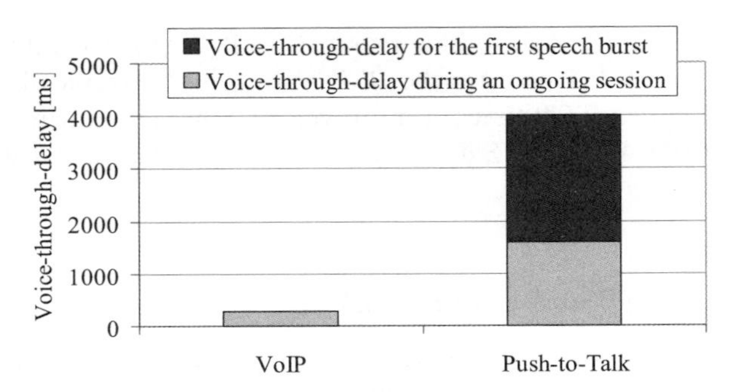


**Figure 9.23** Voice-through-delay requirement for VoIP and push-to-talk.

- The 'start-to-speak' response time after push-to-talk session establishment – this is the time it takes from the instant that the user requests to speak during an ongoing session to when the user receives feedback from the terminal that he/she can start talking. The start-to-speak time should be less than 1.6 s.
- 'End-to-end channel delay' – this is the time from when a user, who has received a right-to-speak indication from the terminal, starts talking until the receiving user hears the speech. End-to-end channel delay should be less than 1.6 sec, but for the first speech burst in a session it may be up to 4 sec.

The voice-through-delay requirements for push-to-talk and for VoIP are illustrated in Figure 9.23. Push-to-talk can be implemented on Release 99 GPRS/EDGE, which provide typical RTTs of 200–700 ms while full duplex VoIP requires a low-latency radio network like WCDMA Release 99 or HSPA.

### *9.3.4 Real time gaming*

There are various categories of network gaming applications, and these categories impose different requirements on the mobile network. The requirements are in terms of radio connection setup times, latency, and battery lifetime. The following are examples of gaming categories [10]:

- real time action games;
- real time strategy games;
- turn-based strategy games.

The tightest performance requirements in terms of RTT come from real time action games. While the peak bit rate of such action games rarely exceeds 100–200 kbps and the average bit rate is often around 10–30 kbps, RTT requirements are typically in the region of 125–250 ms – or even below – for the most demanding games [10]. Hence, HSPA will be able to support action-based gaming with good end user performance as long as network load is controlled. Note, however, that to deliver these short RTTs the mobile station must constantly be in DCH or in FACH state, which is challenging for battery consumption.

The required data rate in action games varies rapidly. HSPA has the advantage over Release 99 that the data rate can be instantly adapted. This leads to efficient radio resource usage.

For strategy games, RTT is not the most challenging property. However, it is important to keep the radio connection setup time short in order to maximize the interactive gaming experience for the end user. With efficient use of radio resource states (Cell_PCH, URA_PCH) it should be straightforward to provide good strategy gaming experience over both HSPA and WCDMA Release 99.

Turn-based games, like chess, require always-on connectivity but they can sustain relatively longer delays than real time games or strategy games.

### *9.3.5 Mobile-TV streaming*

Providing good-quality streaming video to a mobile screen using the latest video codecs requires on the order of 32 to 128 kbps depending on the content. For most types of content 64 kbps is enough to yield good quality with the H.264 codec. WCDMA networks can provide 64–128 kbps with a very good coverage probability. From the end user point of view no significant improvement is expected from HSDPA at first glance. However, what HSPA brings is more capacity, which in turn enables mass-market delivery of high bit rate clips to end users.

Radio networks may contain areas without third-generation (3G) coverage where EDGE data rates of 50–200 kbps are available, while 3G networks with HSDPA capability can provide data rates beyond 1 Mbps. Hence, the streaming application must adapt the media rate not only to network load but also network technology.

Adaptation of the streaming media rate is already partially supported in some terminals, while the full support of a 3GPP standardized media rate adaptation scheme is included in 3GPP Release 6. To select the appropriate media rate the streaming server should know:

1. What kind of mobile station it is streaming towards – in case it is a terminal with limited bit rate capabilities then the media rate should take those limitations into account.
2. What initial media rate should be used when starting the streaming session – sometimes the mobile may roam in a 2G network, and sometimes it may roam in a 3G network.
3. When to increase or decrease the media rate.

In practice, the mobile station and the streaming server exchange information about their capabilities before the streaming session starts – this is how Step 1 is solved. Selection of the initial media rate is more challenging. In today's networks it is, in practice, based on the phone model. When guaranteed bit rate quality of service solutions are in place in terminals and networks for WCDMA and HSPA then the agreed bit rate may be used to guide selection of the initial media rate. From the streaming application point of view, media rate selection based on guaranteed bit rate is supported in 3GPP Release 6 streaming specifications. Once the server has started playing a clip, it needs to constantly monitor the connection. It is already possible for the streaming server to adapt the media rate based on feedback information from the terminal. However, in 3GPP Release 6 streaming specification improved feedback information is introduced which will simplify continuous media rate adaptation. With all the improvements in place, streaming should work well in EDGE, WCDMA and HSPA networks.

In case streaming becomes a mass-market service, the capacity in the cellular network may not be enough to provide point-to-point connections for all users that would like to watch a streaming clip. In that case it is possible to introduce the point-to-multipoint streaming service called the 'Multimedia Broadcast and Multicast System' (MBMS). With MBMS many users can receive the same video clip concurrently and the number of radio resources consumed is reduced compared with point-to-point streaming.

### *9.3.6 Push e-mail*

The delay values in HSPA networks are generally low enough for push e-mail applications. Another performance aspect – mobile power consumption – is considered in this section. Even if the user does not actively download attachments or send e-mails, the push e-mail application generates the following messages to the mobile:

- The mail header and the first few kilobytes of each received e-mail are pushed to the terminal.
- Keep alive messages are exchanged between the server and the terminal with a frequency of a few minutes. The size of keep alive messages is very small.

Figure 9.24 gives an estimate of mobile power consumption on the basis of a keep alive message every 4 min and receipt of between 0 and 50 e-mail messages per hour. The keep

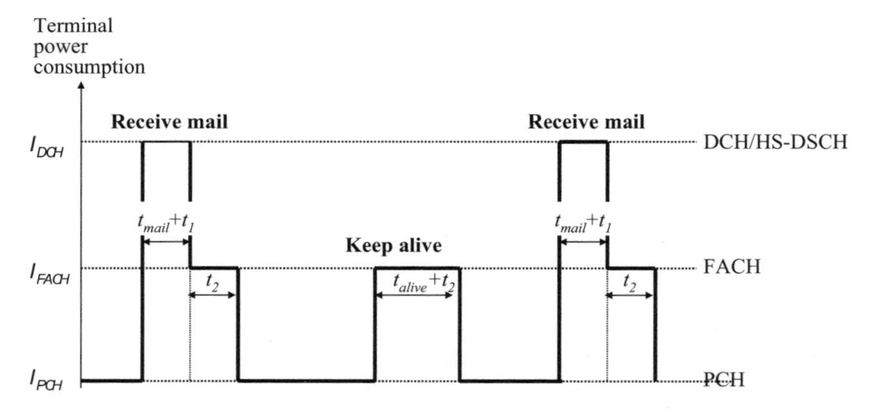


**Figure 9.24** Estimated mobile power consumption with push e-mail: $t_{mail}$ = mail notification time = 2 sec; $t_{alive}$ = keep alive message reception time = 2 sec; $t_1$ = DCH release timer = 2–10 sec; $t_2$ = FACH timer = 2–10 sec; $I_{DCH}$ = mobile current consumption in DCH state = 250 mA; $I_{FACH}$ = mobile current consumption in FACH state = 125 mA; $I_{PCH}$ = mobile current consumption in PCH state = 3 mA.

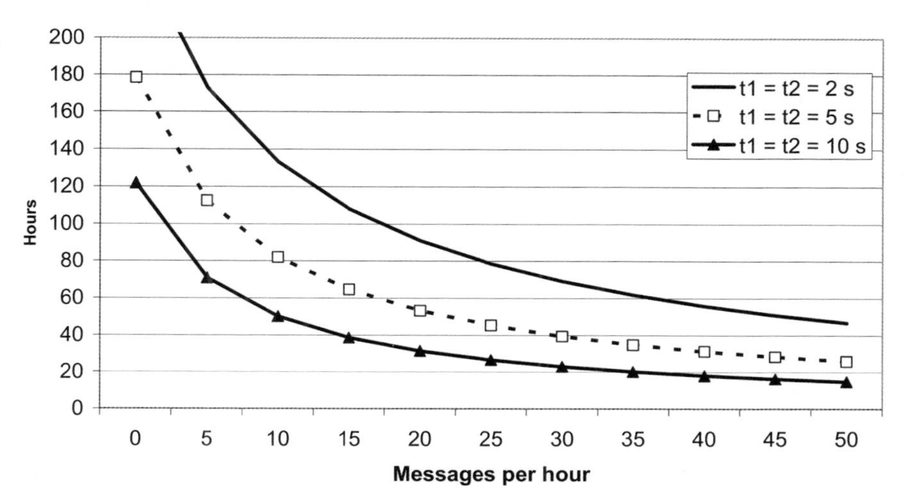


**Figure 9.25** Push e-mail mobile standby time with a 1000-mAh battery.

alive messages are carried on RACH/FACH channels while the message parts – of a few kilobytes – are carried on the HS-DSCH.

Mobile power consumption is illustrated in Figure 9.25. Power consumption depends on the number of messages received and on the channel release timers defined in the radio network parameter settings. If we assume 5-sec DCH and FACH timers and 20 messages per hour, the mobile standby time is 53 h using a 1000-mAh battery. If the release timers are 2-sec, the standby time is over 100 h while longer 10-sec release timers decrease the standby time to below 40 h. These calculations demonstrate that parameters in the push e-mail application (keep alive frequency) as well as in the radio network (release timers) need to be carefully considered to maintain satisfactory standby times.

If PCH state is not used by the network, the UE is moved from FACH to idle state, and the RRC connection is released. When data arrive in the downlink from the 3G core network, the RRC connection needs to be established. The RRC connection setup procedure can increase terminal power consumption considerably and decrease standby time. The use of PCH state is highly beneficial to achieving long standby times.

## 9.4 Application performance vs network load

The aim of this section is to illustrate how network load and cell throughput impact end user experience in the case of laptop-based Internet browsing. For non-real time data services, performance degradation is typically gradual with load increase. In most cases a high network load means that the end user needs to wait a little longer because of the low bit rate. This section shows – by using a simple Internet browsing scenario – that large bandwidth and high cell capacity are beneficial by providing better trunking gains and leading to faster download times and higher network hardware utilization. The scenario also shows that the download time for one Internet page may be two to four times longer in a loaded network than in a lightly loaded network.

For plain voice services the number of blocked calls is an important indicator of how the end user experiences network quality. The number of blocked calls may be computed analytically by means of the well-known Erlang formulas. For data services, where blocking does not occur, an analytical approach to quantify network quality is more challenging. Instead, simulations are typically used to evaluate the performance of cellular data services. To keep it simple the simulation approach taken here does not consider radio interface properties in detail. Instead, the users in one cell are assumed to share a bit pipe characterized by the following two properties:

- The throughput of the bit pipe is constant and independent of the number of users.
- The throughput of the bit pipe is shared equally between all users.

Note that with this simple model some well-known cellular effects – like lower bit rates at the cell border than at the cell centre – are not considered. However, if we take the average cell throughputs of Release 99 DCH and HSDPA from Chapter 7, then this simple model already provides us with good insights into the load vs performance trade-off for WCDMA/HSPA networks. Active browsing users are assumed to have the following simplified usage pattern:

- Each active user downloads a 300-kB-large Internet page every second minute.
- Internet browsing is the only service used in the cell.

The page size of 300 kB is chosen to represent a medium-sized Internet page. The download time of the Internet page is – in the simulations – computed by dividing the page size by the instantaneous bit rate offered by the shared bit pipe. This approach excludes effects like bearer setup times and rendering times and the download times from the simulations are hence to be interpreted as best-case examples. In addition to the active browsing users considered in the simulations, there may be a large number of inactive users within the coverage area of one cell. These inactive users are not considered in the simulations.

The simulation results are shown in Figure 9.26. Internet page download times are given as functions of the number of active browsing users in the cell, and results for 1-, 2-, and 4-Mbps-large bit pipes are shown. The average downlink cell throughput of a Release 99 DCH cell can be close to 1 Mbps, while HSDPA improves the capacity to 2 Mbps with single-antenna terminals and up to 4 Mbps with advanced terminals.

Figure 9.26 shows that capacity is 50, 150, and 360 active browsing users per carrier whenever end users are happy to wait the average 5 sec for a 300-kB Internet page. How many browsing users may be supported per carrier depends heavily on the usage pattern. When users download an Internet page every minute instead of every second minute, then the number of supported users is 50% lower. Calculations show that HSDPA can support a large number of browsing users. The number of users is higher than is the case with WCDMA voice capacity, which is typically 60–130 users per cell [12].

It can also be observed from Figure 9.26 that the cell capacity in terms of the number of browsing users triples when the maximum total cell throughput doubles from 1 to 2 Mbps. This shows that there is a 50% trunking gain available when increasing the cell throughput from 1 to 2 Mbps. For the operator this means that three times as many

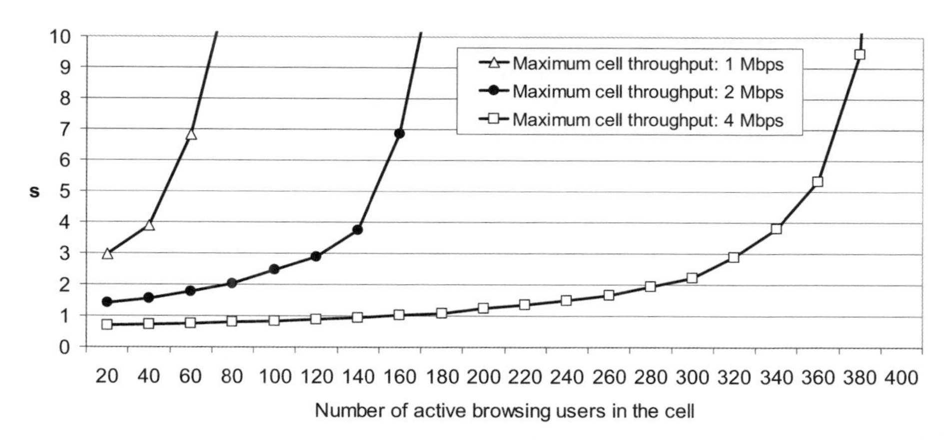


**Figure 9.26** Page download time for a 300-kB Internet page as a function of number of active users.

users can be supported by doubling the cell capacity while maintaining the same download times. From an economical point of view it is hence important to continue developing radio technologies that increase the total cell throughput for high bit rate data services.

As can be seen from Figure 9.26 the network can support more users per cell when the operator allows the time it takes to download the Internet page to be increased. Next, we try to find an approximate formula describing the load vs download time trade-off. Let's assume that we measure the load level by comparing the actual amount of data transmitted during the busy hour with the maximum amount of data the network would be able to transmit during the busy hour if it was full of users all the time:

$$\text{Load level} = \frac{\text{Processed data during busy hour}}{\text{Maximum possible data production during 1 hour}} \tag{9.1}$$

Figure 9.27 again shows the download time for the 2-Mbps throughput case, but now as a function of load level introduced. In the figure the following intuitive formula is also shown:

$$\text{Download time} = \frac{\text{Empty network download time}}{1 - \text{Load level}} \tag{9.2}$$

where empty network download time is given by the time it takes to download 300 kB using a 2-Mbps data rate – that is, 1.2 sec. From the figure it is clear that the intuitive formula predicts the time to download the 300-kB WWW page well. If it is assumed that the load level is around 50% to 75%, then download times during the busy hour are around two to four times larger than the download time in an empty network.

Assume now, finally, that we view a HSPA cell as an entity that maximally can transmit $X$ megabytes to the users in the cell during the busy hour. From the Internet page example, we now know that – when we load the cell so that it produces $0.75 \times X$ megabytes during the busy hour – the end users need to wait some four times longer for the page during the busy hour than during hours with very low network load. With

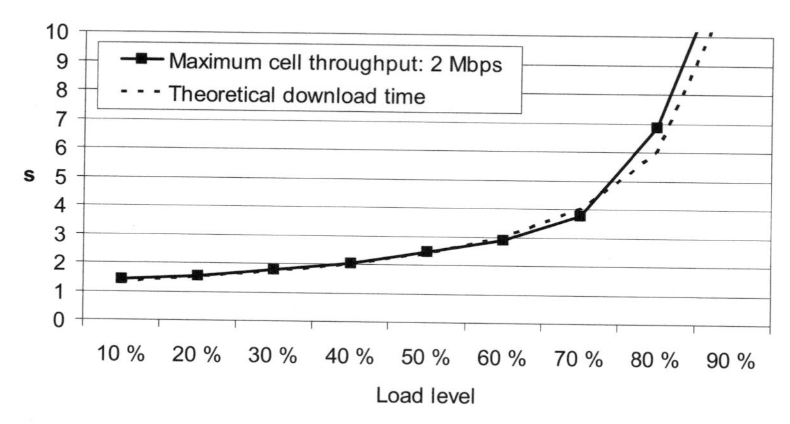


**Figure 9.27** Average download time for a 300-kB Internet page as a function of load level.

this approach and by knowing the number of megabytes that each data user consumes on average per month, it is straightforward to obtain a first understanding for how many data users a HSDPA cell can support while still maintaining adequate service levels. This calculation method is demonstrated in Section 7.5.

## 9.5 References

[1] ITU-T Recommendation G.114 (2003), One-way transmission time.

[2] ITU-T Recommendation P.800 (1996), Methods for subjective determination of transmission quality.

[3] 3GPP Technical Specification, TS 24.008, Mobile Radio Interface Layer 3 Specification, Core Network Protocols, available at *www.3gpp.org*

[4] 3GPP Technical Specification, TS 25.331, Radio Resource Control (RRC), Protocol Specification, available at *www.3gpp.org*

[5] 3GPP Technical Specification, TS 25.433, UTRAN Iub interface Node B Application Part (NBAP) Signalling, available at *www.3gpp.org*

[6] 3GPP Technical Specifications, TS 25.413, UTRAN Iu Interface Radio Access Network Application Part (RANAP) Signalling, available at *www.3gpp.org*

[7] C. Johnson, H. Holma, and I. Sharp (2005), Connection setup delay for packet switched services, *IEE International Conference on 3G and Beyond (3G 2005), London, November.*

[8] C. Johnson, R. Cuny, G. Davies, and N. Wimolpitayarat (2005), Inter-system handover for packet switched services, *IEE International Conference on 3G and Beyond (3G 2005), London, November.*

[9] OMA (Open Mobile Alliance), Push to Talk over Cellular Requirements, Version 1.0, 29 March 2005.

[10] Forum Nokia, Multiplayer Game Performance over Cellular Networks, Version 1.0, 20 January 2004.

[11] IETF RFC 2988, Computing TCP's Retransmission Timer, November 2000.

[12] H. Holma and A. Toskala (2004) *WCDMA for UMTS* (3rd edn.). Wiley, Chichester, UK.

# 10

# Voice-over-IP

Harri Holma, Esa Malkamäki, and Klaus Pedersen

Voice over IP (VoIP) has turned out to be an attractive solution for carrying voice over the packet-switched domain in the fixed network. A number of computer-based VoIP clients are emerging (e.g., Skype [1]) which allow packet-based voice calls between computers and handheld devices to be made over the public Internet. VoIP is also emerging as an add-on feature to Internet Protocol (IP) applications like messaging or netmeeting.

Wideband code division multiple access (WCDMA) and high-speed packet access (HSPA) make it possible to carry good quality VoIP over wide area cellular networks as well. The implementation of VoIP requires a proper understanding of VoIP radio performance, but other aspects of VoIP also need to be considered including multi-operator service level agreements, international roaming, termination fees, and regulatory aspects. This chapter focuses on the radio performance of VoIP. The main motivations for running VoIP over cellular networks are discussed, IP header compression is introduced, and system capacity results are presented. The radio performance of VoIP on Release 99 dedicated channels has previously been studied ([2]–[4]).

## 10.1 VoIP motivation

Circuit-switched voice has been the main source of revenue for cellular operators and still constitutes >70% of their revenue. So far, cellular networks have not been able to support good-quality voice over packet-switched channels, but with WCDMA/HSPA radio performance will be good enough for VoIP. VoIP can be implemented either to support rich call services or just because it will provide the same plain vanilla voice service but with lower cost than traditional circuit-switched voice. This section briefly introduces the different drivers necessary to run VoIP over WCDMA/HSPA and we differentiate between the following three cases:

*HSDPA/HSUPA for UMTS* Edited by Harri Holma and Antti Toskala

1. Consumer-rich calls, in which voice is one component of a multimedia session containing, for example, video or peer-to-peer gaming.
2. Corporate-rich calls, in which the corporation's own private multiservice data network is expanded to cover wireless access networks as well.
3. Plain vanilla voice.

The current advanced second-generation (2G) enhanced data rate for global evolution (EDGE) and third-generation (3G) WCDMA networks allow simultaneous circuit-switched voice and packet-switched data connection. This configuration is well suited for content-to-person services, like WAP browsing and e-mail download, where the destination of the voice call and the packet-switched connection are different. It is also possible to run person-to-person packet services, like real time video sharing, together with circuit-switched voice calls. VoIP, however, could make the implementation of rich call services simpler since both voice and the data service would be carried via packet-switched networks to the same destination. This aspect is considered to be important, especially in cases where the other end of the rich call connection is not a mobile terminal but an ADSL/WiFi-connected computer with a similar VoIP client.

A VoIP scenario with rich call services is illustrated in Figure 10.1.

Business users can access their corporate intranet using virtual private networks (VPNs). Intranet services, including netmeeting, may also carry voice. In order to use those VoIP-based intranet services outside the office, the wide area mobile solution must be able to support VoIP. In this case cellular VoIP is required to extend corporate services to wide area coverage.

Plain vanilla mass-market VoIP puts a lot of pressure on radio efficiency since it can be justified only if it is more efficient than the circuit-switched alternative. VoIP can also be

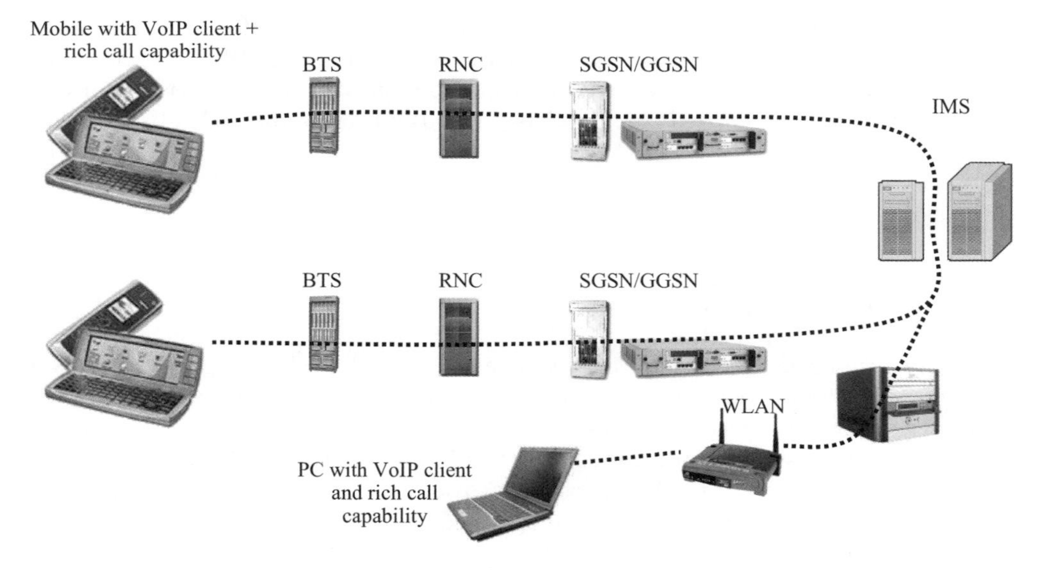


**Figure 10.1** VoIP with rich call capabilities.

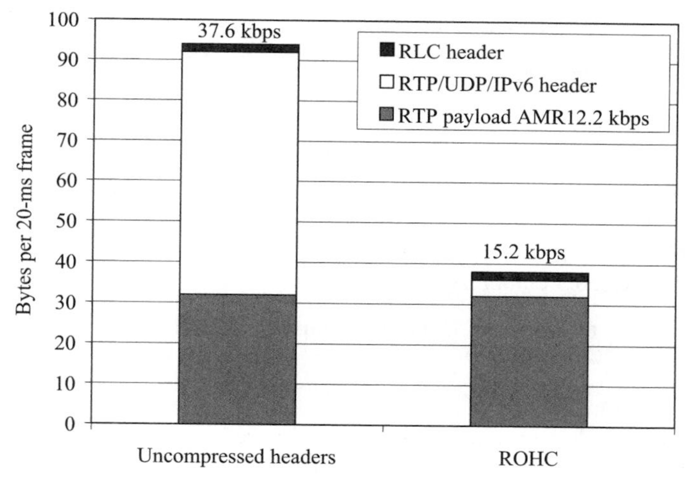


**Figure 10.2** Benefit from robust IP header compression (ROHC) in 12.2-kbps VoIP.

justified for plain voice services if circuit-switched voice is not supported. A number of other packet-based radio systems, including WLAN or Wimax, are not able to carry circuit-switched voice. If the voice service is required over those radio systems, VoIP is the only option – even for a simple voice service.

## 10.2 IP header compression

The size of a full IPv6 header together with a Real Time Protocol/User Datagram Protocol (RTP/UDP) header is 60 bytes, while the size of a typical voice packet is 30 bytes. Without header compression two-thirds of the transmission would be just headers. IP header compression can be applied to considerably improve the efficiency of VoIP traffic in HSPA. We assume usage of robust header compression (ROHC), which is able to push the size of the headers down to a few bytes [7]. ROHC in the Third Generation Partnership Project (3GPP) is part of Release 4. Figure 10.2 illustrates the required data rate with full headers and with compressed headers. The required data rate is reduced from close to 40 kbps down to below 16 kbps.

Header compression with HSPA is done on the Layer 2 Packet Data Convergence Protocol (PDCP) in the user equipment (UE) and in the radio network controller (RNC); therefore, it saves not only air interface capacity but also Iub transmission capacity. The header compression location is illustrated in Figure 10.3.

## 10.3 VoIP over HSPA

High-speed downlink/uplink packet access (HSDPA/HSUPA) was originally designed for high bit rate non-real time services while VoIP is a low bit rate service with strict or

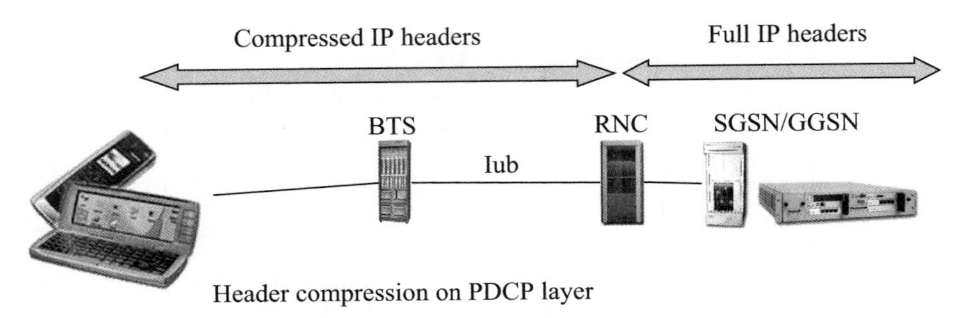


**Figure 10.3** IP header compression with VoIP.

tough requirements. The simulation results in this section show, however, that 3GPP HSPA can still provide an attractive performance for VoIP.

## *10.3.1 HSDPA VoIP*

### 10.3.1.1 Packet scheduling and delay budget

HSDPA VoIP simulations assume proportional fair packet scheduling (for details see Chapter 6). Code-multiplexing of users ($M_{users}$) is assumed. The scheduler selects those $M_{users}$ with highest priority from the scheduling candidate set for transmission in the next 2-ms transmission time interval (TTI). The scheduling candidate set includes users that fulfill the following criteria [6]:

1. Users that have at least $M_{pkts}$ of VoIP packets buffered in the Node B. The value for $M_{pkts}$ depends on the maximum allowed VoIP delay and was between 3 and 4 in the following simulations.
2. Users whose head-of-line packet delay is equal to or larger than $(M_{pkts} - 1) \times 20$ ms.
3. Users with pending retransmissions in their hybrid automatic repeat request (HARQ) manager.

Using these criteria, we try to avoid scheduling users with low amounts of buffered data in the Node B, which might cause a loss of system capacity. Note that a single VoIP packet with ROHC is typically only 38 bytes or 304 bits while the HSDPA transport block size with, say, three high-speed downlink shared channel (HS-DSCH) codes can be clearly beyond 1500 bits. Therefore, a single transport block can carry multiple VoIP packets.

According to the International Telecommunication Union (ITU) model [7], one-way mouth-to-ear delay should be less than 250 ms to achieve a good to acceptable voice quality rating. We estimate that the available VoIP packet delay budget for Node B scheduling, air–interface transmission, and UE reception roughly equals 80–150 ms, depending on whether the VoIP call is between two mobiles or between a land-line and a mobile user.

#### 10.3.1.2 Channelization codes and power allocation

The simulations assume a 3GPP Release 5 solution where an associated dedicated physical channel (DPCH) is used to carry signalling. The spreading factor (SF) for the associated DPCH is assumed to be SF512. The associated DPCH can be in soft handover mode. Assuming an average soft handover overhead of 30%, each user occupies on average 1.3 DPCH channelization codes. Furthermore, the channelization codes for transmission of common channels are reserved. Assuming code-multiplexing of $M_{users}$ per TTI, $M_{users}$ high-speed shared control channel (HS-SCCH) channelization codes with SF128 should also be allocated. The remaining channelization codes can be used for high-speed physical downlink shared channel (HS-PDSCH) transmission with SF16. Figure 10.4 shows the number of available HS-PDSCH codes per cell as a function of number of users, assuming $M_{users} = 4$ for cases with more than 60 VoIP users per cell. For less than 60 VoIP users, it is assumed that one HS-SCCH code is allocated per group of 15 VoIP users. The number of available HS-PDSCH codes decreases as a function of the number of users due to the channelization code overhead from having an associated DPCH for each user. As an example, for a 30% soft handover overhead, an associated dedicated channel (DCH) SF of 512 and 100 users, there are 10 HS-PDSCH codes available out of a total of 15 for VoIP transmission to users.

3GPP Release 6 allows usage of a fractional DPCH (F-DPCH) where multiple users – up to ten – can share an associated DPCH allowing more codes to be allocated for the HS-DSCH. Figure 10.4 also shows the available HS-PDSCH codes using a fractional DPCH. Those users with a DPCH in soft handover still require a dedicated DPCH for the other branch. With 100 users we can still allocate 14 HS-PDSCH codes for data

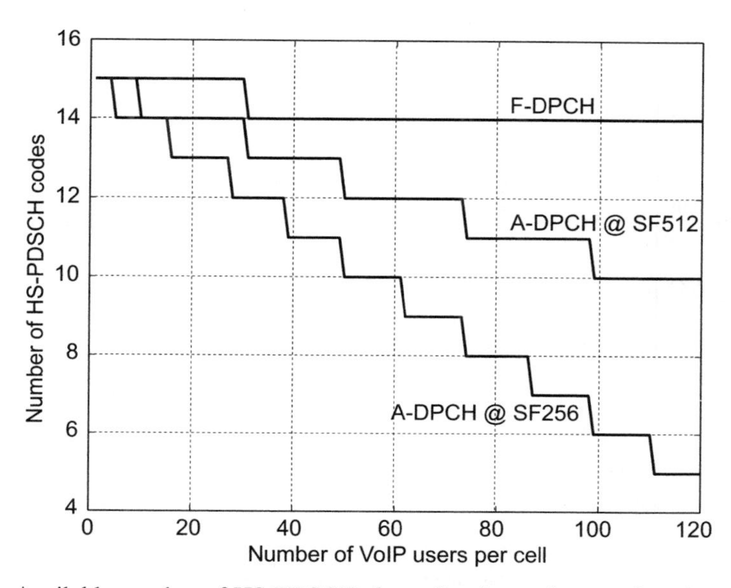


**Figure 10.4** Available number of HS-PDSCH channelization codes as a function of number of users per cell.

transmission, a clear improvement over only 10 codes with a 3GPP Release 5 associated DCH. When more HS-PDSCH codes are available, stronger channel coding and more robust modulation can be applied, thus improving spectral efficiency.

In terms of power allocations the simulations assume that common channels take 3 W, associated DCHs 1 W, the HS-SCCHs 2 W and the HS-DSCH 10 W, leading to an average Node B power of 16 W, which leaves some room for fluctuations in associated DCH power levels.

#### 10.3.1.3 Capacity results

The downlink network simulator described in [6] has been used to investigate the performance of VoIP on HSDPA. Each newly arrived VoIP packet in the Node B is associated with a discard timer. Whenever a buffered packet has been transmitted, it is moved to the HARQ manager and its discard timer is de-activated. Hence, whenever a packet has been transmitted, it can be dropped only if it has not been successfully received after the maximum number of allowed transmissions. For more details of the simulator, see [6].

Figure 10.5 shows the macro-cell simulation results with different delay values for transmission from the RNC to the UE's play-out buffer. The maximum capacity with 80-ms, 100-ms, and 150-ms maximum delays is 73, 87, and 105 users with 5% cell outage. There is clearly a trade-off between delay and capacity with VoIP: if more delay can be tolerated, voice capacity increases.

These VoIP capacity figures can be compared with the estimated Release 99 voice capacity of 64 users [8]. HSDPA can improve voice capacity as a result of advanced L1

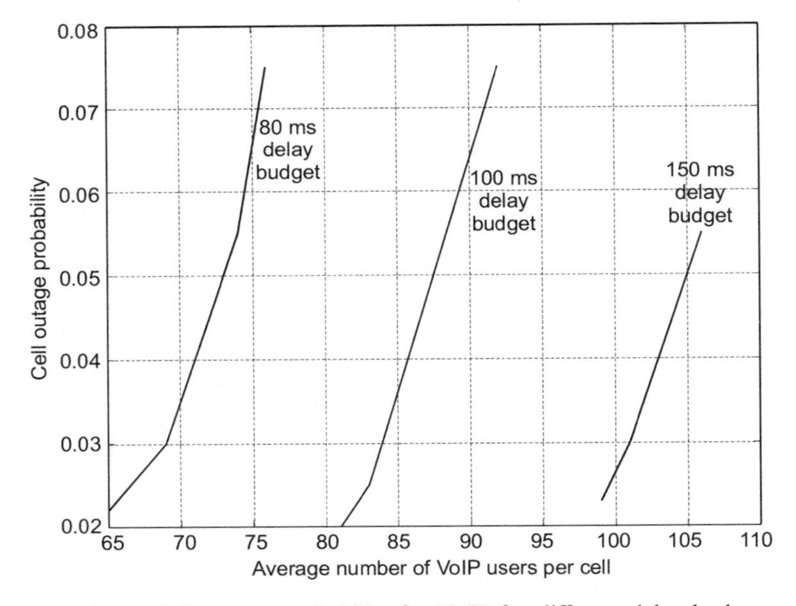


**Figure 10.5** Cell outage probability for VoIP for different delay budgets.

features – including fast HARQ retransmissions, link adaptation, and turbo-coding – compared with the Release 99 DCH that uses convolutional channel coding without any link adaptation or HARQ.

## *10.3.2 HSUPA VoIP*

### 10.3.2.1 Algorithms

VoIP over HSUPA can be implemented in several different ways. The HSUPA specification defines two TTI lengths for the enhanced uplink dedicated channel (E-DCH): 10 ms and 2 ms. A 10-ms TTI is mandatory for all UEs and support of a 2-ms TTI depends on the UE capability. Furthermore, two different scheduling modes are defined for HSUPA: Node B scheduling mode with L1 medium access control (MAC) signalling in the uplink and downlink, and RNC controlled non-scheduled mode.

For a 10-ms TTI, four HARQ processes are specified, which implies a round trip time of 40-ms for fast HARQ. Thus, only one retransmission is possible in order to keep transmission delay below 80 ms. Figure 10.6 illustrates the transmission of VoIP packets over the E-DCH. A new VoIP packet is received from the speech codec every 20 ms. Thus, every second TTI is occupied by a new VoIP transmission. If retransmission is needed, then transmission of the next VoIP packet is delayed by 10 ms and worst-case transmission delay increases to 60 ms.

For a 2-ms TTI, eight HARQ processes are specified and the HARQ round trip time is 16 ms. The 80-ms transmission delay limit allows using up to four retransmissions. Figure 10.7 illustrates VoIP transmission with a maximum of three retransmissions resulting in a worst-case delay of 50 ms. For a 2-ms TTI, it is possible to limit the

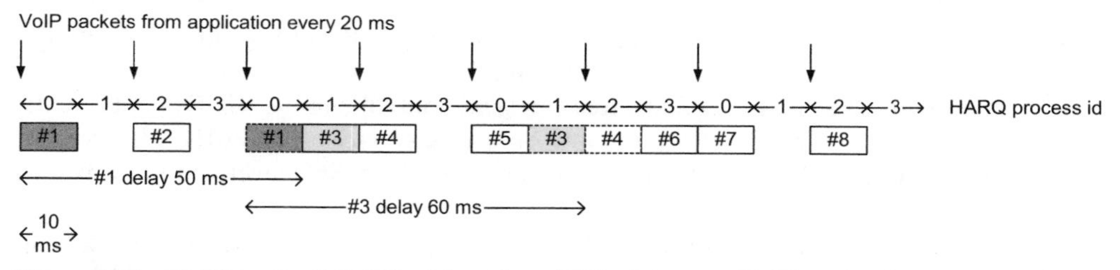


**Figure 10.6** VoIP on the E-DCH with a 10-ms TTI – here, one VoIP packet is transmitted every 20 ms.

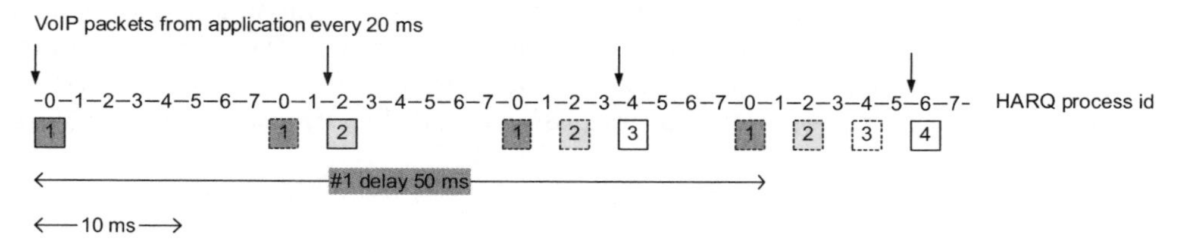


**Figure 10.7** VoIP on the E-DCH with a 2-ms TTI – here, one VoIP packet is transmitted every 20 ms.

number of HARQ processes used by one VoIP user. This can be used for time-multiplexing of different users into separate HARQ processes. However, if three retransmissions per packet and no extra delay due to HARQ process allocation are allowed, then four HARQ processes out of eight need to be allocated for each user.

The advantage of a 10-ms TTI is that all UEs support it, it requires lower peak rates and performs better at the cell edge and in soft handover than is the case for a 2-ms TTI. With a 2-ms TTI, higher cell capacity is achieved since more HARQ retransmissions can be allowed. Also, time-multiplexing of different users is possible.

Non-scheduled transmission of the E-DCH is specified for guaranteed bit rate services and is therefore suitable for VoIP. The maximum number of bits per MAC-e payload data unit (PDU) per MAC-d flow is configured by the serving RNC (SRNC) through radio resource control (RRC) signalling. The allowed bit rate should take into account speech codec rate, header compression efficiency and variations, and the existence of Real Time Control Protocol (RTCP) packets. The non-scheduled data rate can be changed via RRC signalling.

Node B scheduling with uplink rate requests and downlink rate grants is also possible for VoIP. The Node B can send an absolute grant to the UE and activate only some HARQ processes (for a 2-ms TTI). This enables time-multiplexing of users. The advantage of scheduled transmission is that the UE can request a higher grant if needed – for instance, due to RTCP packets. The scheduled grant is, however, a power allocation that does not guarantee a minimum bit rate and in soft handover other non-serving Node Bs can lower the serving grant of the UE. Therefore, RNC controlled non-scheduled transmission is more attractive for the VoIP service.

#### 10.3.2.2 Capacity results

HSUPA is expected to provide some capacity gain over the DCH for a variety of services as a result of fast L1 HARQ and faster scheduling than in Release 99. On the other hand, the overhead from the enhanced dedicated physical control channel (E-DPCCH) 'eats' part of the capacity gain, especially for low data rate services like VoIP. This section provides example HSUPA VoIP capacity results based on link level simulations and system level load equations.

Link level throughput for both a 10-ms and a 2-ms TTI are shown in Figure 10.8. For the 10-ms TTI, the peak data rate is 32 kbps and for the 2-ms TTI it is 160 kbps. Both curves assume a maximum of four transmissions. Due to delay limitations and channel usage, only two transmissions are allowed for VoIP with the 10-ms TTI, which implies a maximum block error rate of 50–70% for the first transmission and single-link throughput of approximately 60% of the maximum value. For the 2-ms TTI, more transmissions can be allowed and capacity is calculated for 50% and 33% single-user throughput, which correspond to an average of two and three transmissions per VoIP packet, respectively.

Uplink capacity can be estimated by using the following load formula:

$$NR_{dB} = -10\log 10\left(1 - \frac{\rho}{W/R} N \cdot v \cdot (1+i)\right) \tag{10.1}$$

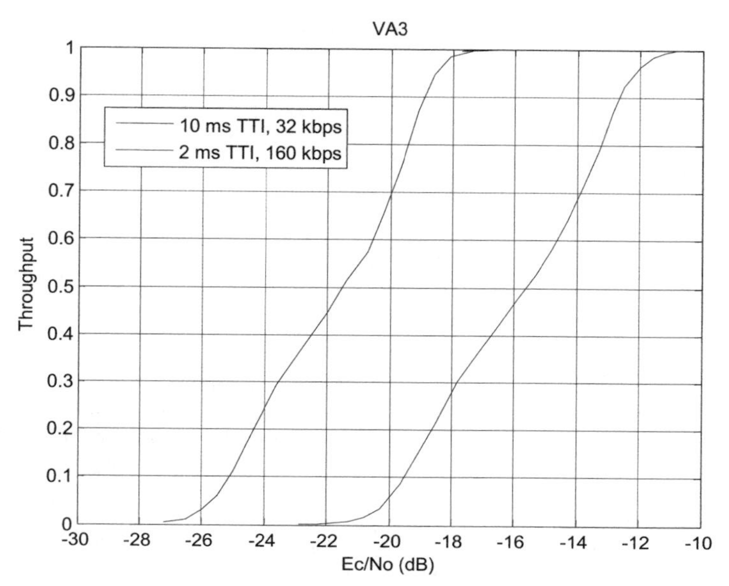


**Figure 10.8** Single-link throughput of 32 kbps with 10-ms TTI transmission and 160 kbps with 2-ms TTI transmission as a function of total $E_c/N_0$ in the Vehicular A channel.

where $\rho$ is the $E_b/N_0$ target, $W$ is the chip rate, $R$ is the enhanced dedicated physical data channel (E-DPDCH) bit rate, $N$ is the number of users, $v$ is the equivalent activity factor, $i$ is the other-cell-to-own-cell-interference ratio, and $NR_{dB}$ is the noise rise in decibels. When the uplink load formula is used, the overhead from the DPCCH, E-DPCCH, HS-DPCCH and retransmissions on the E-DPDCH are included in the equivalent activity factor $v$.

Uplink capacity calculations assume $i = 0.65$ and voice activity of 50%. The channel quality indication (CQI) for the downlink is assumed to be sent on the HS-DPCCH once every 10 ms. The E-DPCCH is only transmitted with E-DPDCH. The DPCCH is sent continuously as it carries mandatory pilot bits and power control bits.

Uplink noise rise is shown in Figure 10.9 as a function of the number of VoIP users. For both TTI lengths two curves are shown. The curves clearly show the power of HARQ: the number of VoIP users can be increased by allowing more retransmissions – that is, by transmitting initially at a lower power level. For a 10-ms TTI, the restriction of only two transmissions per VoIP packet has the effect of limiting capacity. Higher capacity is achievable with a 2-ms TTI since more retransmissions can be allowed. The results are shown for 50% single-user throughput (on average, two transmissions per VoIP packet) and for 33% throughput (on average, three transmissions per VoIP packet).

Due to faster Node B scheduling and HARQ, higher noise rise can be tolerated with HSUPA than with Release 99. We assume a maximum 6-dB noise rise with HSUPA.

Continuous DPCCH transmission causes a quite significant overhead for VoIP traffic where a new packet arrives every 20 ms, although transmission can be as fast as that in a 2-ms TTI. The DPCCH carries pilot bits for channel estimation and power control bits

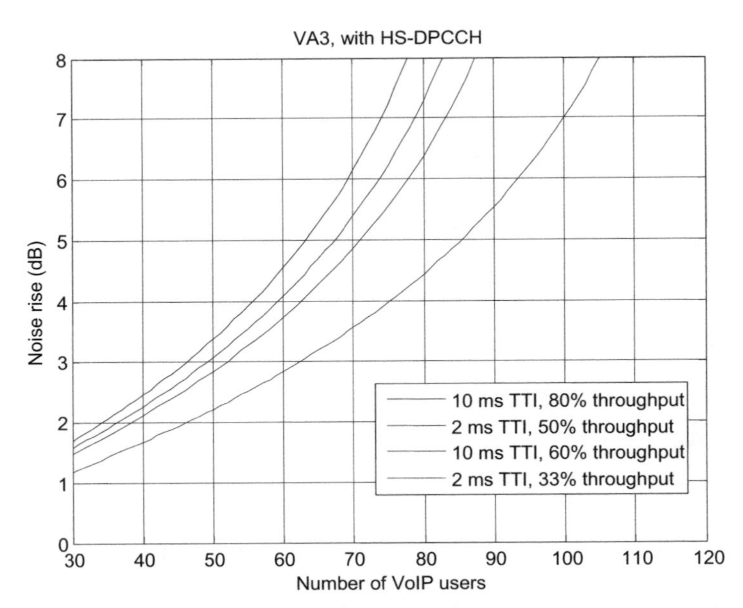


**Figure 10.9** Uplink noise rise as a function of number of VoIP users for different TTI lengths and single-user throughputs in the Vehicular A channel using a 3-km/h channel.

for downlink power control. When no data are transmitted between VoIP packets, no channel estimation is needed either. Possibilities for gating DPCCH transmission when there is no other uplink transmission are being studied in 3GPP for Release 7. The target is to reduce interference and improve capacity. HSUPA uplink capacity is addressed in 3GPP [9].

The spectral efficiency gain from HSUPA can be achieved with several retransmissions. Each retransmission needs to be decoded by the Node B receiver, and the Node B baseband needs to be dimensioned with twice to three times as much processing power than is the case where the number of retransmissions is kept small.

### *10.3.3 Capacity summary*

HSDPA and HSUPA VoIP simulation results are summarized in Figure 10.10. HSDPA results are based on 3GPP Release 5 with an associated DCH, single-antenna Rake receiver, and a maximum 80-ms transmission delay. HSUPA results are based on 3GPP Release 6 with a maximum 60-ms transmission delay. Achieved capacity in the downlink and uplink is similar to WCDMA circuit-switched voice capacity.

HSDPA capacity can be increased by using a fractional DPCH and the advanced terminal receivers that are part of 3GPP Release 6. HSUPA capacity can be increased with 3GPP Release 7 based gating. When all these enhancements are included, VoIP capacity is expected to exceed 120 users with an adaptive multi-rate (AMR) 12.2-kbps codec.

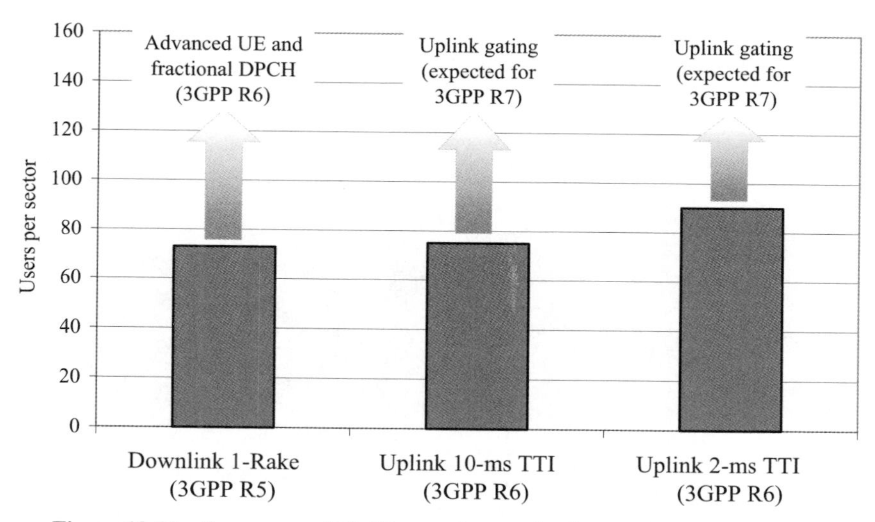


**Figure 10.10** Summary of VoIP capacity results for an AMR of 12.2 kbps.

## 10.4 References


[1] *www.skype.com*

[2] F. Poppe, D. de Vleeschauwer, and G. H. Petit (2000), Guaranteeing quality of service to packetized voice over the UMTS air interface, *Eighth International Workshop on Quality of Service, June*, pp. 85–91.

[3] F. Poppe, D. de Vleeschauwer, and G. H. Petit (2001), Choosing the UMTS air interface parameters, the voice packet size and the dejittering delay for a voice-over-ip call between a umts and a pstn party, *IEEE INFOCOM*, **2**, 805–814, April.

[4] R. Cuny and A. Lakaniemi (2003), VoIP in 3G networks: An end-to-end quality of service analysis, *IEEE Proc. Vehicular Technology Conference, Spring, Vol. 2, pp. 930–934.*

[5] IETF RFC 3095, Robust Header Compression (ROHC), Framework and four profiles: RTP, UDP, ESP, and uncompressed, July 2001.

[6] W. Bang, K. I. Pedersen, T. E. Kolding, and P. E. Mogensen (2005), Performance of VoIP on HSDPA, *IEEE Proc. Vehicular Technology Conference, Stockholm, June.*

[7] ITU, One Way Transmission Time, ITU-T Recommendation G.114.

[8] H. Holma and A. Toskala (eds) (2004), *WCDMA for UMTS* (3rd edn), John Wiley & Sons, Chichester, UK.

[9] 3GPP Technical Report, TR 25.903, Continuous Connectivity for Packet Data Users, Release 7.

# 11

# RF requirements of an HSPA terminal

Harri Holma, Jussi Numminen, Markus Pettersson, and Antti Toskala

This chapter presents the principal parts of a Third Generation Partnership Project (3GPP) terminal's radio frequency (RF) performance requirements with an emphasis on the new aspects introduced using high-speed downlink/uplink packet access (HSDPA/HSUPA). Section 11.1 presents the transmitter requirements and Section 11.2 the receiver requirements. The different frequency bands are introduced in Section 11.3. For detailed requirements, the reader is referred to [1].

## 11.1 Transmitter requirements

### *11.1.1 Output power*

Commercial wideband code division multiple access (WCDMA) and HSDPA terminals are Power Class 3 with 24 dBm maximum output power or Power Class 4 with 21 dBm power. Power Class 4 has a tolerance of +1/−3 dB – that is, terminal output power must be in the range of 21–25 dBm. Tolerance in Power Class 3 is +2/−2 dB. If terminal output power is 22 dBm, the terminal could be classified either as a Class 3 or a Class 4 terminal due to the overlap in class definition. The power classes are summarized in Table 11.1. Higher terminal output power can improve uplink data rates in the weak coverage area.

HSDPA introduces a new uplink channel for L1 feedback called the 'high-speed dedicated physical control channel' (HS-DPCCH). The transmission of HS-DPCCH

**Table 11.1** UE power classes.

| | *Power Class 3* | *Power Class 4* |
|---|---|---|
| Maximum power | +24 dBm | +21 dBm |
| Tolerance | +1/−3 dB | +2/−2 dB |

*HSDPA/HSUPA for UMTS* Edited by Harri Holma and Antti Toskala

**Table 11.2** Release 5 UE output power with HS-DPCCH.

| *Amplitude radio DPCCH/DPDCH* | *Corresponding typical bit rate* (kbps) | *Maximum allowed output power reduction* (dB) |
|---|---|---|
| $1/15 < \beta_c/\beta_d < 12/15$; | >8–16 | — |
| $13/15.5 < \beta_c/\beta_d < 15/8$ | ≤8–16 | −1 |
| $15/7 < \beta_c/\beta_d < 15/0$ | 0 (i.e., no data transmission) | −2 |

takes place in parallel with the normal DPCCH creating a multicode transmission. Multicode transmission sets higher linearity requirements for the user equipment (UE) transmitter's RF parts since the peak-to-average ratio increases. 3GPP specifications allow the UE to lower its maximum output power for those time slots when the HS-DPCCH is transmitted. The allowed power reduction depends on the relative amplitude of the uplink DPDCH $\beta_d$ and DPCCH $\beta_c$. If the relative power of the DPCCH is low compared with the DPDCH, no power reduction is allowed. If the uplink data rate is approximately 16 kbps or higher, no power reduction is required. A power reduction of 1 dB can be used for lower data rates. The maximum 2-dB power reduction is only relevant when the uplink connection has no data transmissions. UE output power limits using the HS-DPCCH are summarized in Table 11.2. Power reduction is not expected to have any effect on link budget dimensioning since the networks are typically dimensioned to provide at least 64 kbps in the uplink using DPDCH.

For Release 6 the solution devised to arrive at a definition of power reduction has been slightly modified to embrace a simple definition for all combinations, including use of both HSUPA and HSDPA. The term 'cubic metric' (CM) is introduced as a metric for allowed power reduction. The specifications allow reduction of the maximum output power when the CM is increased due to the use of parallel code channels over the reference CM value of 1 ($CM = 1$ for $\beta_c/\beta_d = 12/15$, $\beta_{hs}/\beta_c = 24/15$). Thus, maximum power reduction is calculated against the CM value of 1, and the maximum CM value is 3.5, equal to the maximum allowed 2-dB power reduction.

The CM is defined in [1] with upwards rounding with 0.5 steps as:

$$\mathrm{CM} = \mathrm{CEIL}\left\{ \frac{20 * \log 10((v\_norm^3)_{\mathrm{rms}}) - 20 * \log 10((v\_norm\_ref)_{\mathrm{rms}})}{k}, 0.5 \right\}$$

Where $k$ is 1.85 if channelization codes are taken only from the lower half of the code tree, otherwise 1.56, and with $v\_norm$ representing the normalized voltage waveform of the input signal and $v\_norm\_ref$ being the normalized voltage waveform of the reference signal (12.2-kbps AMR speech). This approach also replaces the Release 5 definitions for HSDPA-only devices without Release 6 HSUPA support.

In addition to maximum output power, minimum output power is also defined. The terminal must be able to go down at least to −50 dBm to ensure protection of base stations when the terminal is very close to the base station antenna – for example, in indoor cells.

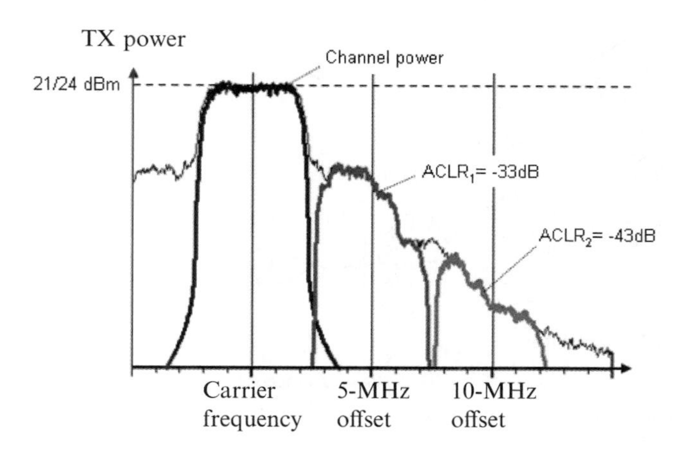


**Figure 11.1** Adjacent channel leakage ratio with terminal transmission.

### *11.1.2 Adjacent channel leakage ratio*

The adjacent channel leakage ratio (ACLR) describes the amount of power allowed to leak into neighbouring carriers. The requirements are neither HSDPA- or HSUPA-specific since output power reduction with HSDPA and HSDPA control channels is defined to allow the same power amplifier to be used for HSPA as for WCDMA while still fulfilling the same ACLR requirements. If there were no reduction in power allowed when the peak-to-average ratio increases, there would be difficulties to maintain ACLR performance without over-dimensioning the amplifier.

The purpose of the ACLR test case is to ensure protection of base station reception performance. This is relevant for base stations that are close to the terminal while the terminal is transmitting to a distant base station with high transmission power. This is especially important for interference cases between operators as terminals obviously cannot connect to the closest base station if that belongs to a different operator.

This case is illustrated in Figure 11.1 for the first and second adjacent carrier. ACLR values do not represent the power level at a single frequency point, but they are integrated over the 3.84-MHz bandwidth with the actual receiver filter modelled in the measurement. Measurement is done at full power, but it is also valid for lower power levels until it gets closer to the minimum power level. At this point the noise floor will start to shape the ACLR.

Also, certain transmitter emission requirements are defined where the individual points in the curve in Figure 11.1 need to be below certain power levels. This is needed for regulatory purposes, and in some countries there are specific requirements for emission mask details (e.g., FCC requirements in the USA).

### *11.1.3 Transmit modulation*

Transmit modulation requirements do not have specific HSDPA-related additions, but with HSUPA there are now similar issues to those of base station error vector magnitude (EVM), which describes how much a particular base station transmitter chain leaks

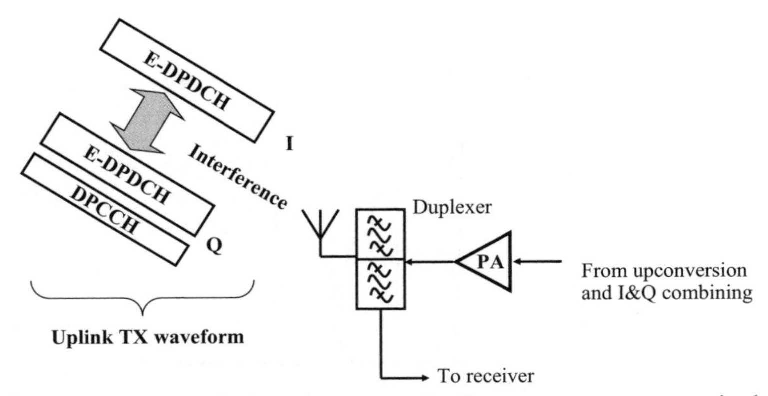


**Figure 11.2** Power leakage between codes due to error vector magnitude.

power between downlink channelization codes. This has been made tighter for base stations using HSDPA due to the introduction of 16-quadrature amplitude modulation (16QAM).

No new modulation was introduced in the uplink direction with HSUPA, so dual-channel quadrature phase shift keying (QPSK) from Release 99 is still being used. EVM with multicode transmission describes how much power is leaking from one code to the other due to the phase inaccuracy of the transmitter, even if in an ideal channel the code channels remain orthogonal. Figure 11.2 shows the example case when two codes are being used with the maximum data rate with a single enhanced dedicated physical data channel (E-DPDCH). The channels are in theory totally orthogonal as they are on different branches (in-phase and quadrature) of the dual-channel QPSK signal. Note that the power level between the dedicated physical control channel (DPCCH) and the E-DPDCH is not to scale in Figure 11.2 – the DPCCH power level with a spreading factor (SF) of 256 is much below the power level of SF4 or SF2 of the E-DPDCH. The smaller SF makes interference between parallel E-DPDCHs more critical, as processing gain is small and does not help in suppressing interference. Existing EVM requirements are valid also for HSUPA transmission and there is work ongoing in 3GPP to set the minimum performance requirements also for code domain power stability, which would guarantee that the UE transmits all the channels with correct weighting.

Phase discontinuity is important even in the case of a single-code channel (DPDCH or E-DPDCH), as modulation carries the information in the phase of the signal. Thus, excessive phase discontinuity will also degrade system performance with data on a single code. Therefore, 3GPP Release 5 specifications contain a test case for phase discontinuity with single DPDCH code as well.

## 11.2 Receiver requirements

### *11.2.1 Sensitivity*

Receiver sensitivity is testing terminal receiver performance at low signal power levels (and in the presense of thermal noise), modelling the case at the edge of the system

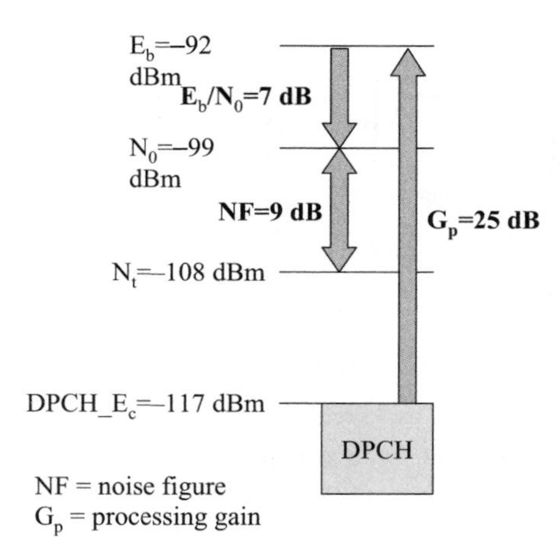


**Figure 11.3** Receiver sensitivity test case.

coverage. Figure 11.3 shows the test settingfor sensitivity measurement. The required signal power level before de-spreading is −117 dBm. Sensitivity is defined for 12.2-kbps voice with a processing gain of 25 dB, thus bringing the signal to −92 dBm after de-spreading. Assuming an $E_b/N_0$ requirement of 7 dB for a block error rate (BLER) of 1%, the noise level has to be −99 dBm. Since the thermal noise level with 3.84 Mbps is −108 dBm, the required terminal noise figure has to be below 9 dB. The required sensitivity of −117 dBm applies for Band I. For other bands the sensitivity varies between −114 and −117 dBm, corresponding to noise figures from 9 to 12 dB. Band-specific sensitivity requirements are defined since the size of the band and the duplex distance between the uplink and downlink is different in the bands. The different bands are introduced in Section 11.11.3.

The sensitivity test is done with the terminal transmitter at full power (21 dBm or 24 dBm), as would most likely be the case at the edge of cell coverage. This allows taking into account leakage of transmitter power to the receiver band. The sensitivity test is defined only for the 12.2-kbps voice reference test channel; there are no HSDPA-specific or HSUPA-specific tests related to receiver sensitivity.

To achieve the required performance in the test case, quite a large attenuation is needed between the transmitter and receiver. The signal sent to the duplex filter in the terminal is at even a higher power than the actual output power, due to attenuation in the duplex filter itself. Separation between transmitter and receiver needs to be achieved with both duplex filter separation available and band pass filters in the transmitter chain, as indicated in the example transmitter chain in Figure 11.4 (PA denotes 'power amplifier' and AGC denotes 'automatic gain control'). Note the example transmitter shown in Figure 11.4 is only one of many possible solutions, this one using intermediate frequency in the transmitter section.

### *11.2.2 Adjacent channel selectivity*

The requirement of adjacent channel selectivity (ACS) in 3GPP Release 99 is valid for HSDPA and HSUPA as well. ACS describes how much higher the power level of the

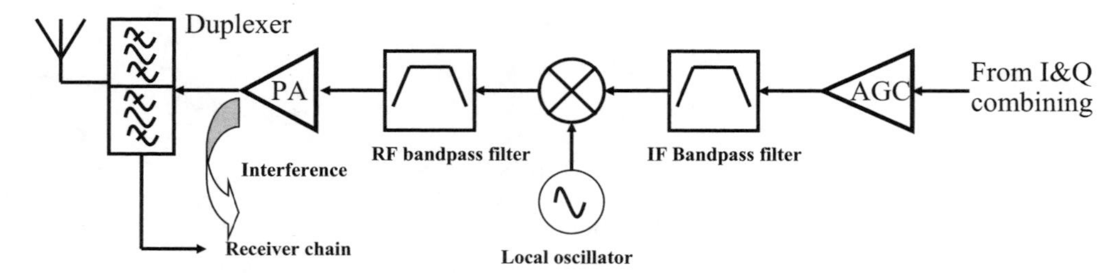


**Figure 11.4** Transmitter chain with interference in the receiver section.

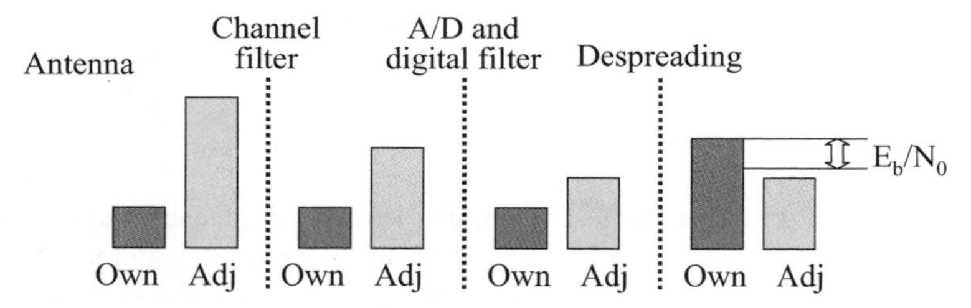


**Figure 11.5** Example of ACS partitioning in the receiver chain: 'Own' = own carrier and 'Adj' = adjacent carrier.

adjacent carrier can be while the terminal can still operate at its current frequency. Such a situation may happen between operators in real networks. 3GPP specifications require an ACS of 33 dB. Release 5 does contain a new test case for ACS as well, but that is only to broaden test coverage and is not directly linked to HSDPA/HSUPA operation. In the terminal design, ASC is obtained by the channel filter and by baseband digital filtering (as illustrated in Figure 11.5).

### *11.2.3 Blocking*

In-band blocking defines how high signal levels from carriers need to be for the terminal to receive signals in the same frequency band. There are requirements for different frequency offsets of 10 and 15 MHz. The 5-MHz offset case is covered in ACS measurement in the previous section. Figure 11.6 illustrates the situation with the 10-MHz offset where the blocker is at the level of −56 dBm. The own-cell signal is 3 dB above the sensitivity level, so the signal level is −114 dBm for Band I. Including processing gain brings the signal level to −89 dBm. With an $E_b/N_0$ of 7 dB, noise plus interference power has to be below −96 dBm. With the 10-MHz offset the selectivity requirement is 40 dBs, and with the 15-MHz offset the corresponding value is 52 dBs.

Narrowband blocking is another set of requirements covering situations in which a second-generation narrowband system has been deployed in the same frequency band. The narrowband system could be the Global System for Mobile Communications (GSM) or Interim Standard 95 (IS-95). The requirement is valid, for example, for UMTS 850, UMTS 1800 (GSM 1800 band), or UMTS 1900 (PCS 1900 band). The

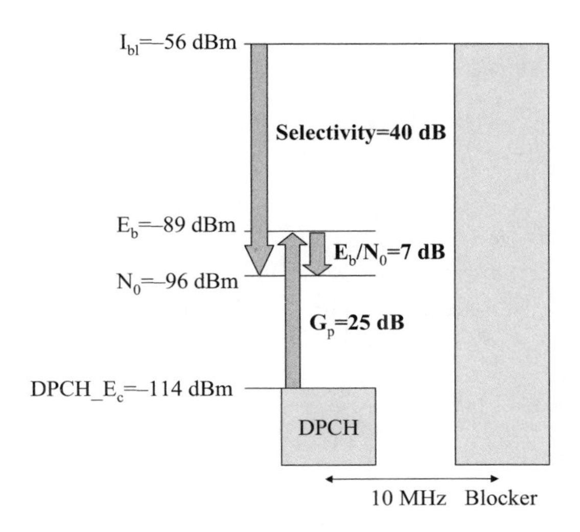


**Figure 11.6** In-band blocking with a 10-MHz offset.

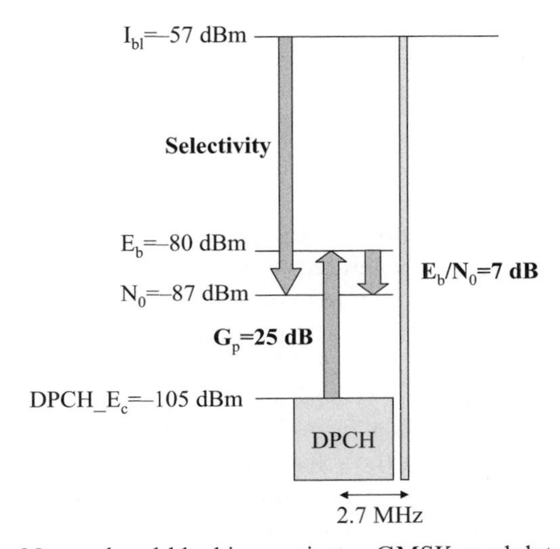


**Figure 11.7** Narrowband blocking against a GMSK-modulated interferer.

test signal is a GMSK-modulated signal with a central frequency that is either 2.7 MHz or 2.8 MHz away from the WCDMA central frequency (as illustrated in Figure 11.7). In Figure 11.7 we show the 2.7-MHz case, where the absolute power level of the narrow-band signal has been set to −57 dBm at the receiver input to represent the case for deployment in the field when there is a GSM carrier next to a WCDMA carrier. The power level of the desired signal is −105 dBm, corresponding to a 10-dB relaxation compared with the sensitivity test case of −115 dBm in those bands.

If GSM and WCDMA base stations are co-sited, the signals received at the terminal are on the same level, thus avoiding any blocking problems. The blocking requirement is

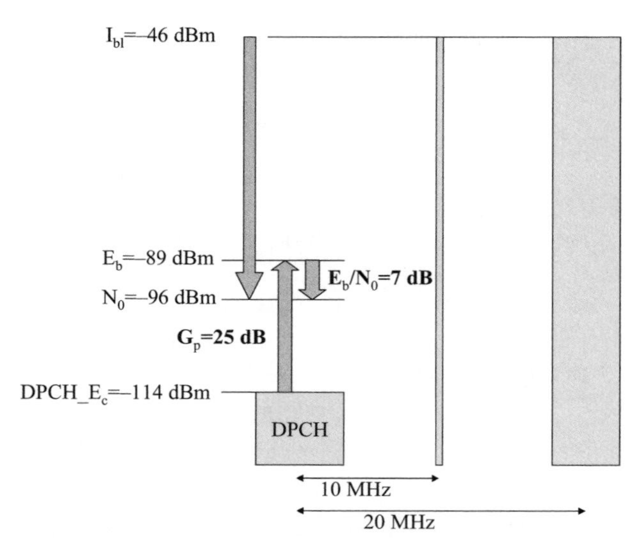


**Figure 11.8** Inter-modulation test case.

only relevant when GSM and WCDMA are deployed uncoordinated – for example, by different operators using different sites.

### *11.2.4 Inter-modulation*

The inter-modulation case is intended to test the terminal receiver tolerance of a third-order inter-modulation product generated by two high-power signals that are 10 MHz and 20 MHz apart. This requirement is to maintain performance in the case where several systems co-exist in the area. The test signal that is 10 MHz apart is a narrowband continuous wave signal while one that is 20 MHz apart is a wideband signal. The test setting is shown in Figure 11.8, where both signals are on at the same time. Test signals have a power level of −46 dBm while the desired signal is at −114 dBm, corresponding to a 3-dB relaxation compared with the sensitivity test case of −117 dBm.

Additionally, there is a narrowband inter-modulation test case for bands where deployment with narrowband systems is most likely. In this additional case there are two narrowband signals with continuous wave signals that are 3.5 or 3.6 MHz apart and GMSK-modulated signals that are 5.9 or 6.0 MHz apart.

### *11.2.5 Receiver diversity and receiver type*

The use of receiver diversity and advanced baseband algorithms in the terminal was raised in connection with the HSDPA introduction. 3GPP Release 5 has HSDPA performance requirements that can be fulfilled using a single-antenna Rake receiver. Release 6 includes additional requirements for HSDPA terminals that have receiver diversity and an equalizer receiver. Currently, the specifications contain requirements for the following cases:

- A single Rake receiver (3GPP Release 5).
- A Rake receiver with receiver diversity (Enhanced Type I in 3GPP Release 6).
- A single equalizer receiver (Enhanced Type II in 3GPP Release 6).
- An equalizer receiver with receiver diversity (Enhanced Type III in 3GPP Release 7).

Note that Type III control channel performance is still to be discussed in 2006. The gain from advanced receivers in terms of data rates and system capacity is analysed in Chapter 7.

Actual antenna performance is independent of these performance requirements. The test cases with receiver diversity assume the ideal situation in which antennas are totally uncorrelated. Obviously, for real life implementation this is not the case, but the antennas will have a correlation that depends just as much on terminal design as on the frequency band being used: the lower the frequency, the higher the correlation between the antennas. Additionally, antennas are not likely to have identical gain, which will reduce the benefits achieved. If receiver diversity is assumed in network capacity planning and dimensioning, an additional margin should be taken into account on top of 3GPP performance requirements to take antenna correlation into account. An example antenna correlation impact for system throughput during 800-MHz and 2-GHz operation is shown in Figure 11.9. The same physical antenna separation is assumed for both bands. The effective antenna separation in terms of wavelengths $\lambda$ is smaller during 800-MHz operation than during 2-GHz operation. A smaller, effective antenna separation leads to higher antenna correlation and lower gain from antenna diversity. The topmost curve is the ideal case without antenna correlation. The middle curve is for 2-GHz operation with an antenna separation of $0.5\lambda$, and the lowest curve is for 800-MHz operation when the wavelength is increased and effective antenna separation is reduced to $0.2\lambda$. The effect of antenna correlation with $0.5\lambda$ is small, clearly less than 5% compared with the ideal case, while with $0.2\lambda$ the effect is approximately 10–15%. Capacity gains are still clear with antenna diversity. Chapter 7 shows that capacity gains are 50–60% in macro-cells. Practical terminal design will dictate final performance, where power azimuth spectrum (PAS) characteristics – including antenna gains – will determine effective performance in the field.

### *11.2.6 Maximum input level*

With the introduction of 16QAM there is the need to preserve more accurate phase and amplitude information through the receiver chain. Otherwise, 16QAM performance would be severely degraded. To avoid this, a specific test case has been defined to test terminal performance at maximum input signal. This corresponds to the case in which the terminal is close to the base station, in the area where 16QAM would be used in the network. The test case measures the throughput to ensure proper HSDPA receiver chain operation at maximum input level. This makes the test case applicable to all devices supporting 16QAM. The test cases for the maximum input level for Release 99 and for HSDPA are shown in Figure 11.10. For HSDPA the case has been modified to accommodate the larger signal envelope variations with 16QAM. All terminals in Categories 1 to 10 can use this test case to validate tolerance of a high input signal level. Additionally,

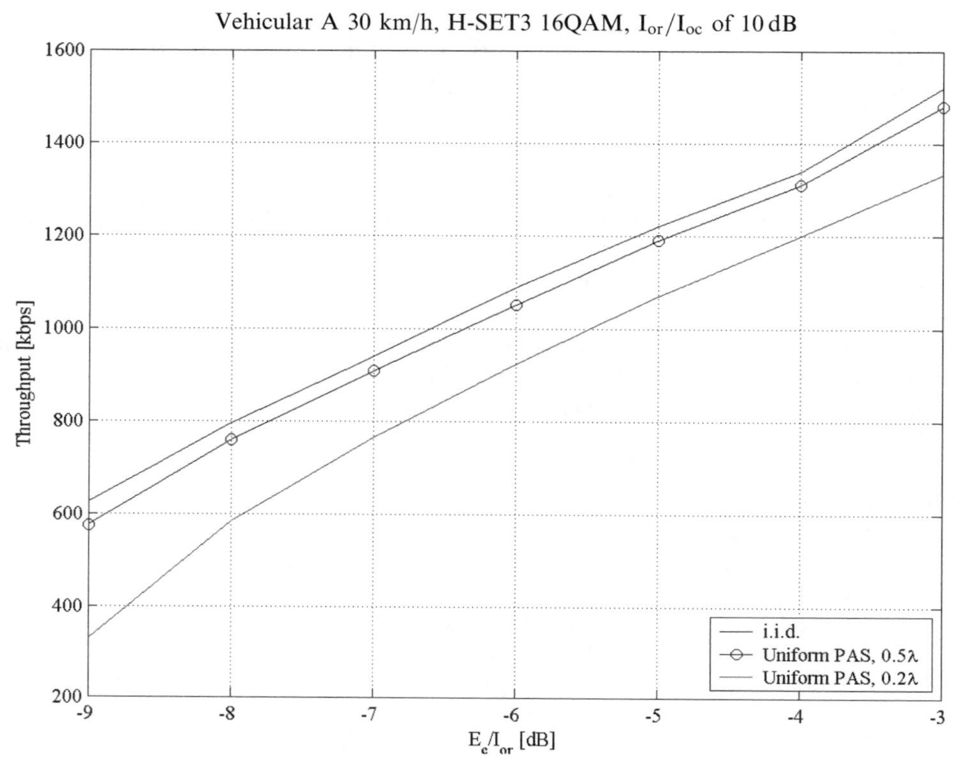


**Figure 11.9** Impact on system performance of an example case with antenna correlation in an ideal case (i.i.d) and in the 2-GHz (0.5λ) and 800-MHz cases (0.2λ).

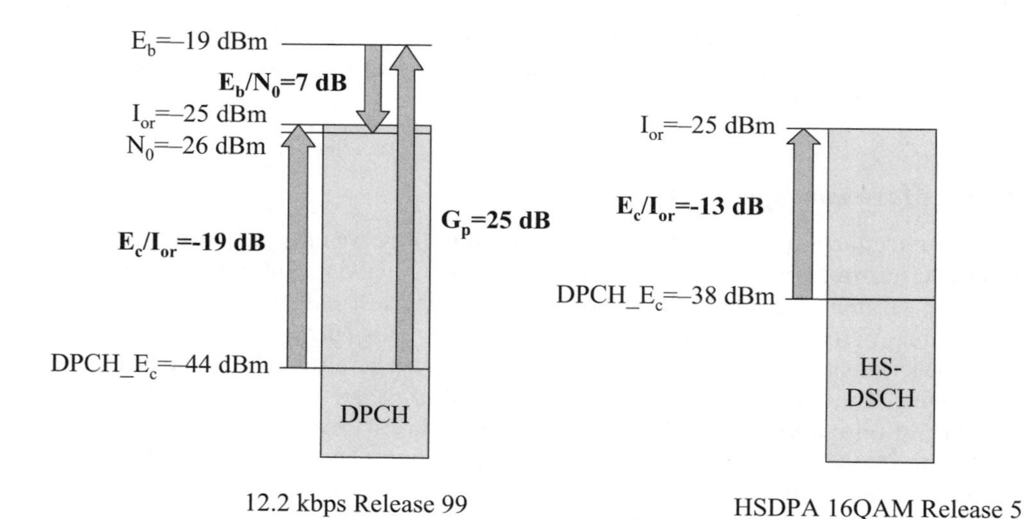


**Figure 11.10** Terminal maximum input signal level test with DCH and with HSDPA 16QAM.

| Operating band | 3GPP name | Total spectrum | Uplink [MHz] | Downlink [MHz] | |
|---|---|---|---|---|---|
| Band VII | 2600 | 2x70 MHz | 2500-2570 | 2620-2690 | New 3G band |
| Band I | 2100 | 2x60 MHz | 1920-1980 | 2110-2170 | Mainstream WCDMA band |
| Band II | 1900 | 2x60 MHz | 1850-1910 | 1930-1990 | PCS band in USA and Americas |
| Band IV | 1700/2100 | 2x45 MHz | 1710-1755 | 2110-2155 | New 3G band in USA and Americas |
| Band III | 1800 | 2x75 MHz | 1710-1785 | 1805-1880 | Europe, Asia and Brazil |
| Band IX | 1700 | 2x35 MHz | 1750-1785 | 1845-1880 | Japan |
| Band VIII | 900 | 2x35 MHz | 880-915 | 925-960 | Europe and Asia |
| Band V | 850 | 2x25 MHz | 824-849 | 869-894 | USA, Americas and Asia |
| Band VI | 800 | 2x10 MHz | 830-840 | 875-885 | Japan |

**Figure 11.11** WCDMA frequency bands in 3GPP.

there is a separate test case with QPSK-only being used to test terminal Categories 11 and 12.

The total received power level is −25 dBm while the designed signal is 19 dB below the total power at the level of −44 dBm for Release 99 and 13 dB below at the level of −38 dBm for the HS-DSCH. The HSDPA test case requires a throughput of 700 kbps with four codes and transmission in every third TTI. For reference, the maximum throughput with four codes and every third TTI is 960 kbps.

## 11.3 Frequency bands and multiband terminals

3GPP has produced WCDMA specifications for all relevant cellular bands that have room for the WCDMA carrier. Frequency variants and their typical usage areas globally are listed in Figure 11.11. Frequency variants are 3GPP release independent, which means that – even if the frequency variant is added with a 3GPP Release 7 schedule – the products for that band can use an earlier 3GPP release as the design basis. Only the new RF requirements must be met in addition to the support of new band-specific signalling elements.

WCDMA deployment has started in Europe and Asia in the mainstream 2.1-GHz Band I with a total $2 \times 60$-MHz allocation. WCDMA terminals typically include WCDMA 2100 together with a number of GSM bands. WCDMA networks in the USA first started at the 1.9-GHz Band II and expanded to the 850-MHz Band V. WCDMA terminals in the USA need, in practice, to support a dual-band WCDMA 1900 + 850 MHz. Once the new 3G Band IV at 1.7/2.1 GHz is available in the USA, frequency variants will be required in the market place. There may be other dual-band frequency variants required in those markets in Asia and in Brazil where the operators have both 850-MHz and 2100-MHz bands. WCDMA deployment at 900-MHz and 1800-MHz frequencies calls for those bands to be included in multiband WCDMA terminals together with the mainstream 2.1-GHz frequency.

The different frequency variants use exactly the same 3GPP WCDMA/HSPA speci-

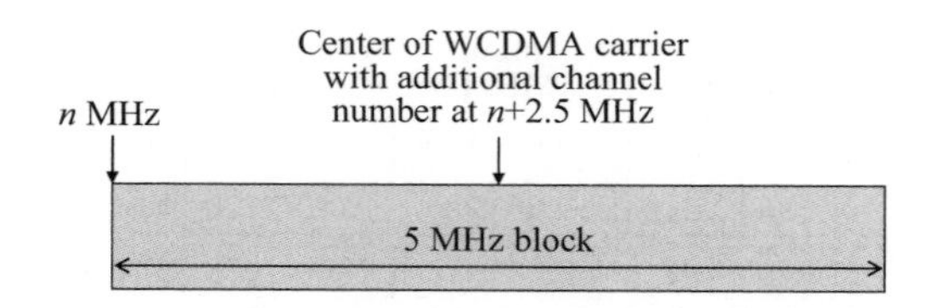


**Figure 11.12** Additional channel numbers allow locating the WCDMA carrier in the middle of the 5-MHz block.

fications except for differences in RF parameters and requirements. The differences-between frequency variants are summarized below:

1. Additional channel frequencies with a 100-kHz offset are included for locating the WCDMA carrier exactly in the middle of a 5-MHz block for Bands II, IV, V, and VI (see Figure 11.12). The normal channel numbers are multiples of 200 kHz.
2. Narrowband blocking requirements for those bands (II, III, IV, VI, and 900) where GSM may be deployed in the same band. Carrier separation between the WCDMA signal and narrowband interference is 2.7 MHz (see Figure 11.7). Such a separation is the minimum possible when WCDMA is located in the middle of a 5-MHz block and the first GSM carrier is 0.2 MHz from the edge of the block; so, in total, $5.0/2 + 0.2 = 2.7$ MHz. For Band III the channel raster is 200 kHz without the 100-kHz offset and, thus, the narrowband blocking distance is 2.8 MHz. A narrow-band intermodulation test case is also included for these bands (as covered in Section 11.11.2.4).
3. Relaxed terminal sensitivity requirements for those bands (II, III, V, and VIII) where separation between the uplink and downlink is only 20 MHz or less, and for Band VII. Those requirements allow a high enough Duplex attenuation to be achieved between transmission and reception in a small terminal. The relaxation is 2–3 dB compared with the other bands.

## 11.4 References

[1] 3GPP, Technical Specification Group RAN, User Equipment (UE) Radio Transmission and Reception (FDD), 3GPP TS 25.101 version 6.11.0, March 2006, available at *www.3gpp.org*

# Index